D1485715

Principles of Oil Well Production

Principles of Oil Well Production

T. E. W. Nind
Trent University, Ontario, Canada

Second edition

McGraw-Hill Book Company
New York St. Louis San Francisco Auckland Bogotá
Singapore Johannesburg London Madrid Mexico Montreal
New Delhi Panama São Paulo Hamburg Sydney
Tokyo Paris Toronto

Library of Congress Cataloging in Publication Data

Nind, T E W
Principles of oil well production.

Includes indexes.
1. Petroleum engineering. I. Title.
TN870.N5 1981 622'.3382 80-17523
ISBN 0-07-046576-2

4567890 KPKP 89876543

The editors for this book were Jeremy Robinson and Olive Collen, the designer was Caliber Design Planning, Inc. and the production supervisor was Teresa F. Leaden. It was set in Baskerville by Progressive Typographers.

Printed and bound by The Kingsport Press.

To W. E. (Wally) Gilbert

Contents

Preface
to the Second Edition

The first edition of this book appeared in 1964. Conscious, over the intervening years, of the developments that have taken place in the oil industry and flattered by steady sales, I have become increasingly aware that the "balance" of the first edition was out of kilter, that some of the stresses were inappropriate to today's industry, and that some of the original material needed amendment or expansion to incorporate new ideas and advances that have been made. I hope that this new edition may correct some of those shortcomings.

It has never been my intention to present an all-inclusive handbook, but rather to stress an approach and to develop a framework to assist the production engineer not only in solving specialized problems, but with day-to-day operations as well. A fairly extensive reorganization of the original work has been undertaken, and some new material incorporated: the criterion that has been uppermost in my mind in the selection of topics has been that of general and lasting applicability. Where appropriate, necessary, and possible, a mathematical analysis is presented, but in such presentations an attempt has been made to stress underlying assumptions and approximations rather than a strict formalism.

It is hoped that the text may prove of value to the practicing engineer as well as to the student: indeed, I am of the opinion that the book may serve best those with some background and experience for the simple reason that production engineering involves a "feel" for practical situations and for the possibilities that exist in the field. W. E. (Wally) Gilbert, who introduced me to the fascination of the subject almost thirty years ago, used to say, "There are no miracles in the oil business." It is the unexpected result that can pave the way for further developments and new approaches—if this book goes some distance toward explaining well behavior it will have served its purpose.

I wish to thank the many petroleum engineers who, over the years, have helped me to formulate and clarify my thoughts. In particular, and recently, I acknowledge Elmo Blount and all those who work with him at the Field Research Laboratory of Mobil Oil Company. Shirley Hartwick undertook the typing of this edition, and her patience, speed, and accuracy—particularly given the challenges of the formulas and the tables—were outstanding. My appreciation too to McGraw-Hill Book Company for its interest and support in publishing the first edition and now this revision.

Finally, as I said in the preface to the first edition, it is to the many authors listed in this volume, and to many associates and friends, that credit belongs for any worthwhile ideas I have presented. I reserve to myself all misinterpretations and mistakes.

T. E. W. Nind

Peterborough, Ontario, Canada

Excerpts from the Preface to the First Edition

In order to make sound recommendations concerning the manner in which an oil well should be produced, the production engineer needs a thorough and clear grasp of the principles governing the movement of oil, gas, and water from the formation to the wellhead. Only with such understanding can the engineer correctly apply the available engineering techniques and decide on the precise specifications of the production equipment to be used in any particular well. A survey of the literature shows that, although a considerable volume of technical material has been published, the principles of production engineering have received far less attention. There are some excellent papers devoted to the fundamental aspects of the subject, but they are scattered and each concentrates on one particular phase of individual well performance, for example, productivity indexes, pressure losses under two-phase vertical flow, gas/oil ratio behavior, or the problems of deep-well pumping.

It is the purpose of this book to gather together between two covers some of the principles of oil well production techniques and to indicate how these may be used in deciding how best to produce—that is, to optimize the profits from—a particular well. In order to keep the book within reasonable limits, at least four major omissions have been made, as well as many minor ones. These are, first, any discussion of gas or condensate well production; second, an outline of gathering systems and the problems connected with the surface separation of gas and water from oil; third, a treatment of formation stimulation, which is now a subject of such magnitude that it warrants a book to itself; and, fourth, a detailed account of production equipment.

No conscious attempt has been made to avoid the use of advanced mathematics in the treatment of the various topics covered in this book; nevertheless, it is the author's opinion that many of the subjects touched upon are of such complexity that they are not amenable to rigid mathematical analysis. Rather than trying to construct a simplified model in such instances, empirical methods have largely been resorted to, illustrated wherever possible by examples. Despite this, it is hoped that the material is sufficiently mathematical to interest, if not to satisfy, those who prefer the strictly analytical to the empirical approach.

The problems that are included at the end of the book are intended to serve a double purpose. It is hoped, of course, that they will be of value to the instructor if the book should be used as a class text; at the same time, some of the problems are intended as illustrations of new, or relatively unproved, techniques of production engineering, which may in fact be found to be useful in the future.

In closing, I would like to express my gratitude to all those who have made the writing of this book possible and in particular to Mr. W. E. Gilbert, of Princeton, New Jersey, for his guidance and for the benefit of his experience, so freely given over the past eight years, and to Shell Oil Company for the opportunity for discussion and for access to data that have together provided much of the stimulus so necessary to the continuance of a project of this kind. I would also like to thank my wife for her continued interest, my sister for reading the first draft and for making many suggestions in connection with style, and Mr. Cy Pannell for his unceasing willingness to type and retype the manuscript.

To the many authors listed in this volume and to the many associates and friends who have contributed by personal discussion belong any worthwhile ideas presented, but I take this opportunity to claim as my own any misinterpretations or mistakes.

T. E. W. Nind

Principles of Oil Well Production

Reservoir Performance

1

1-1 INTRODUCTION

The theme of this book is that a producing oil well is only one part of a complex system, involving the reservoir, the wells themselves, and the surface facilities. Each element in the system affects the others, and in order to achieve an efficient operation it is essential to ensure mutual compatibility.

It follows that the successful production engineer will have a sound knowledge of reservoir engineering and will be fully conversant with advances in oil well and surface equipment technology. It is not, however, the author's purpose to produce a handbook for reference, but rather to stress the principles that need to be kept in mind in the design and operation of producing wells. It is hoped that an understanding of such principles will assist the engineer in adapting to the various constraints and opportunities afforded by reservoir conditions, by equipment advances and availability, and by the overall economic factors guiding the company's operations. An overview of the major categories of well completion designed to lift oil, water, and gas from the formation to the surface may form an appropriate introduction.

There are, essentially, four ways in which wells may be produced: they may be flowed, gas-lifted, pumped, or produced intermittently. Each of these methods has its own variations, and overlap occurs between the categories. *Flowing wells* may be simple flowing completions with production taking place through the tubing; there may or may not be a tubing-casing packer installed; chokes may be set in the tubing, at the wellhead, or in the surface flow lines;

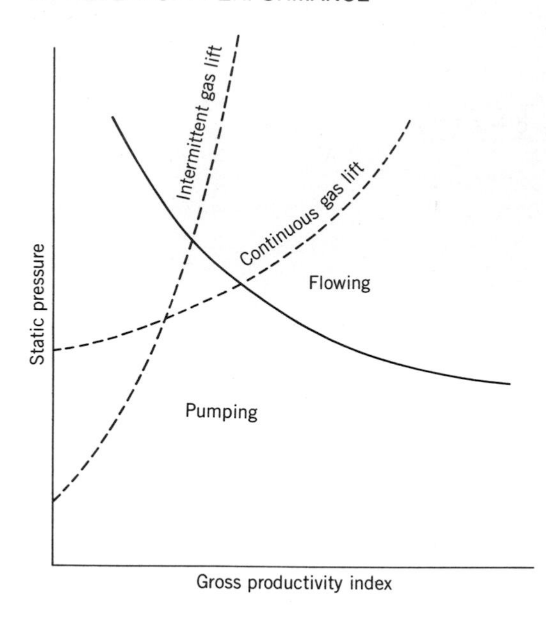

Fig. 1-1 Relative ranges of application of the four essential production methods. (*After Babson, Ref. 1. Courtesy* API Drill. Prod. Practice.)

the completion may be slim-hole; and a time-clock intermitter may be used to control heading or to meet a restricted allowable. *Gas lifting* may be continuous or intermittent (overlapping the fourth category); various types of gas-lift mandrels may be employed; and the technique may be combined with other types of lift, for example, in conjunction with a free-fall plunger. *Pumping* can take a variety of forms (sucker-rod, hydraulic, and centrifugal pumping are all common), and within any one category various techniques are available—for instance, conventional or air-balanced units, long-stroke pumping, and so on. *Intermittent production* (or production by slugs) may result from the use of a free-fall plunger, from a chamber-lift installation, from an intermittent gas-lift operation, and so on.

Each of these basic production techniques has a broad range of operation, and an attempt to illustrate this is shown in Fig. 1-1, which is an adaptation and simplification of a diagram due to Babson (Ref. 1)[1]. Such a broad-brush approach must, however, be applied with a realization that not only are the indicated lines of demarcation vague and subject to wide zones of overlap, but there are many factors that must be taken into account in deciding on the preferable production technique for a particular well or group of wells. Some obvious parameters that need to be considered are:

1. Well depth
2. Present and anticipated gas/liquid ratios (GLRs)

[1] Numbers in parentheses pertain to references at the end of the chapter.

3. Sand and wax problems
4. Hole deviation
5. Casing diameter
6. Inflow performance relationship (or IPR; a productivity index—the well's potential) now and in the future
7. Formation pressure and pressure decline
8. Secondary and tertiary recovery plans
9. Future life and cumulative production anticipated
10. Availability of (high-pressure) gas
11. Work-over difficulties (for example, offshore completions)
12. Oil viscosity
13. Present and future water/oil ratios (WORs)
14. Economic and profitability criteria in light of a company's policies and requirements

The balance of this first chapter is taken up with summarizing some aspects of reservoir performance to which the production engineer must always be alert. No attempt has been made to cover the items fully or rigorously. The objective has been to present them in a reasonably scientific manner, but more importantly in such a way that the production engineer can grasp the underlying principles involved and apply them when thinking through problems.

1-2 PERMEABILITY

Permeability is a measure—under nonturbulent flow conditions—of the ease with which fluid flows through a porous rock and is a function of the degree of interconnection between the pores; for example, in Fig. 1-2 are illustrated two samples that, it is assumed, have the same porosity. Evidently fluid will flow from north to south much more readily through sample *a* than through sample *b*, where the flow is restricted by the fine capillaries.

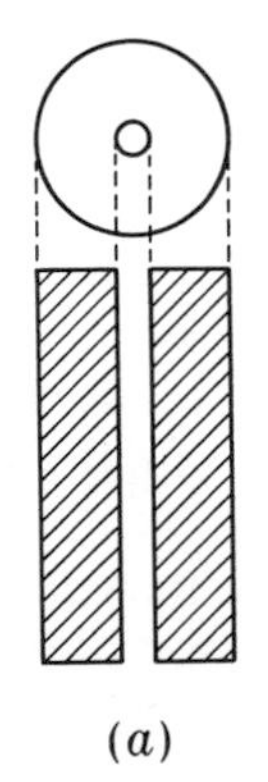

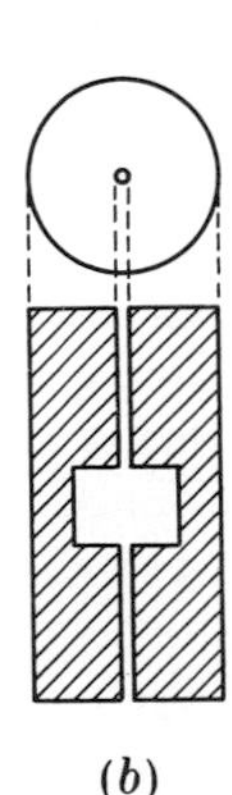

Fig. 1-2 Rocks with same porosity may have widely differing permeabilities.

Henry Darcy carried out the pioneer work on permeability when investigating the flow of water through filter sands (Ref. 2), and for that reason the unit of permeability in the oil industry is known as the *darcy*. The *millidarcy* (md) is one-thousandth of a darcy and is the commonly used unit in the industry, the permeabilities of pay sands varying from about 1 to 1000 md (0.001 to 1 darcy) or higher.

Darcy's equation, which holds for viscous flow through a rock of constant permeability k, may be written

$$\frac{Q}{A} = -\frac{k}{\mu}\frac{dp}{dl} \tag{1-1}$$

where Q/A = rate of flow per unit cross-sectional area across a rock face of area A

$-dp/dl$ = rate of pressure drop in the overall direction of flow

μ = viscosity of the fluid

It can be shown by considering the dimensions of the factors in the equation that k has the dimensions of area. Unfortunately, the system of units that must be employed to obtain k in oil field terms is a nonstandard one.[2] If Q is measured in cubic centimeters per second, μ in centipoises, l in centimeters, A in square centimeters, and p in atmospheres, the resultant value of k is in *darcies;* multiplying this result by 1000 gives the answer in *millidarcies* (md).

The relationship between darcies and area is

$$1 \text{ darcy} = 10^{-8} \text{ sq cm (approx)}$$

that is, a darcy equals ten billionths of a square centimeter, and a millidarcy is one-thousandth of this.

The permeability at different places in the same reservoir rock may vary over wide limits, and there is probably little point in trying to measure the permeabilities of individual core samples too accurately. Roughly speaking, it may be said that when values are less than 50 md, the wells draining the reservoir will be relatively poor producers on the basis of daily production per foot of net pay (unless some formation-stimulation treatment such as fracturing or acidizing is applied); when values are between 50 and 250 md, the wells will be average to good; permeabilities greater than 250 md will result in good wells, other things being equal. However, such sweeping generalizations make no allowance for individual well problems, such as high water cut, high gas/oil ratio (GOR), and sand problems. In addition to varying from place to place, permeability may vary directionally. In many fields, because the pay-zone beds were initially deposited in roughly horizontal layers, the vertical permeability (permeability perpendicular to the bedding planes) is considerably less than the horizontal permeability (that parallel to the bedding planes). Then again, it may have happened during deposition that the sand grains

[2] Some conversion factors to the metric system of units are listed on pages 345–346.

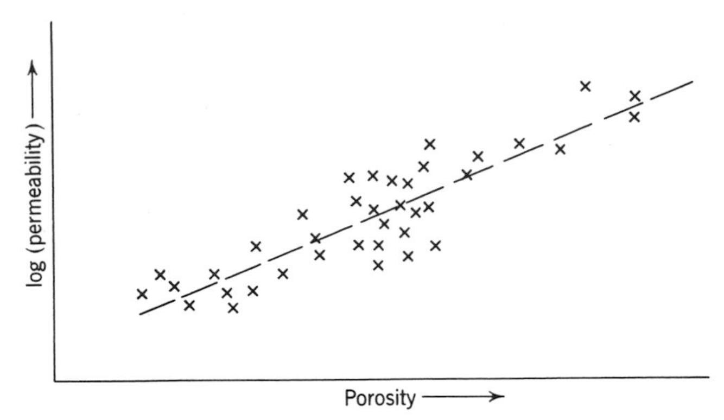

Fig. 1-3 Type of correlation between porosity and permeability noted in many fields.

were oriented with their long axes in the same general direction (for example, by current action); in such a case there will be differences in the permeabilities in different horizontal directions. It should always be remembered, too, that Eq. (1-1) holds only for viscous, or laminar, flow. Laboratory tests combined with calculations based on field data indicate that, in general, oil velocities in the formation in a producing oil field satisfy this condition, except possibly for the region within a foot or two of the well bore itself (see Secs. 3-4 and 3-5). However, for some gas wells at high production rates, turbulent flow may have considerable effect (see, for example, Katz et al., Ref. 3, pages 46–51, 436–437).

Although there is no direct connection between permeability and porosity, it seems reasonable to expect that in sandstones laid down under similar conditions, the permeability should, in general, increase with the porosity. It has in fact been frequently found that within a field, or perhaps within a single stratigraphic unit over a fairly wide area, it is possible to obtain a reasonable straight-line correlation when the logarithm of permeability is plotted against porosity (Fig. 1-3).

The great majority of oil reservoirs in practice contain at least two fluids, namely, connate water and oil; if free gas is also present, there will be three fluids in the reservoir. Evidently there will be a greater resistance to the flow of oil through a rock containing, say, 20 percent connate water than through the same rock in which connate water is absent, since the connate water will block some of the flow channels. This problem has been approached by a number of workers (Refs. 4 through 8), the first organized work on the subject being carried out by Wyckoff and Botset (Ref. 4).

If for example an oil-water system is considered, then, by analogy with Eq. (1-1), *effective permeabilities* to oil and to water, k_o and k_w, respectively, may be defined by the equations

$$k_o = -\frac{Q_o \mu_o}{A} \bigg/ \frac{dp}{dl} \tag{1-2}$$

and
$$k_w = -\frac{Q_w \mu_w}{A} \bigg/ \frac{dp}{dl} \tag{1-3}$$

In these equations the rate of pressure drop in the oil may differ slightly from that in the water owing to the effect of capillary forces which come into play when fluids flow through small-diameter tubes and pores. This difference will here be neglected.

It is found experimentally that oil viscosity, water viscosity, total throughput rate (provided that viscous flow conditions are maintained), back pressure, and length and cross-sectional area of the core have relatively little effect on the plots of k_o and k_w against the fluid saturation in the rock sample.[3] Roughly speaking, then, the curves of k_o and k_w for any particular core are dependent only on the oil and water saturations S_o and S_w within the core (or only on one of them, since the sum of S_o and S_w is unity).

The factors k_o and k_w are called the *effective permeabilities to oil and water*, respectively, and three important points should be noted about the effective permeability curves of an oil-water system.

1. The factor k_o drops very rapidly as S_w increases from zero. Similarly, k_w drops sharply as S_w decreases from unity. That is to say, a small water saturation will markedly reduce the ease with which oil moves through the rock, and vice versa.
2. The factor k_o drops to zero while there is still considerable oil saturation in the core (point C of Fig. 1-4). In other words, below a certain minimum saturation the oil in the core will not move; this minimum saturation is called the *residual oil saturation* (S_{or}) or the *critical oil saturation* (S_{oc})[4]; similarly for water, with a *residual* (S_{wr}) or *critical* (S_{wc}) *water saturation* (point D of Fig. 1-4).
3. The values of both k_o and k_w are always less than k (except at points A and B). In fact, it appears to be true that, except at A and B, the sum of k_o and k_w at any particular oil saturation is always less than k, that is,

$$k_o + k_w \leq k \tag{1-4}$$

Suppose now that two-liquid experiments are run on two cores of different permeabilities k_1 and k_2. Curves like those shown in Fig. 1-5 might result. Such curves give no direct way of comparing the k_w-k_o curves of the two cores, since they start from different points, k_1 and k_2. If, however, instead of

[3] However, it should be noted that some regular trends are discernible, if not marked. Moreover, it appears that the curves are dependent on the direction in which the liquid saturations are varied (for example, oil saturation increasing or decreasing). See Pirson, Ref. 9, pp. 81–83, for a discussion of this phenomenon.

[4] The term *residual oil saturation* is also used to define the oil saturation remaining in the pay zone at the end of the life of the field. This may be considerably in excess of the critical oil saturation.

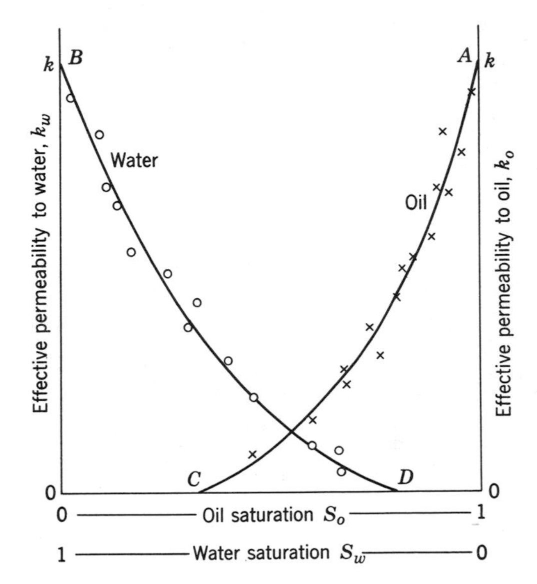

Fig. 1-4 Typical effective-permeability curves (oil-water system).

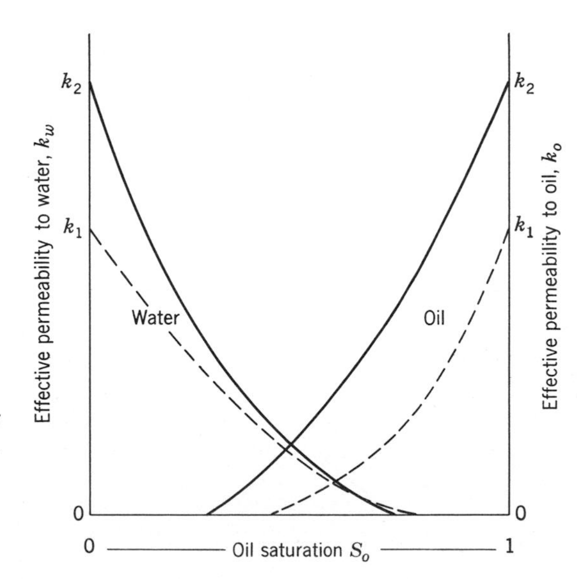

Fig. 1-5 Effective-permeability curves for two different cores.

plotting k_w-k_o against fluid saturation, the ratios

$$k_{rw} = \frac{k_w}{k} \qquad k_{ro} = \frac{k_o}{k} \tag{1-5}$$

are plotted, then in each case the k_{rw}-k_{ro} curve starts from the point unity so that a comparison between the k_{rw} (and similarly between the k_{ro}) curves can be made (Fig. 1-6). The quantities k_{rw} and k_{ro} are called the *relative permeabilities* to water and to oil, respectively. It should be noted that k_{rw} and k_{ro} always lie in the range 0 to 1 and that, from Eq. (1-4) above,

$$k_{ro} + k_{rw} \leq 1 \tag{1-6}$$

Unfortunately, relative-permeability curves are found to depend markedly on the formation; if there is occasion to use these curves (for example, in a reservoir study), the curves for the particular formation under study should be determined experimentally. If this is not possible, recourse should be made to those experimental curves published in the literature.

Up to this point a water-oil system has been under consideration. It is also possible to determine curves for gas-oil and gas-water systems. For these systems the relative-permeability curves have the general shape shown in Fig. 1-7. The important points in this case are the following:

1. Although k_{ro} drops rapidly as S_g increases from zero, small saturations of oil frequently have relatively little effect on k_{rg}.
2. Critical oil S_{oc} and critical gas S_{gc} saturations exist. The critical oil saturation in a gas-oil system is not necessarily the same as the critical oil satura-

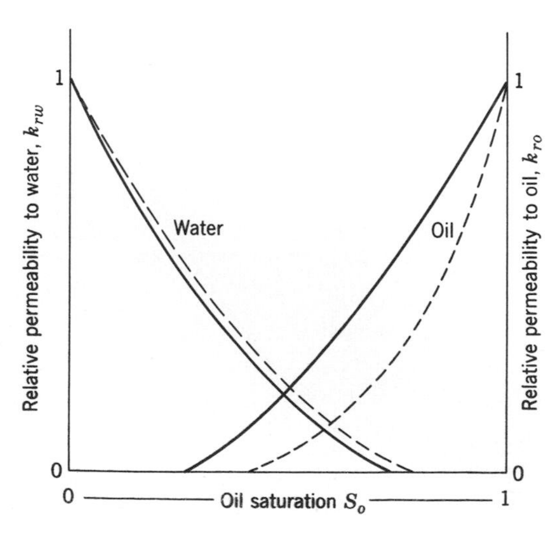

Fig. 1-6 Relative-permeability curves for two different cores.

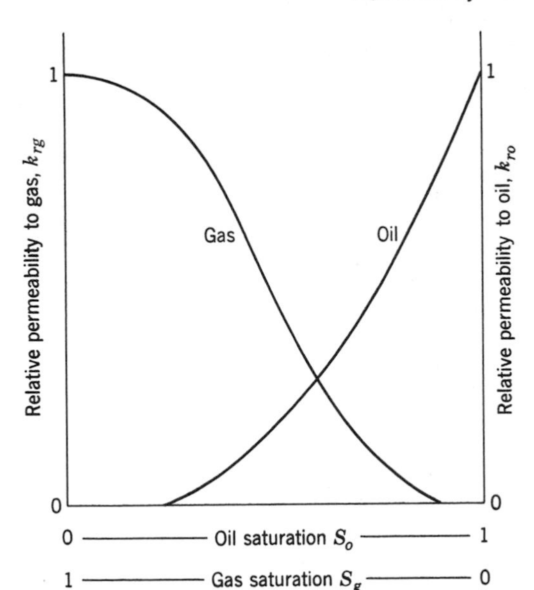

Fig. 1-7 Typical relative-permeability curves (gas-oil system).

tion in a water-oil system, even though the same core has been used. The critical gas saturation S_{gc} is generally of the order of 5 to 10 percent.

3. Both k_{rg} and k_{ro} are less than or equal to unity, and

$$k_{rg} + k_{ro} \leq 1 \tag{1-7}$$

4. It appears that the ratio k_{rg}/k_{ro} tends to increase with the degree of consolidation of the rock. Thus, in general, the less porous and permeable the rock, the higher will be the relative permeability to gas compared with that to oil at a given gas saturation (see Pirson, Ref. 9, pages 68–74.)
5. One point arises in connection with the gas-oil curves that was not apparent for the water-oil system, and that is the question of connate water. It has been found that in order to obtain gas-oil relative-permeability curves which give results agreeing with field practice, it is necessary to run the experiments with the connate water saturation present in the core. In this case the connate water may be regarded as part of the rock, and single-fluid oil or gas permeability measurements should be made in the presence of connate water also. This is the reason that gas and oil saturations are sometimes given in terms of the hydrocarbon-filled pore space, on the assumption that the connate water is an immobile phase, the only characteristic of which is to reduce the effective porosity.[5]

[5] This argument assumes that the connate (or interstitial) water saturation is less than or equal to the critical water saturation, so that no movement of water occurs. However, in many fields the connate water saturation is greater than the critical, so that some water is produced with the oil from the very first barrel of a well's production. A characteristic of such connate water production is that the well's water cut remains reasonably constant throughout its life (see Sec. 1-6).

In connection with relative- or effective-permeability experiments generally, it should be noted that the experimental technique used may have a considerable effect on the results, so that while qualitative arguments based on the general shape of the curves may be in order, quantitative use of the curves is always a source of possible error.

Computer simulations of reservoir behavior frequently involve use of three-phase relative-permeability inputs, but the dependability of experimentally derived relative-permeability curves for three fluids flowing simultaneously is debatable.

1-3 RADIAL FLOW EQUATIONS

Suppose a well is producing liquid at the rate of q bbl/day (stock-tank oil) from a horizontal, homogeneous reservoir of net pay thickness h ft and infinite areal extent; and suppose that the flow conditions do not change with time (that is, steady-state flow prevails). Under such circumstances, and on the assumption that the liquid produced has a low and constant compressibility, it is possible to derive a formula relating the pressure in the formation at a particular point to the distance of the point from the well bore and to the liquid production rate. (For a proof of the relationship for the case in which the liquid is incompressible, see, for example, Pirson, Ref. 9, page 392.)

Let the radius of the well bore be r_w ft and let the pressure at the sand face be p_{wf} psi. If the liquid has viscosity μ cP, the pressure p (psi) in the formation at the radius r ft from the centerline of the well bore (see Fig. 1-8) is approximately

$$p = p_{wf} + \frac{qB_o\mu}{0.007082kh} \ln\left(\frac{r}{r_w}\right) \tag{1-8}$$

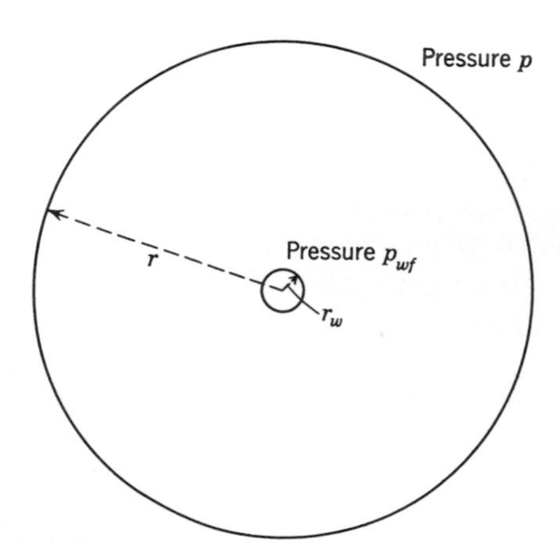

Fig. 1-8 Single well in an infinite homogeneous reservoir.

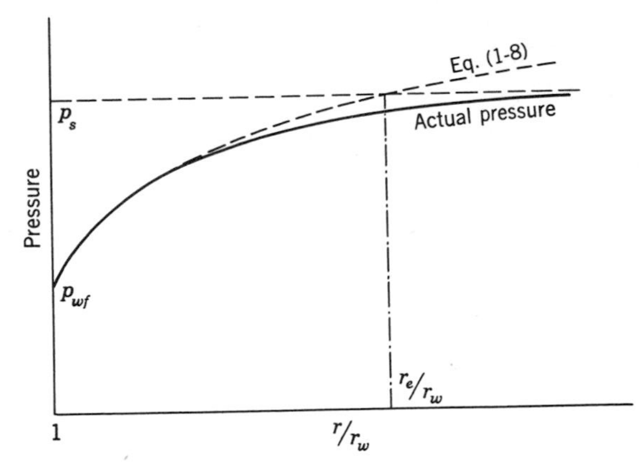

Fig. 1-9 Pressure distribution in the formation.

where B_o is the oil formation volume factor and the formation permeability k is measured in millidarcies.

Equation (1-8) is evidently unrealistic for large values of r, since it implies that p becomes very large as r increases, whereas in practice p tends to p_s, the static pressure of the reservoir. If r_e is the value of r, which makes the right-hand side of Eq. (1-8) equal to p_s, then the equation gives a reasonably good approximation to the actual pressure distribution for values of r less than r_e (Fig. 1-9).

The value r_e is called the *drainage radius* of the well; evidently it has no physical significance for one well in an infinite reservoir.

Example 1-1 A field is drilled up on a rectangular 80-acre spacing. The reservoir pressure (p_s) is 1000 psi, the permeability (k) 50 md, the net sand thickness (h) 20 ft, the oil viscosity (μ) 3 cP, and the oil formation volume factor (B_o) 1.25. The wells are completed with 7-in. casing. What is the production rate per well when the producing pressure at the bottom of the well is 500 psi?

Let x (ft) be the distance between adjacent wells, as shown in Fig. 1-10. Since 1 acre is 43,560 sq ft,

$$x^2 = 80 \times 43{,}560$$

or

$$x = 1864 \text{ ft}$$

That is, as a first approximation, each well drains a circle of 932-ft radius.

Since r_w is $3\frac{1}{2}$ in., or $\frac{7}{24}$ ft, r_e/r_w is 3200, and

$$\ln\left(\frac{r_e}{r_w}\right) = 8.06$$

Use of Eq. (1-8) results in

$$q = \frac{0.007082}{1.25} \times \frac{50 \times 20 \times (1000 - 500)}{3 \times 8.06}$$

$$= 117 \text{ bbl/day stock-tank oil}$$

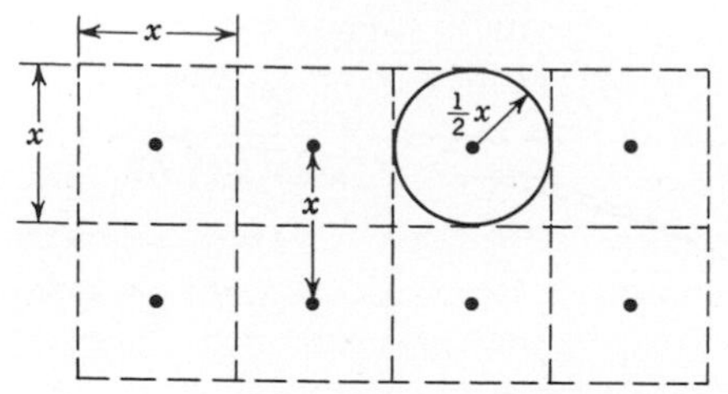

Fig. 1-10 Regular 80-acre spacing pattern.

In this example it should be noted that the exact value of r_e is relatively unimportant. For instance, if a value of 700 ft is used in place of 932 ft, ln (r_e/r_w) is 7.78 and q becomes 121 bbl/day. If a value of 1200 ft is used, ln (r_e/r_w) is 8.31 and q becomes 113 bbl/day. Thus, an exact knowledge of the drainage radius (r_e) of the well is not necessary to obtain a reasonably accurate value for the production rate at a known value of the producing pressure at the bottom of the well (or conversely, to obtain a reasonably accurate value of the producing pressure at the bottom of the well at a known value of the production rate).

When this steady-state radial-flow equation is used, its limitations should always be clearly remembered, namely, that it is valid (approximately) only for values of r less than r_e, and then only for single-liquid flow (the liquid having small constant compressibility) through a homogeneous, horizontal formation, and that in theory it is applicable only to the case of a single well producing from an infinite reservoir (where the concept of drainage radius lacks physical significance).

The model may be made more realistic by assuming that the well is located at the center of a bounded, circular reservoir of radius r_e, no flow taking place across the external boundary. A pseudo steady state is one in which the pressure is taken to decline at the same rate throughout the reservoir, this rate of decline being assumed to be not only independent of position but also of time (during periods of constant production rate q). Under such circumstances it can be shown (see, for example, Craft and Hawkins, Ref. 10, pages 285–287) that Eq. (1-8) should be modified to read

$$p_e = p_{wf} + \frac{qB_o\mu}{0.007082\,kh}\left[\ln\left(\frac{r_e}{r_w}\right) - \frac{1}{2}\right] \tag{1-9}$$

where p_e is the pressure at the boundary.

This model, however, raises the problem of the average pressure in the drainage area of a well. If a well is to produce oil, which implies a flow of fluids through the formation to the well bore, the pressure in the formation at the foot of the well must be less than the pressure in the formation at some distance from the well. The pressure in the formation at the foot of a producing well is known as the *bottom-hole flowing pressure* (flowing BHP, p_{wf}). If all the wells in a producing field were closed in, eventually (perhaps after a few weeks to, possibly, many months, depending on the field characteristics) the pressure throughout the field would equalize. This equalization is caused by the flow of fluids from areas of high pressure to areas of low, lowering the pressure in the former and raising it in the latter. This equilibrium pressure is known as the *field static pressure* p_s. The field static pressure is normally defined at some datum level (for example, so many feet below sea level), and an applicable pressure change with depth is worked out for each field to permit pressures not recorded at the datum level to be corrected to this level.

In practice, because of the long shut-in time that would be required for the field and the attendant loss in income, field static pressures are rarely, if ever, recorded. Instead, individual well *bottom-hole static pressures* (static BHP, p_{ws}) are measured. This can be, and sometimes is, done by the simple expedient of closing in the individual well (leaving all other wells in the field producing) and recording the bottom-hole pressure (BHP) continuously or at intervals[6] until it stabilizes. There are, however, two drawbacks to this procedure: the first is that a considerable length of time may be required before the pressure stabilizes, and the second is that the pressure may never stabilize, interference from surrounding producers causing the BHP of the closed-in well to fall slowly after a certain shut-in time. For these reasons, methods have been developed (Refs. 13 through 16) by which the variation in BHP recorded for the first few days[7] after shut-in can be extrapolated to obtain a figure for the static BHP of the individual well.

A brief description of one of these methods of extrapolation is given below (Sec. 1-4), but in introduction it should be said that the resulting answer is not very meaningful in a physical sense. Consider a single well at the center of a bounded circular reservoir (Fig. 1-11). From the point of view of the flow of oil to the well bore while the well is producing, the pressure differential under which the oil is moving is the pressure at the boundary minus the flowing BHP. From the point of view of the reservoir itself (and the pressure that should be used in applying the material balance equation, for instance, see Pirson, Ref. 9, pages 473–481), the field static pressure p_s that is obtained under equilibrium conditions is somewhere between the value of p_{wf} and that of p_{boundary}. The extrapolation procedures, however, based on an analysis applicable to an infinite reservoir, result in yet another pressure (denoted as p^* by Horner, Ref. 13). Methods of correcting this extrapolated value to p_s have been suggested, (Refs. 13, 16), but none appears to be completely satisfactory. Finally, the question of well interference in a field containing more than one producing well will further complicate the picture.

On the credit side, it may be said that the extrapolation method will give

[6] For example, with a permanently installed BHP transmitter (Refs. 11, 12) or with a wire-line instrument (Pirson, Ref. 9, pages 358–363).

[7] The recording time varies from field to field and from well to well. It may be as low as 1 day or as high as 2 weeks or more.

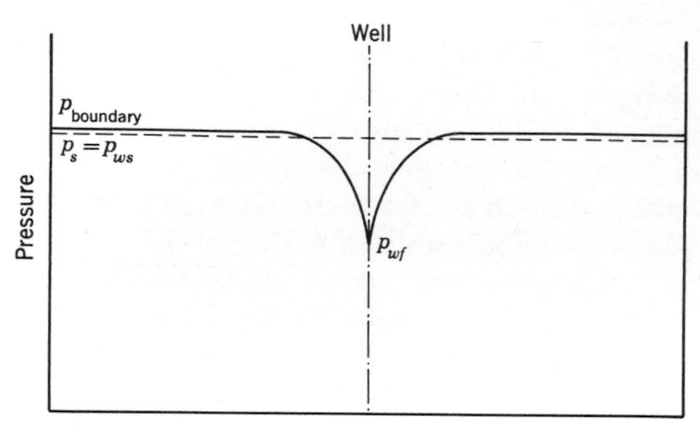

Fig. 1-11 Pressure distribution around a producing well located at center of circular reservoir.

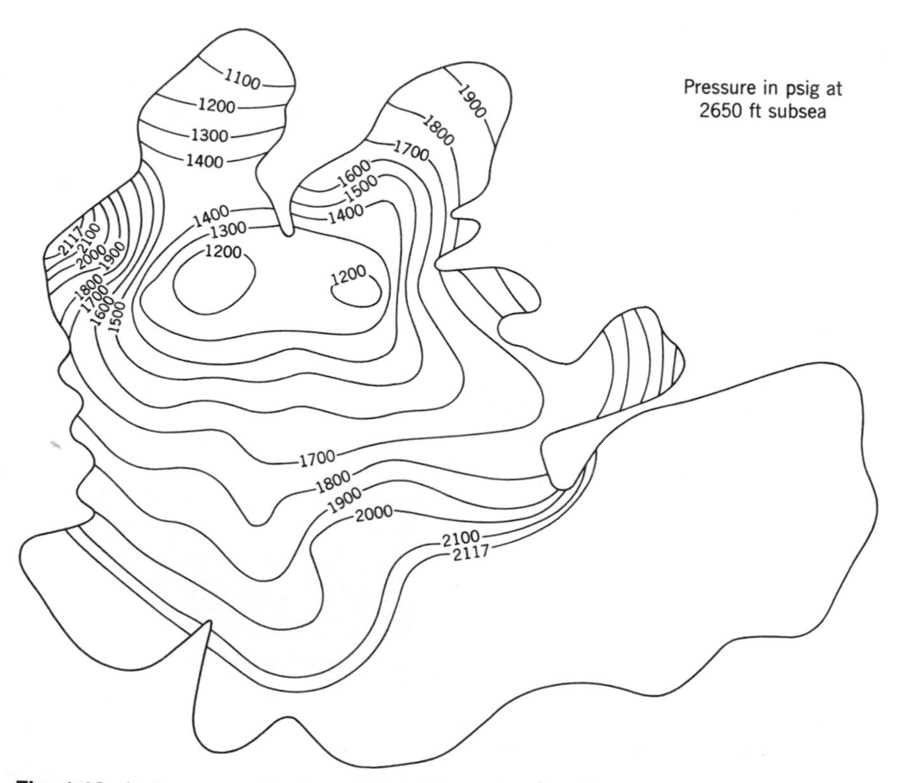

Fig. 1-12 Isobar map, Weyburn Field, Saskatchewan, March 1958. (*After Benn and Tosi, Ref. 17. Courtesy Dept. Mineral Resources, Province of Saskatchewan.*)

reliable and meaningful results in a new well that is not yet drawing oil from its entire drainage area. Furthermore, in many older wells the method gives results that are usable even if a degree of inaccuracy is known to exist, and in a qualitative way these results can be extremely informative. Figures 1-12 and 1-13, for example, show lines of equal pressure, based on individual-well static pressure surveys for the Weyburn Field in the province of Saskatchewan, Canada, as of March 1958 and September 1959 (Ref. 17). It will be noted that a marked pressure low has developed in the northwestern corner of the field in the course of 18 months, indicating that withdrawals from this section of the field have been, relative to the remainder of the field, too high. The sharp pressure gradients existing within the field in September 1959 imply that there was considerable movement of formation fluids from the southern to the northern end of the field (Darcy's law) at about that time.

If the average drainage area pressure, $\bar{p}$ say, were to be used in place of p_e in Eq. (1-9), that equation should be further modified to read

$$\bar{p} = p_{wf} + \frac{qB_o\mu}{0.007082\,kh}\left[\ln\left(\frac{r_e}{r_w}\right) - \frac{3}{4}\right] \tag{1-10}$$

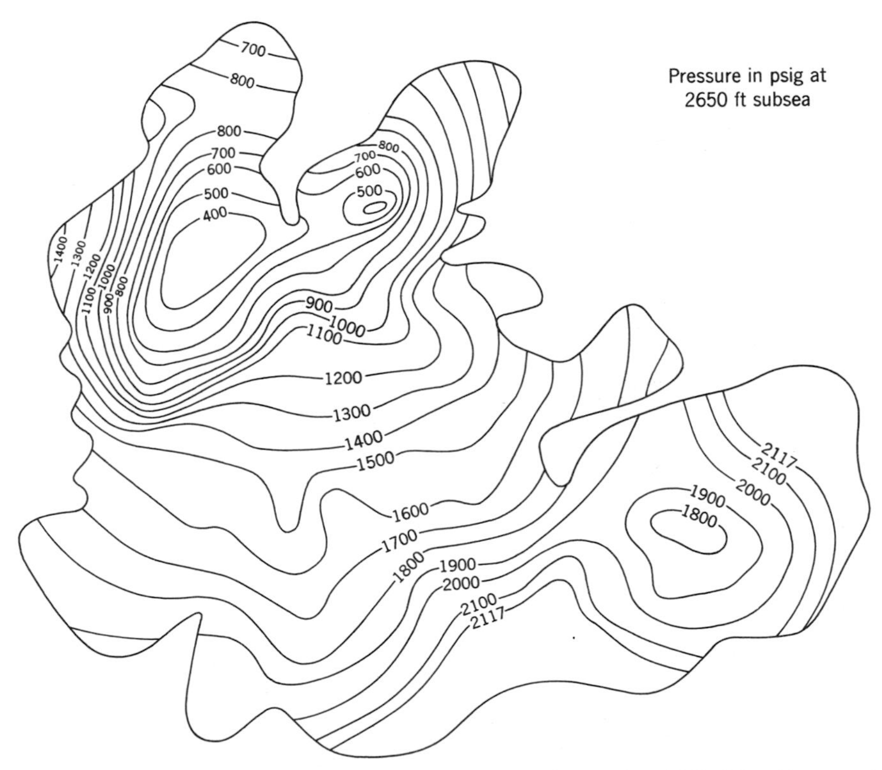

Fig. 1-13 Isobar map, Weyburn Field, Saskatchewan, September 1959. (*After Benn and Tosi, Ref. 17. Courtesy Dept. Mineral Resources, Province of Saskatchewan.*)

In what follows in this book, differences between Eqs. (1-8), (1-9), and (1-10) will be ignored, and the symbol p_s will be used for the *static pressure,* which may be looked upon as the pressure at the boundary, the volumetric average pressure, the built-up pressure as discussed in the next section, or the built-up pressure modified as recommended by Horner (Ref. 13).

1-4 PRESSURE BUILDUP ANALYSIS: HORNER'S METHOD

Suppose a producing well is closed in by lowering a blank liner so that all flow of oil, gas, and water into the well bore is stopped. Oil and gas in the formation will continue to move into the region of lower pressures around the well bore, and this process will continue until the pressure throughout the formation is constant. It can be shown that, for the case of a single well producing from an infinite homogeneous reservoir containing only oil (so that there is no free gas movement), a plot of log $[(T + \theta)/\theta]$ against p is a straight line, where T is the production time of the well prior to shut-in, θ is the time since the well was shut in (measured in the same units as T), and p is the BHP recorded at time θ (Fig. 1-14). In practice, T is usually defined as the well's cumulative production divided by its production rate immediately prior to shut-in.

When the closed-in time θ is infinite, $(T + \theta)/\theta$ is equal to unity and log $[(T + \theta)/\theta]$ is zero. Thus, an extrapolation of the above line until it cuts the axis along which log $[(T + \theta)/\theta]$ is zero will give a value of the pressure at infinite closed-in time, which in the case of a single well in an unlimited reservoir will be the value of p_{ws} (or p_s). Moreover, the slope of this straight line is dependent on the formation permeability, the relationship being

$$\text{Slope of line} = 162.6\,\frac{B_o q \mu}{kh} \qquad \text{psi/cycle } \log_{10} \tag{1-11}$$

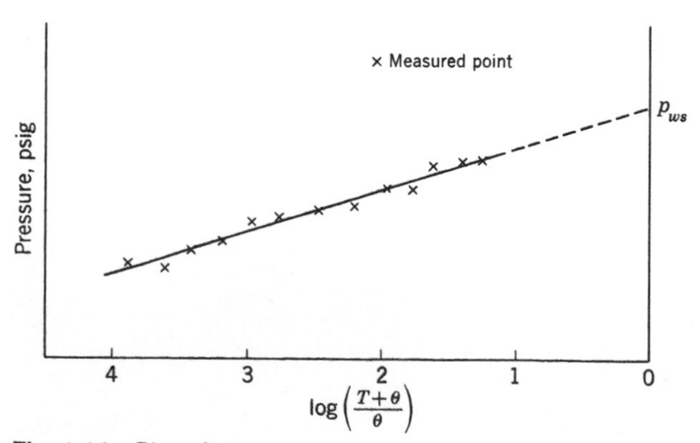

Fig. 1-14 Plot of log $(T + \theta)/\theta$ against pressure.

where q = production rate, stock-tank bbl/day
k = permeability, md
h = formation net thickness, ft
μ = oil viscosity, cP
B_o = oil formation volume factor

The following example, taken from Horner (Ref. 13) and using measurements made on well CB-161 in Casabe Field, Columbia, illustrates the method.

Example 1-2 Well CB-161, Casabe Field, was completed to the "A" sands on February 7, 1959 and closed in from February 16 to March 8 for a bottom-hole survey. Its cumulative production at the instant of closing in was 5847 bbl, and its production rate prior to closing in was 641 bbl/day. The BHP after closing in was recorded at intervals, and the readings shown in Table 1-1 resulted. The net pay thickness h was 349 ft (from electric log), the viscosity μ was 40 cP (from PVT), and the oil formation volume factor B_o was 1.075 (from PVT). The problem is to determine the static BHP and the formation permeability.

The first step in the calculation is the determination of the average production time of the well, and this is given by

$$T = \frac{5847}{641} \quad \text{days} = 219 \text{ hr}$$

Next a tabulation of measured BHP against log $[(219 + \theta)/\theta]$ is set up (Table 1-2), and the results are plotted (Fig. 1-15). From Fig. 1-15 it can be seen that a straight-line plot results and that the value of p_{ws} is 1280 psig. To calculate the permeability Eq. (1-11) is used. The slope of the line in Fig. 1-15 can be obtained by taking the value of the pressure when log $[(T + \theta)/\theta]$ is equal to zero less the value of the pressure when log $[(T + \theta)/\theta]$ is equal to unity; that is, 1280 − 1198, or 82 psi/cycle $\log_{10}$.

TABLE 1-1 Pressure Buildup Data from Well CB-161, Casabe Field (after Horner, Ref. 13)

BHP (p), psig	Closed-In Time (θ), hr	BHP (p), psig	Closed-In Time (θ), hr
1192	19	1239	97
1200	25	1241	103
1206	31	1242	109
1212	37	1241	115
1216	43	1243	121
1220	49	1244	127
1223	55	1245	133
1227	61	1247	139
1230	67	1249	145
1232	73	1249	151
1235	79	1250	157
1236	85	1267	477
1237	91		

TABLE 1-2 Determination of $\log_{10}\left(\frac{219+\theta}{\theta}\right)$: Well CB-161 Casabe Field (after Horner, Ref. 13)

p	θ	$\frac{(219+\theta)}{\theta}$	$\log_{10}\left(\frac{219+\theta}{\theta}\right)$
1192	19	12.53	1.0980
1200	25	9.760	0.9894
1206	31	8.065	0.9066
1212	37	6.919	0.8400
1216	43	6.093	0.7848
1220	49	5.469	0.7379
1223	55	4.982	0.6974
1227	61	4.590	0.6618
1230	67	4.269	0.6303
1232	73	4.000	0.6021
1235	79	3.772	0.5766
1236	85	3.576	0.5534
1237	91	3.407	0.5324
1239	97	3.258	0.5130
1241	103	3.126	0.4950
1242	109	3.009	0.4784
1241	115	2.904	0.4630
1243	121	2.810	0.4487
1244	127	2.724	0.4352
1245	133	2.647	0.4228
1247	139	2.576	0.4109
1249	145	2.510	0.3997
1249	151	2.450	0.3892
1250	157	2.395	0.3793
1267	477	1.459	0.1641

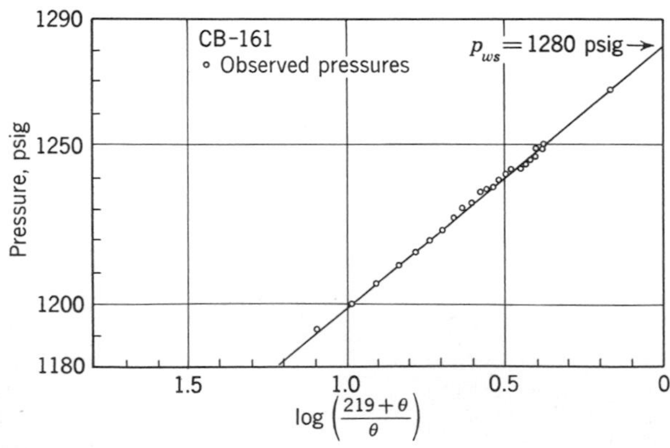

Fig. 1-15 Observed pressure buildup curve, Casabe CB-161. (*After Horner, Ref. 13.*)

Substituting in Eq. (1-11)

$$82 = \frac{162.6 \times 1.075 \times 641 \times 40}{k \times 349}$$

from which $k = 156$ md.

1-5 COMPLETION EFFICIENCY

Up to this point it has been assumed that the well is closed in at the sand face. In reality, the situation is complicated by the fact that the closing in of a well takes place at the surface. Fluids continue to move into the bore itself, and this movement (the *afterproduction* to give it the name by which it is commonly known) affects the redistribution of pressure in the reservoir. After the well has been closed in for a few hours, the greater part of the pressure buildup will in general have been completed; from that time on the BHP will rise only very slowly, so that the rate of fluid movement into the well bore becomes small and may, to all intents and purposes, be neglected. In other words, after a few hours the shape of the buildup curve is determined solely by the pressure redistribution in the formation and is not affected by the afterproduction. A typical example of the influence of afterproduction on the shape of the log $[(T + \theta)/\theta]$ versus p plot is shown in Fig. 1-16.

The oil which flows into the well during the afterproduction period comes from the immediate surroundings of the well bore. If the formation close to the well bore has been damaged in any way by the drilling of the well (for example, by some mud filter cake remaining in place, by invasion of the sand with filtrate, or by resistance of the screen-liner slots to flow) or if it has been improved in some way (for example, by acidization, fracturing, or flushing), this damage or improvement should show up in the shape of the early part of the buildup curve. Thus a comparison of the early and later parts of the

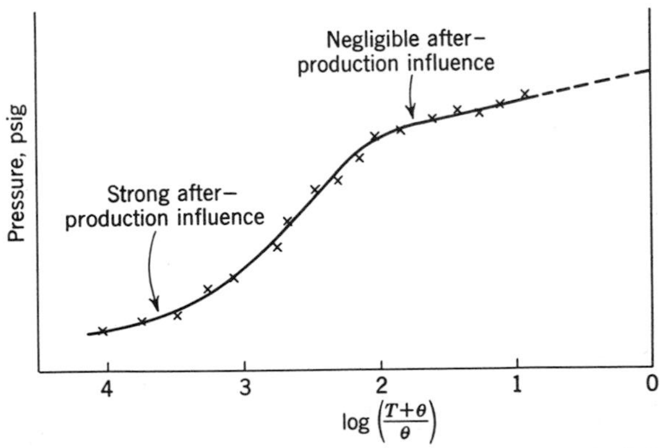

Fig. 1-16 Influence of afterproduction on pressure buildup curve.

buildup curve should give an indication of the degree of formation damage (or improvement) in the vicinity of the well bore. The results of such calculations can be expressed in terms of the skin factor S (Ref. 18), which has the properties that

If S is negative, formation improvement has occurred.
If S is zero, no improvement or damage has occurred.
If S is positive, formation damage has occurred.

Other measures that are in use to express the degree of formation damage or improvement in the neighborhood of the well bore are:

The *condition ratio CR* (Ref. 19) or *productivity ratio PR* (Ref. 9), q/q_I
The *completion factor CF* (Ref. 20), $q/q_I \times 100$
The *damage factor DF* (Ref. 21), $1 - q/q_I$

where q is the production rate (at a fixed drawdown; see Sec. 3-2) of the actual well and q_I is the production rate (at the same drawdown) of an ideal well having no zone contamination.

It can be shown that the condition ratio and the skin factor are linked by the equation

$$\frac{q}{q_I} = \frac{p_{ws} - p_{wf} - (q\mu S/2\pi kh)}{p_{ws} - p_{wf}} \tag{1-12}$$

where the other symbols have their usual meanings (Ref. 16). It follows that a nonzero skin may be taken into account quantitatively in, for example, Eq. (1-10) by introducing the factor S:

$$\bar{p} = p_{wf} + \frac{qB_o\mu}{0.007082kh}\left[\ln\left(\frac{r_e}{r_w}\right) - \frac{3}{4} + S\right] \tag{1-13}$$

in oil field units.

It will be seen in Sec. 3-3 that a considerably higher free gas saturation may build up in the immediate vicinity of the well bore of a producing well than is present further back in the formation. This higher gas saturation results in a lower effective permeability to oil in the neighborhood of the well bore and may in many cases show up on the pressure buildup curve as formation damage, whereas it is in fact just a normal part of the well's history. For this reason among others, considerable care and experience are required when interpreting formation-damage parameters calculated from pressure buildup surveys.

1-6 WOR BEHAVIOR

Instantaneous WOR Formula

Consider a horizontal, homogeneous formation producing only oil and water (no free gas). Then the volume of oil crossing a unit cross-sectional area per unit time in the direction of decreasing pressure is, from Eq. (1-1),

$$q_o = \frac{k_o}{\mu_o}\frac{dp}{dl} \tag{1-14}$$

Similarly, the volume of water crossing a unit cross-sectional area per unit time in the direction of decreasing pressure is (neglecting the effect of capillary forces)

$$q_w = \frac{k_w}{\mu_w}\frac{dp}{dl} \tag{1-15}$$

Dividing Eq. (1-15) by Eq. (1-14) results in

$$\frac{q_w}{q_o} = \frac{k_w}{k_o}\frac{\mu_o}{\mu_w} \tag{1-16}$$

The expression on the left-hand side is the ratio of the rates at which water and oil, respectively, flow through the formation. But oil shrinks when it is produced (because of the gas released from solution) so the stock-tank oil rate will be q_o/B_o. Gas, on the other hand, has a low solubility in water and water has a low compressibility; hence, to a close degree of approximation, q_w may be taken as equal to the surface water rate. Thus, the WOR, measured at the surface, is

$$\frac{q_w}{q_o/B_o} = \frac{B_o q_w}{q_o}$$

or, from Eq. (1-16),

$$\text{(Surface) WOR} = B_o \frac{k_w}{k_o}\frac{\mu_o}{\mu_w} \tag{1-17}$$

$$= B_o \frac{k_{rw}}{k_{ro}}\frac{\mu_o}{\mu_w} \tag{1-18}$$

from Eq. (1-5). This is the *instantaneous WOR formula.*

Water Cut History

When oil and water are flowing through a formation in which no free gas is present, Eq. (1-17) shows that

$$\text{Produced WOR} = B_o \frac{\mu_o}{\mu_w}\frac{k_w}{k_o}$$

provided gravity and capillary-pressure terms are neglected. Moreover, if q_{oi} was the initial production rate of oil from a pay zone under a certain drawdown (Sec. 3-2), the production in these early stages having been water-free, and if q_o and q_w are the current production rates of oil and water, respectively, from the zone for the same value of the drawdown, then

$$\frac{q_o + q_w}{q_{oi}} = \frac{(1/B_o)(k_o/\mu_o)(dp/dl) + (k_w/\mu_w)(dp/dl)}{(1/B_{oi})(k/\mu_{oi})(dp/dl)}$$

$$= \frac{(k_o/\mu_o) + (k_w/\mu_w)}{k/\mu_{oi}} \tag{1-19}$$

the oil formation volume factor and the variation of oil viscosity with pressure being ignored. If the numerator and denominator of Eq. (1-19) are multiplied by μ_o, it follows that, as a first approximation,

$$\frac{\text{Current liquid production rate for zone}}{\text{Initial liquid production rate for zone}} = \frac{k_o + (\mu_o/\mu_w)k_w}{k} \tag{1-20}$$

No Water Influx

In this case the water saturation S_w will remain constant (at the interstitial, or connate, water saturation), and the produced WOR will either be zero (if S_w is less than the critical value) or remain roughly constant at some nonzero value (if S_w is greater than the critical value). There may be some variation in the WOR due to variations in B_o and μ_o with pressure, but this will probably not be marked in practice.

Water Influx

Suppose water is moving into the field via a permeable stringer. As production proceeds, the water saturation in the stringer gradually increases and the oil saturation decreases. That is, for the stringer (see Fig. 1-4), the ratio k_w/k_o increases; hence, from Eq. (1-17), the produced WOR increases for the stringer and also for the whole formation. The abruptness with which this increase occurs will depend on the magnitude of the stringer's gross production compared with that of the entire pay, on the shape of the relative-permeability curves, and on the value of μ_o/μ_w. The larger μ_o/μ_w (other things being equal), that is, the more viscous the oil, the more rapidly will the WOR build up in terms of the cumulative production of the well. It is, however, to be expected that the WOR will eventually flatten out and only increase slowly. The reason for this is that once the water has flooded the permeable stringer, it has established a "pipeline" into the well bore; from this time forward the encroaching water will travel by this path of least resistance rather than flood out other oil-saturated zones in the net pay. Evidently the value at which the WOR levels out will increase as the thickness of the flooded stringer increases, as the permeability of the stringer relative to the remainder of the pay increases, and also as the oil viscosity increases. In practice, the WOR may be expected to continue to rise slowly with increasing cumulative production owing to a gradual increase in percentage of pay that is watered out.

From the above analysis it appears that a typical curve of WOR against cumulative production from the well will be as shown in Fig. 1-17.

Finally, it should be noted that the gross productivity index (PI) (Sec. 3-2) of a wet well may increase with cumulative production, in contrast to the PI

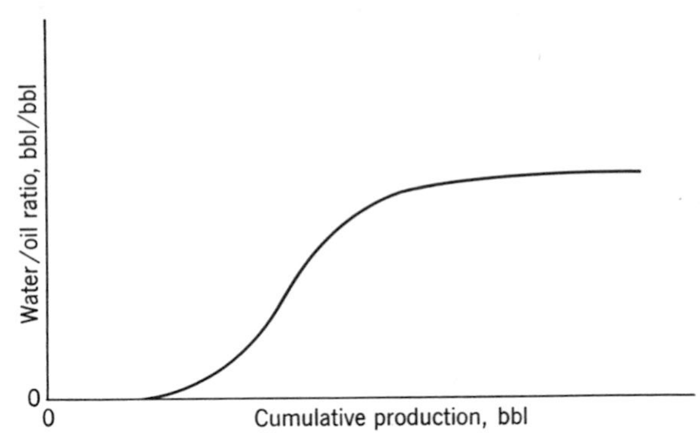

Fig. 1-17 Typical WOR behavior: encroaching edge water.

behavior of a well producing with a zero cut (Sec. 3-3). This result follows from Eq. (1-20), from which it is seen that the current gross liquid production rate of the permeable stringer will be greater than the initial gross liquid production rate (at the same drawdown) if

$$k_o + \frac{\mu_o}{\mu_w} k_w > k$$

that is, if

$$k_w > \frac{\mu_w}{\mu_o}(k - k_o) \tag{1-21}$$

This inequality may hold, provided that the ratio μ_w/μ_o is sufficiently small, that is, provided that the oil in question is sufficiently viscous. (It will be recalled from Eq. (1-4) that $k_w \leq k - k_o$, the equality holding only for the flow of a single liquid.)

1-7 GOR BEHAVIOR

Instantaneous GOR Formula

Consider a horizontal, homogeneous formation producing only oil and free gas (no water production, although connate water may be present in the formation; see Sec. 1-2). Then the volume of oil crossing a unit cross-sectional area per unit time in the direction of decreasing pressure is

$$q_o = \frac{k_o}{\mu_o}\frac{dp}{dl}$$

and that of gas is

$$q_g = \frac{k_g}{\mu_g}\frac{dp}{dl}$$

where the pressure drop dp across the distance dl is the same for both the oil and the gas if capillary forces are neglected. Dividing,

$$\frac{q_g}{q_o} = \frac{k_g}{k_o}\frac{\mu_o}{\mu_g} \tag{1-22}$$

This is the ratio of the rates at which gas and oil flow through the formation. The stock-tank oil rate will be q_o/B_o bbl, and the surface free-gas rate will be q_g/B_g cu ft.[8] However, in addition to the free gas produced from the formation, each barrel of stock-tank oil will release a volume R_s cu ft of gas when it is brought from formation to stock-tank conditions, so the total surface GOR (scf/bbl) is

$$R_s + \frac{q_g/B_g}{q_o/B_o} = R_s + \frac{B_o}{B_g}\frac{q_g}{q_o}$$

Thus from Eq. (1-22),

$$\text{(Surface) GOR} = R_s + \frac{B_o}{B_g}\frac{k_g}{k_o}\frac{\mu_o}{\mu_g} \qquad \text{scf/bbl} \tag{1-23}$$

$$= R_s + \frac{B_o}{B_g}\frac{k_{rg}}{k_{ro}}\frac{\mu_o}{\mu_g} \qquad \text{scf/bbl} \tag{1-24}$$

This is the *instantaneous GOR formula.*

GOR History: Depletion-Drive Field

As long as the reservoir pressure is still above the bubble point, there is no free gas in the reservoir, except possibly in the vicinity of the well bore. In the case of a well being produced with very high drawdown, this free gas close to the well bore may cause a considerable rise in producing GOR, as outlined in Sec. 3-3, and it may erroneously be concluded that free-gas movement is general throughout the formation. However, setting this possibility to one side, it appears that when the reservoir pressure is greater than the saturation pressure, the producing GOR is equal to R_{si}, the initial volume of gas in solution per unit volume of stock-tank oil.

When the reservoir pressure is below, but close to, the bubble point, there will be a low free-gas saturation in the formation. However, if this is less than the critical gas saturation, there is no overall movement of free gas; that is, the effective permeability to gas, k_g, is still zero. Equation (1-23) implies that the producing GOR, when this condition holds, is equal to the gas saturation at the current reservoir pressure, which is somewhat less than the initial gas saturation R_{si}. Thus the producing GOR falls steadily with increasing cumulative production until the free gas saturation in the formation attains the critical

[8] *The gas formation volume factor Bg* is defined as the volume (in barrels) occupied by 1 cu ft of gas (at standard conditions) when subjected to reservoir temperature and pressure.

value. (In practice, this effect is frequently masked by the flow of gas from the low-pressure region surrounding the well bore.)

When the formation pressure is dropped still further and the free-gas saturation builds up above the critical value, free gas will move in the formation. Since the gas formation volume factor is usually small and the viscosity ratio μ_o/μ_g in Eq. (1-23) is large, the term

$$\frac{B_o}{B_g}\frac{\mu_o}{\mu_g}\frac{k_g}{k_o}$$

will become important compared with R_s even when k_g is still very small. As the pressure decreases further, the producing GOR increases as a result of the increase in k_g/k_o (Fig. 1-7). However, if the reservoir pressure is decreased sufficiently, the GOR starts to drop again. This can be seen from the following considerations.

The gas formation volume factor B_g is defined as the volume (in bbl) occupied by 1 scf of gas when subjected to reservoir temperature and pressure, and in symbols by

$$\frac{pB_g}{ZT} = \frac{14.7}{520}\frac{1}{5.614}$$

or

$$B_g = \frac{14.7}{5.614 \times 520}\frac{ZT}{p} \tag{1-25}$$

where Z is the gas compressibility factor and T is the (absolute) temperature.

If the compressibility factor Z and the reservoir temperature T are assumed to remain constant, and if A is defined as below

$$\frac{14.7}{5.614 \times 520} ZT = A$$

A then being a constant, Eq. (1-23) becomes

$$\text{Instantaneous GOR} = R_s + \frac{p}{A} B_o \frac{\mu_o}{\mu_g}\frac{k_g}{k_o} \tag{1-26}$$

When p is, say, 2000 psia, a drop of 100 psi will have little effect on the value of p/A, and the increase in $B_o(\mu_o/\mu_g)(k_g/k_o)$ will dominate Eq. (1-26). When p is only 200 psia, a drop of 100 psi will halve the value of p/A; thus it is apparent that, dependent on the rate of change of $B_o(\mu_o/\mu_g)(k_g/k_o)$ with pressure, a stage will be reached at which the term p/A begins to control the right-hand side of Eq. (1-26) and the GOR starts to drop with increasing cumulative production.

If it is remembered that in the final stages of a well or field the production rates will be low, so that only a small increase in field cumulative may be obtained in a considerable length of time, the producing GOR history can be expected to be somewhat as shown in Figs. 1-18 and 1-19.

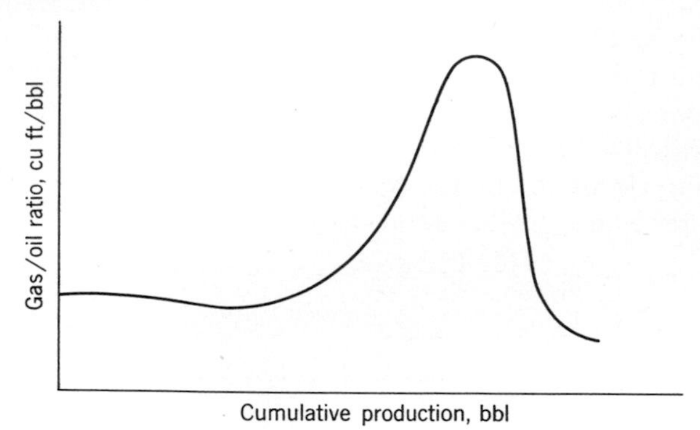

Fig. 1-18 Typical GOR behavior with cumulative production: depletion-drive field.

1-8 RESERVOIR-PERFORMANCE CURVES

As a working tool most reservoir engineers maintain and keep up-to-date reservoir-performance curves. Such curves are not made for individual wells as a rule; they are based on all the wells draining a particular pool, tract, or lease. The information, which is plotted against time, usually by months, generally includes most of the following at least (Fig. 1-20) (Ref. 22):

Pressure at a certain datum level, based on individual well surveys
Oil production rate, bbl/day or bbl/month
Cumulative oil production
GOR, scf/bbl
Cumulative gas production
Cumulative GOR (cumulative gas divided by cumulative oil)
WOR, bbl/bbl, or water cut (water rate divided by gross production rate)
Cumulative water produced
Number of productive wells

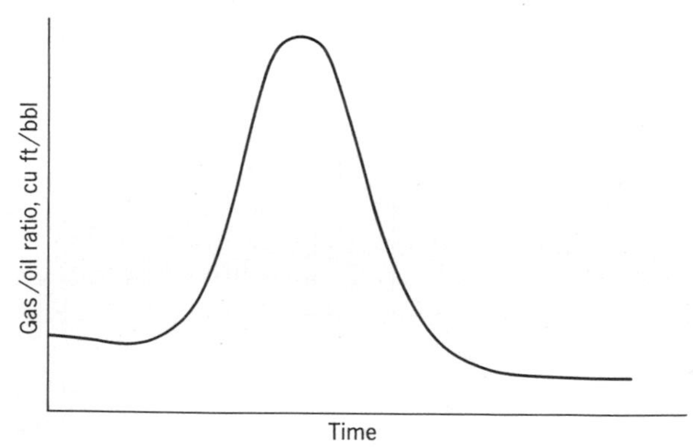

Fig. 1-19 Typical GOR behavior with time: depletion-drive field.

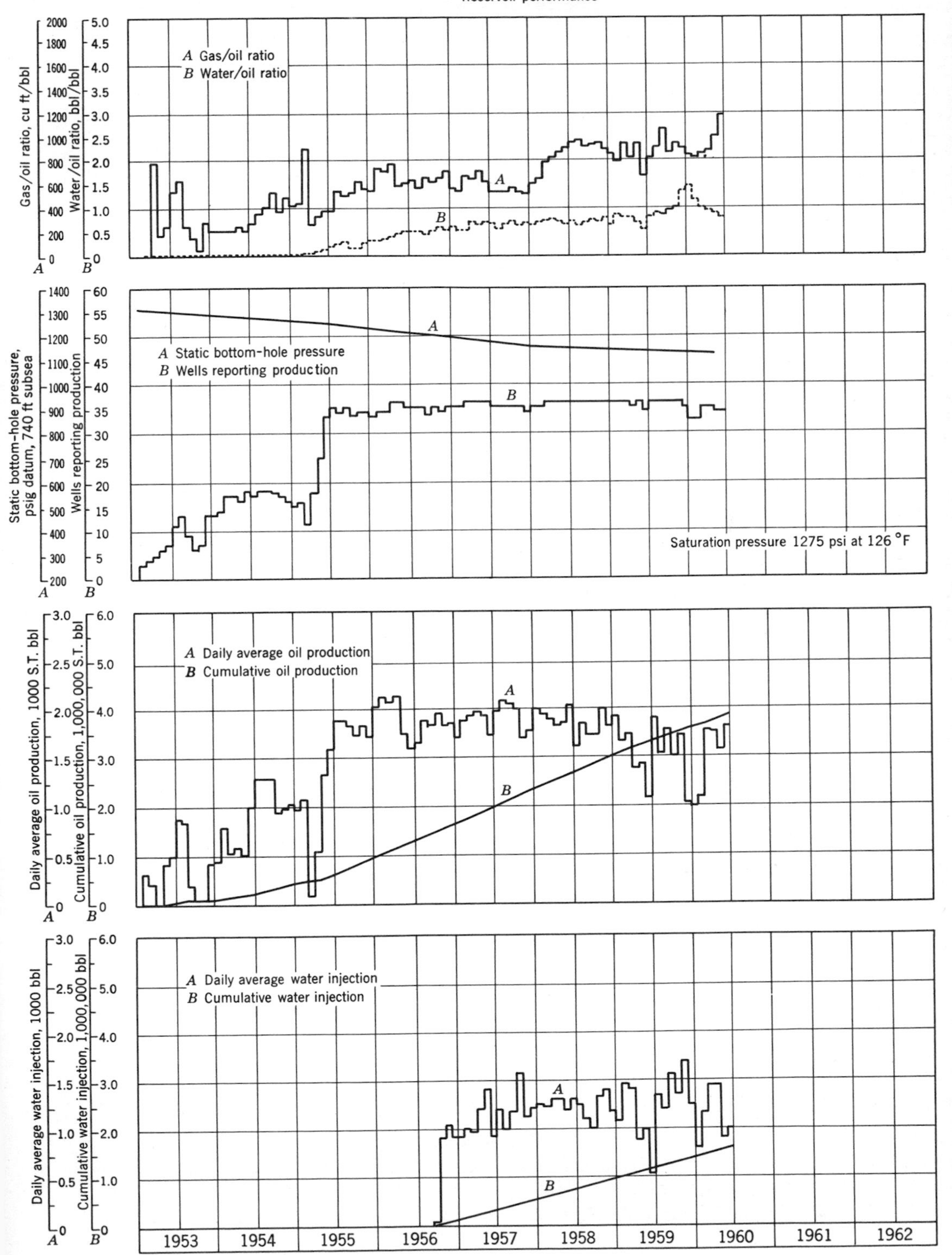

Fig. 1-20 Reservoir-performance curves, Success Roseray Sand Pool, Success Field, Saskatchewan. (*Courtesy Dept. Mineral Resources, Province of Saskatchewan.*)

These performance curves, and the reservoir engineer's analysis of them, can be of very great assistance to the production engineer, both from the point of view of estimating individual well pressures and also from the wider point of view of understanding and appreciating the behavior of the pool as a whole.

In addition to referring to records of past behavior, the production engineer will need to call on the reservoir engineer for predictions of field performance whenever attempts are made to analyze future well production characteristics and to plan changes in production techniques and equipment to match the future depletion behavior of the reservoir as a whole (see Sec. 5-6 for an example of such an exercise).

REFERENCES

1. Babson, E. C.: "The Range of Application of Gas-Lift Methods," *API Drill. Prod. Practice,* 1939, p. 266.
2. Darcy, H.: *Les Fontaines Publiques de la Ville de Dijon,* Victor Dalmont, Paris, 1856.
3. Katz, Donald L., et al.: *Handbook of Natural Gas Engineering,* McGraw-Hill Book Company, Inc., New York, 1959.
4. Wyckoff, R. D., and H. D. Botset: "Flow of Gas-Liquid Mixtures through Unconsolidated Sands," *Physics,* **1**(9):325 (1936).
5. Rapoport, L. A., and W. J. Leas: "Relative Permeability to Liquid in Liquid-Gas Systems," *Trans. AIME,* **192:**83 (1951).
6. Miller, F. G.: "Steady Flow of Two-Phase Single Component Fluids through Porous Media," *Trans. AIME,* **192:**205 (1951).
7. Leverett, M. C.: "Flow of Oil-Water Mixtures through Unconsolidated Sands," *Trans. AIME,* **132:**149 (1939).
8. Brownscombe, E. R., R. L. Slobad, and B. H. Caudle: "Laboratory Determination of Relative Permeability," *API Drill. Prod. Practice,* 1949, p. 302.
9. Pirson, Sylvain J.: *Oil Reservoir Engineering,* 2d ed., McGraw-Hill Book Company, Inc., New York, 1977.
10. Craft, B. C., and Murray F. Hawkins: *Applied Petroleum Reservoir Engineering,* Prentice-Hall, Inc., Englewood Cliffs, N.J., 1959.
11. Lozano, G., and W. A. Harthorn: "Field Test Confirms Accuracy of New Bottom-hole Pressure Gauge," *J. Petrol. Technol.,* **11**(2):26 (1959).
12. Kolb, R. H.: "Two Bottom-hole Pressure Instruments Providing Automatic Surface Recording," *J. Petrol. Technol.,* **12**(2):79 (1960).
13. Horner, D. R.: "Pressure Build-up in Wells," *Proc. Third World Petrol. Congr., The Hague,* Sec. II:503 (1951).
14. Arps, J. J., and A. E. Smith: "Practical Use of Bottom-hole Pressure Build-up Curves," *API Drill. Prod. Practice,* 1949, p. 155.
15. Miller, C. C., A. B. Dyes, and C. A. Hutchinson: "The Estimation of Permeability and Reservoir Pressure from Bottom-hole Pressure Build-up Characteristics," *Trans. AIME,* **189:**91 (1950).
16. Perrine, R. L.: "Analysis of Pressure Build-up Curves," *API Drill. Prod. Practice,* 1956, p. 482.

17. Benn, P. J., and S. Tosi: *Weyburn Field Study,* Department of Mineral Resources, Province of Saskatchewan, Regina, Saskatchewan, 1960.
18. Van Everdingen, A. F.: "The Skin Effect and Its Influence on the Productive Capacity of a Well," *Trans. AIME,* **198:**171 (1953).
19. Gladfelter, R. E., G. W. Tracy, and L. E. Wilsey: "Selecting Wells Which Will Respond to Production-Stimulation Treatment," *API Drill. Prod. Practice,* 1955, p. 117.
20. Arps, J. J.: "How Well Completion Damage Can Be Determined Graphically," *World Oil,* **140**(5):225 (1955).
21. Thomas, G. B.: "Analysis of Pressure Build-up Data," *Trans. AIME,* **198:**125 (1953).
22. *Reservoir Performance Charts, December 31, 1960,* Department of Mineral Resources, Province of Saskatchewan, Regina, Saskatchewan, 1961.

Production-Rate-Decline Curves

2

2-1 INTRODUCTION

Production-rate-decline curves are widely used throughout the producing side of the oil industry in assessing individual well and field performance and in forecasting future behavior. When estimates are based on the mathematical or graphical techniques of production-rate-decline curve analysis, it should always be remembered that this analysis is merely a convenience, a method that is amenable to mathematical or graphical treatment, and has no basis in the physical laws governing the flow of oil and gas through the formation. Such curves can be drawn for individual wells, for a group of wells within a pool, or for all the wells in a pool taken together; the curve of oil production rate on the reservoir-performance graph (Fig. 1-20) is a typical example.

It will be seen in Sec. 2-4 that forecasts dependent on production-rate-decline curves are not additive, in the sense that if the rates from two wells are separately assumed to be declining according to some mathematical extrapolation, then the rate from the two wells taken together is not in general declining in the same way. This result can lead to confusion in assessing future reserves or potential productivity if it is not recognized, and may be one of the causes of the markedly different conclusions that are sometimes drawn by differing agencies from the same basic data.

On the other hand production-rate-decline curves are easy to use, are generally kept up to date in some manner in the field office (at least), give quick and reliable information on expectations over the next few months, and show graphically which wells (or groups of wells) are producing below expectation so that work-overs and rehabilitation programs may be planned (Ref. 1).

2-2 EXPONENTIAL DECLINE

It will be assumed in what follows that one well only is being considered, but the analysis applies equally to a group of wells considered as a unit.

Because the obvious way to plot production rate is against time, this was the first method used. It was found that, following a period during which the production rate was steady (at or near the well's allowable, or the market demand), there came a time when the well could no longer make its allowable, and the production rate dropped off fairly regularly, or *declined,* month by month. A typical plot of production rate against time is shown in Fig. 2-1, in which an average curve has been shown by means of the dashed line. Evidently, if some regular (mathematical) form can be given to the curved part of the dashed line, it will be possible to extrapolate this into the future and so predict what the well will be producing in, for instance, 1, 2, 5, or 10 years' time. If the data are plotted as production rate versus cumulative oil production, it is found that the declining part of the curve becomes a reasonably straight line, which is, of course, easy to extrapolate (Fig. 2-2). If q is the production rate and Q is the cumulative production, the equation of this straight line may be written

$$q = mQ + c \tag{2-1}$$

where m and c are constants. If the production rate q is maintained for a short time δt, the cumulative production in that time is $q\ \delta t$; hence, the cumulative production is the sum of the products $q\ \delta t$ from the start of production (time zero, say) to the present day (time t, say). In mathematical terms

$$Q = \int_0^t q\ dt \tag{2-2}$$

or

$$q = \frac{dQ}{dt} \tag{2-3}$$

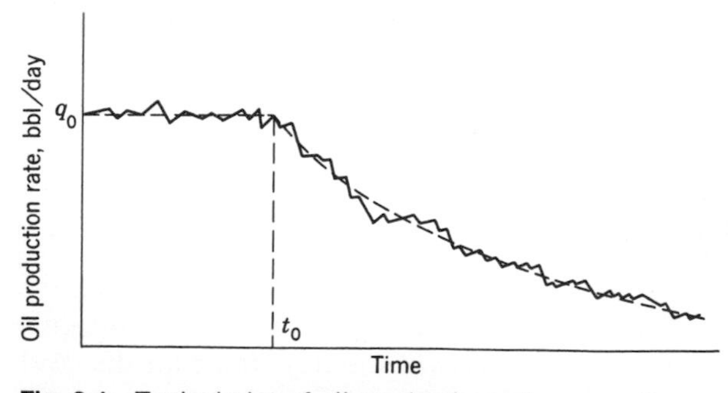

Fig. 2-1 Typical plot of oil-production rate versus time.

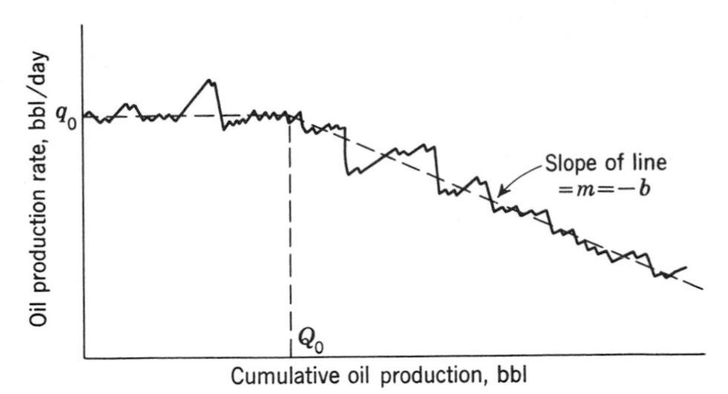

Fig. 2-2 Typical plot of oil-production rate versus cumulative production.

If Eq. (2-1) is differentiated with respect to t,

$$\frac{dq}{dt} = m\frac{dQ}{dt}$$

so that from Eq. (2-3)

$$\frac{dq}{dt} = mq$$

or

$$\frac{1}{q}\frac{dq}{dt} = m \tag{2-4}$$

From Fig. 2-2 it is evident that the slope of the line obtained in the production decline period is negative, and m may be written as $-b$, where b is positive.

Substitution in Eq. (2-4) yields

$$\frac{1}{q}\frac{dq}{dt} = -b \tag{2-5}$$

The positive constant b is called the *continuous* (or *nominal*) *production decline rate.*

In Eq. (2-1)

$$q = -bQ + c \tag{2-6}$$

If the production decline commences when the well's cumulative production is Q_0 (Fig. 2-2) and if the steady production rate prior to that time was q_0,

$$q_0 = -bQ_0 + c$$

or

$$c = q_0 + bQ_0$$

Substituting in Eq. (2-6) and rearranging gives

$$Q - Q_0 = \frac{q_0 - q}{b} \tag{2-7}$$

or, in words, the *cumulative production during the decline period is equal to the difference between the initial and the current production rates divided by the continuous decline rate.*

From Eq. (2-5)

$$\frac{dq}{q} = -b\,dt$$

or, on integration,

$$\ln q = -bt + a \tag{2-8}$$

where a is a constant.

If the decline period commences at time t_0 (Fig. 2-1) and if the steady production rate prior to that time was q_0,

$$\ln q_0 = -bt_0 + a$$

so that

$$a = bt_0 + \ln q_0$$

Substitution in Eq. (2-8) gives

$$\ln q = \ln q_0 - b(t - t_0) \tag{2-9}$$

or

$$q = q_0 \exp\,[-b(t - t_0)] \tag{2-10}$$

Equation (2-9) shows that, for this type of production rate decline, a *plot of production rate versus time on semilogarithmic paper is a straight line, the slope of the line being equal to minus the continuous decline rate* (Fig. 2-3).

Equation (2-10) enables the production rate at any instant to be found once the initial production rate q_0 is known. Suppose for simplicity that the production rate decline commences as soon as the well is put on production, so that t_0 is zero. Equation (2-10) reduces to

$$q = q_0 \exp\,(-bt)$$

The production rate after 1 year will be given by

$$q_1 = q_0 \exp\,(-b)$$

The production rate after 2 years will be given by

$$\begin{aligned} q_2 &= q_0 \exp\,(-2b) \\ &= q_0 \exp\,(-b) \exp\,(-b) \\ &= q_1 \exp\,(-b) \end{aligned}$$

Thus,

$$\frac{q_1}{q_0} = \frac{q_2}{q_1} = \frac{q_3}{q_2} = \ldots = \exp\,(-b) \tag{2-11}$$

which implies that *the ratio of the production rate at the end of any year to that at the beginning of the same year is always the same.* This ratio is frequently written as $1 - d$, and d is called the *annual production decline rate* (it may be expressed as a decimal or as a percentage). Evidently the equation connecting the annual and the continuous decline rates is

$$\exp\,(-b) = 1 - d \tag{2-12}$$

This type of production rate decline, a few of the properties of which have been proved above, is known by various names: *logarithmic* decline [from Eq. (2-9) and Fig. 2-3]; *exponential* decline [from Eq. (2-10) and Fig. 2-1];

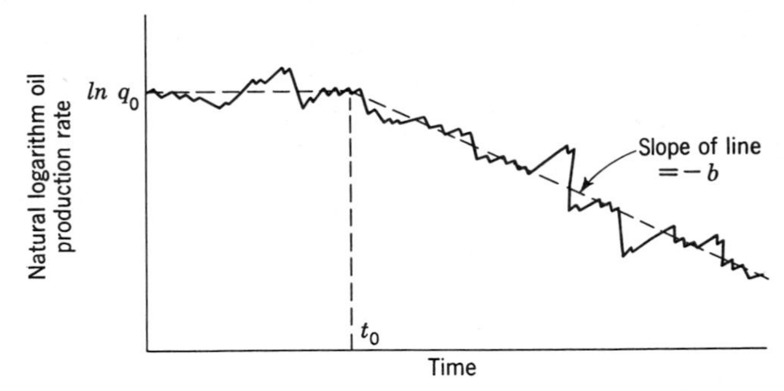

Fig. 2-3 Typical plot of natural logarithm of oil-production rate versus time.

straight decline (from Fig. 2-2 or Fig. 2-3); *constant-rate* decline (from the fact that the decline rate b or d is a constant); or *proportional* decline [from Eq. (2-11)].

Before the use of the equations is illustrated by means of an example, it should be noted that in the analysis leading to Eq. (2-11), 1 year was used as the interval of time. This, of course, is not a limitation on the method. It would have been valid to have employed a day, a week, a month, or a century as the unit of time. It is important, however, in calculations to be consistent in the time units. If a year is chosen, then production rates must be expressed as annual rates, that is, as daily rates × 365; if a month is chosen, then production rates must be expressed as monthly rates, that is, as daily rates × 30.42; and so on. Further, it is worth noting the connection between annual and monthly decline rates d and between annual and monthly continuous decline rates b. If d_m is the monthly decline rate, then, from Eq. (2-11), the production rate at the end of the first month is $q_0(1 - d_m)$; at the end of the second month it is $q_1(1 - d_m)$, which equals $q_0(1 - d_m)(1 - d_m)$, or $q_0(1 - d_m)^2$, and so on. Thus at the end of 12 months the production rate is $q_0(1 - d_m)^{12}$. But at the end of 12 months, the production rate is $q_0(1 - d_a)$, where d_a is the annual decline rate, so

$$1 - d_a = (1 - d_m)^{12} \tag{2-13}$$

Similarly, if b_m is the monthly and b_a the annual continuous decline rate

$$\begin{aligned} \exp(-b_a) &= [\exp(-b_m)]^{12} \\ &= \exp(-12b_m) \end{aligned}$$

so that

$$b_a = 12b_m \tag{2-14}$$

Example 2-1 A well that came in at 100 bbl/day has declined to 80 bbl/day at the end of the first year. Calculate the yearly and monthly decline rates and the yearly and monthly continuous decline rates. If the economic limit of the well is 2 bbl/day, calculate the life of the well and its cumulative production.

Yearly and Monthly Decline Rates

By definition,

$$80 = 100(1 - d_a)$$

so

$$d_a = 0.2$$
$$= 20 \text{ percent/year}$$

From Eq. (2-13)

$$(1 - d_m)^{12} = 1 - 0.2 = 0.8$$

which gives

$$d_m = 0.0184$$
$$= 1.84 \text{ percent/month}$$

Yearly and Monthly Continuous Decline Rates
From Eq. (2-12)

$$\exp(-b_a) = 1 - d_a$$

so

$$b_a = 0.223$$

From Eq. (2-14)

$$b_m = \frac{0.223}{12} = 0.0186$$

Life of the Well

a. Using 1 year as the time unit.
From Eq. (2-10) with

$$t_0 = 0$$
$$q_0 = 100 \times 365$$
$$q = 2 \times 365$$
$$b = 0.223$$

it follows that

$$2 \times 365 = 100 \times 365 \times \exp(-0.223T)$$

where T = life of the well.
This gives T = 17.5 years.

b. Using 1 month as the time unit.
In this case,

$$t_0 = 0$$
$$q_0 = 100 \times 30.42$$
$$q = 2 \times 30.42$$
$$b = 0.0186$$

and Eq. (2-10) becomes

$$2 \times 30.42 = 100 \times 30.42 \times \exp(-0.0186T)$$

so

$$T = 210.0 \text{ months}$$
$$= 17.5 \text{ years}$$

Cumulative Production

a. Using 1 year as the time unit.
From Eq. (2-7) with

$$
\begin{aligned}
Q_0 &= 0 \\
q_0 &= 100 \times 365 \\
q &= 2 \times 365 \\
b &= 0.223
\end{aligned}
$$

it follows that

$$Q = 160{,}000 \text{ bbl}$$

b. Using 1 month as the time unit.
From Eq. (2-7) with

$$
\begin{aligned}
Q_0 &= 0 \\
q_0 &= 100 \times 30.42 \\
q &= 2 \times 30.42 \\
b &= 0.0186
\end{aligned}
$$

it follows that

$$Q = 160{,}000 \text{ bbl}$$

A question that frequently arises is the effect that an increase in production rate might be expected to have on the decline rate. A formal answer to this question may be given on the assumption that there is no change in the future cumulative production resulting from the alteration in production rate. In carrying out the calculation it is sometimes postulated that there will be no change in the cumulative production down to some assigned economic limit. This assumption seems to introduce yet another note of unreality into what is already a rather unrealistic calculation because the well cannot react to what we may regard as some (financial) limitation. Moreover, the introduction of such a limit adds an unnecessary complication to the algebra, as well as giving to it an added—and spurious—air of authenticity.

If it is assumed, then, that the ultimate cumulative production is unaltered and that the current production rate of q_0 changes to $q_0^{(a)}$ while the current decline rate b changes to $b^{(a)}$, Eq. (2-7) gives

$$\frac{q_0}{b} = \frac{q_0^{(a)}}{b^{(a)}}$$

or

$$b^{(a)} = \frac{q_0^{(a)}}{q_0} b \tag{2-15}$$

that is, the original continuous decline rate is multiplied by the ratio of the new to the original production rate.

In order to determine the (economic) life of the well under the new conditions it is necessary to introduce the economic production rate limit q_e; this will be assumed to be the same for both the original and the accelerated project. If N is the future life of the original project,

$$q_e = q_0 \exp(-bN)$$

from Eq. (2-10).

If $N^{(a)}$ is the future life of the accelerated project,

$$q_e = q_0^{(a)} \exp\left[-b^{(a)}N^{(a)}\right]$$

It follows that the future life of the accelerated project is given by either of the equations

$$\exp\left[-b^{(a)}N^{(a)}\right] = \frac{q_0}{q_0^{(a)}} \exp(-bN)$$

or

$$\exp\left[-b^{(a)}N^{(a)}\right] = \frac{b}{b^{(a)}} \exp(-bN) \tag{2-16}$$

2-3 HYPERBOLIC AND HARMONIC DECLINES

It has been found in many of the older producing fields that an assumed exponential production rate decline early in a well's life has led to conservative answers for the ultimate life of the well and the cumulative recovery. One way of overcoming this is to assume that the decline rate (d or b), instead of being constant, is proportional to the production rate; hence, the smaller the production rate, the smaller the decline rate. In symbols, this assumption implies replacing Eq. (2-5)

$$\frac{1}{q}\frac{dq}{dt} = -b$$

by the equation

$$\frac{1}{q}\frac{dq}{dt} = -C^k q^k \tag{2-17}$$

where C and k are positive constants. Decline curves based on this equation are known as *hyperbolic,* and the constant $a = 1/k$ is called the *hyperbolic constant* (Refs. 1, 2).

Integrating Eq. (2-17) and using the initial condition that

$$q = q_0 \text{ when } t = 0$$

gives

$$q^{-k} = kC^k\, t + q_0^{-k}$$

or

$$\frac{q_0^k}{q^k} = kC^k q_0^k\, t + 1 \tag{2-18}$$

From Eq. (2-17), the initial value of the decline rate is $C^k q_0^k$, which may be written as b_0. Substituting this in Eq. (2-18) gives

$$\frac{q_0^k}{q^k} = kb_0t + 1$$

or

$$q = \frac{q_0}{(1 + kb_0t)^{1/k}} \tag{2-19}$$

Introducing the hyperbolic constant $a = 1/k$ gives the final expression for the production rate at time t, namely

$$q = \frac{q_0}{\left(1 + \frac{b_0 t}{a}\right)^a} \tag{2-20}$$

The value b of the decline rate at time t follows from Eqs. (2-17) and (2-19):

$$\begin{aligned} b &= -\frac{1}{q}\frac{dq}{dt} \\ &= C^k q^k \\ &= \frac{C^k q_0^k}{(1 + kb_0 t)} \\ &= \frac{b_0}{1 + kb_0 t} \end{aligned}$$

so, finally,

$$b = \frac{b_0}{1 + \frac{b_0 t}{a}} \tag{2-21}$$

The special case of $a = 1$ is known as *harmonic decline.* From Eqs. (2-20) and (2-21) the results for harmonic decline are

$$q = \frac{q_0}{1 + b_0 t} \tag{2-22}$$

$$b = \frac{b_0}{1 + b_0 t} \tag{2-23}$$

from which it follows that

$$\frac{q}{q_0} = \frac{b}{b_0} \tag{2-24}$$

To obtain the cumulative production when $a \neq 1$, Eq. (2-20) gives

$$\begin{aligned} Q &= \int_0^t \frac{q_0 dt}{\left(1 + \frac{b_0 t}{a}\right)^a} \\ &= \frac{a}{a-1}\frac{1}{b_0}\left[q_0 - q\left(1 + \frac{b_0 t}{a}\right)\right] \end{aligned} \tag{2-25}$$

Under harmonic decline, the cumulative production is given by

$$Q = \int_0^t \frac{q_0\, dt}{1 + b_0 t}$$

$$Q = \frac{q_0}{b_0} \ln (1 + b_0 t) \tag{2-26}$$

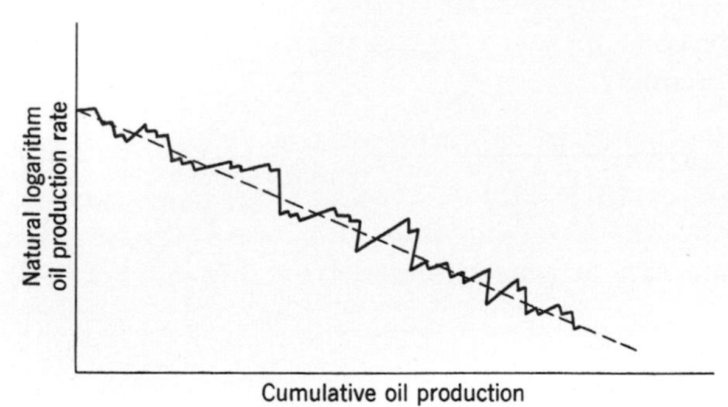

Fig. 2-4 Typical harmonic plot decline.

$$Q = \frac{q_0}{b_0} \ln \left(\frac{q_0}{q}\right) \qquad \text{from Eq. (2-22)} \qquad (2\text{-}27)$$

$$Q = \frac{q_0}{b_0} \ln \left(\frac{b_0}{b}\right) \qquad \text{from Eq. (2-23)} \qquad (2\text{-}28)$$

Finally, Eq. (2-27) may be written in the form

$$\ln q = \ln q_0 - \frac{b_0 Q}{q_0} \qquad (2\text{-}29)$$

It is important to remember, in applying Equations (2-26) through (2-29), that b_0 and b are *instantaneous* decline rates.

Equation (2-29) is the basis for a straight-line plot of oil production rate against cumulative production on semilogarithmic paper, Fig. 2-4, a type of plot that is frequently carried in field offices.

In general, the hyperbolic constant a is given one of the three values, 1, 2, or 3, and it is worth recording that the harmonic decline is the most optimistic of these. Figures 2-5 through 2-8 gives an impression of the relative effects of using an exponential decline or one or another type of hyperbolic decline.

Example 2-2 The production history of the Rotting Horse pool is as follows:

Years	Production Rate, bbl/day
0	5000
1	3730
2	2940
3	2350
4	1900
5	1590
6	1320
7	1160

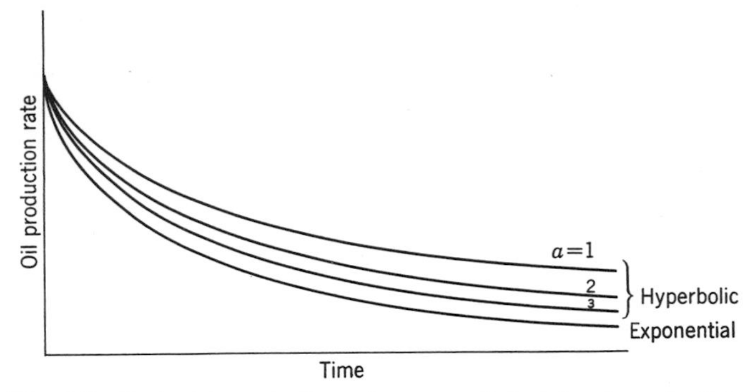

Fig. 2-5 Typical hyperbolic and exponential plots of production rate versus time.

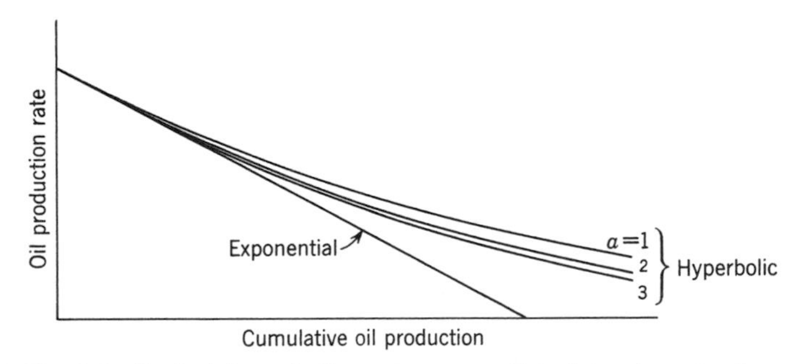

Fig. 2-6 Typical hyperbolic and exponential plots of production rate versus cumulative production.

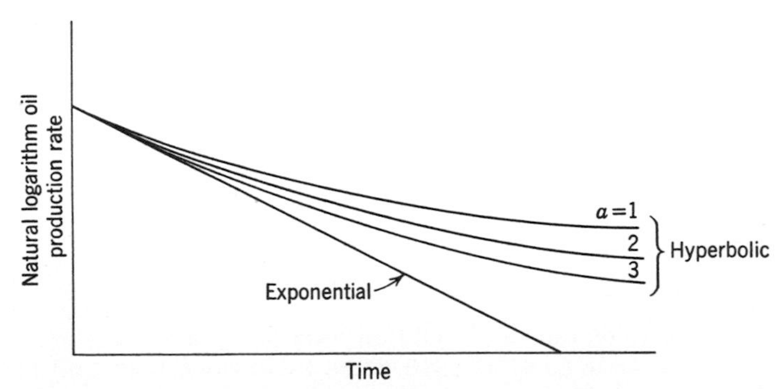

Fig. 2-7 Typical hyperbolic and exponential plots of natural logarithm of production rate versus time.

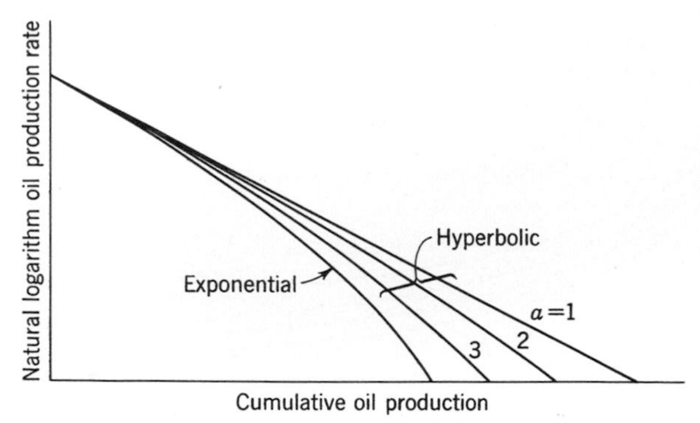

Fig. 2-8 Typical hyperbolic and exponential plots of natural logarithm of production rate versus cumulative production.

If the economic limit of the pool is 200 bbl/day, when will this be reached, and what will be the ultimate production?

First, determine the annual decline rates, which will be taken as the decline rates at the midpoints of each of the years; for example, in the first year

$$3730 = 5000(1 - d_{1/2})$$
$$d_{1/2} = 0.254$$
$$b_{1/2} = 0.293$$

Corresponding values for $b_{3/2}$, $b_{5/2}$, and so forth work out to be 0.238, 0.225, 0.212, 0.178, 0.186, and 0.129.

It is apparent that b is not constant, but declines with time. From Eq. (2-21)

$$\frac{1}{b} = \frac{1}{b_0} + \frac{t}{a}$$

and Fig. 2-9 shows a plot of the reciprocals of the b values against t. Superimposed on this plot are lines of slopes 1, $^1/_2$, and $^1/_3$, corresponding to values of the hyperbolic constant of 1, 2, and 3, respectively. It is clear that a good fit is obtained by using $a = 2$; the corresponding value for $1/b_0$ is 3.17, so that b_0 is 0.315.

Using these values for a and b_0 in Eq. (2-20) gives

$$q = \frac{5000}{(1 + 0.1575t)^2}$$

Production rates calculated from this expression at $t = 1, 2, \ldots, 7$ are generally low when compared with the actual field data.

In an attempt to correct this, a value of $b_0 = 0.31$ is tried, leading to

$$q = \frac{5000}{(1 + 0.155t)^2}$$

Production rates at the ends of successive years from the first through the seventh work out to be 3750, 2910, 2330, 1910, 1590, 1340, and 1150 bbl/day, which are in close agreement with the measured values.

The time to economic limit is calculated from Eq. (2-20) in the form

$$200 = \frac{5000}{(1 + 0.155t)^2}$$

which gives $t = 25.8$ years

The cumulative production at economic limit is obtained from Eq. (2-25) in the form

$$Q = \frac{2}{2-1} \times \frac{1}{0.31} [5000 - 200 \times 5] \times 365$$

since $1 + \frac{b_0 t}{a} = 1 + 0.155t = \sqrt{\frac{5000}{200}} = 5$

Thus $Q = 9{,}420{,}000$ bbl

2-4 IN CONCLUSION—A WARNING

It must be reiterated that the mathematical production-rate-decline curves (exponential, harmonic, or hyperbolic) are conveniences, enabling extrapolations of future well or field performances to be made. There is, however, no physical basis for these curves, and the production engineer must not feel surprised if the wells or pools do not follow the estimated production-rate-decline curves, no matter how carefully these may have been prepared. To illustrate the arbitrary nature of these curves, it will be shown that if two wells, A and B, are each declining exponentially, the sum of their production rates is

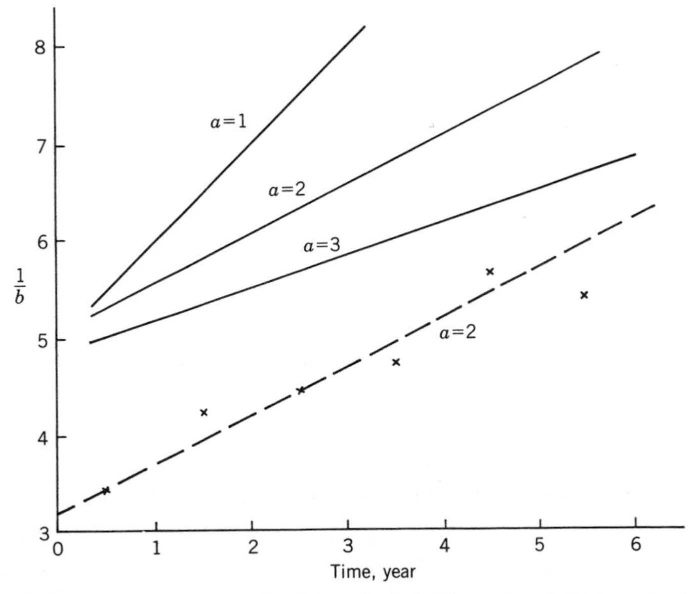

Fig. 2-9 Example 2-2: Reciprocal of decline rate plotted against time.

not, in general, declining exponentially (the same difficulty applies to harmonic and to hyperbolic declines).

Suppose the production rate from well A is declining exponentially with a continuous decline rate b_A. Let the initial production rate of well A be q_{A0}. Then, by Eq. (2-10) with t_0 equal to zero, the production rate at time t is

$$q_A = q_{A0} \exp(-b_A t)$$

Similarly, the production rate of well B (which will be assumed to have come into production at the same time as well A, that is, at time zero) at time t is

$$q_B = q_{B0} \exp(-b_B t)$$

Thus, the combined production rate is

$$(q_A + q_B) = q_{A0} \exp(-b_A t) + q_{B0} \exp(-b_B t)$$

which cannot be written in the form

$$(q_{A0} + q_{B0}) \exp(-bt)$$

unless $b_A = b_B$. So, in general, even though each well is declining exponentially, the two wells considered together cannot be declining exponentially. It follows that any analysis which presupposes exponential (or harmonic, or hyperbolic) declines for wells taken separately, and for those wells treated in groups, is certain to be in error, and estimates based on this type of work will always be incorrect. So, while production-rate-decline curves may be, and are, used, they should always be used with considerable care and circumspection.

REFERENCES

1. Slider, H. C.: *Practical Petroleum Reservoir Engineering Methods,* Petroleum Publishing Company, Tulsa, Oklahoma, 1976.
2. Campbell, John M.: *Oil Property Evaluation,* Prentice-Hall, Inc., Englewood Cliffs, N.J., 1959.

The Performance of Productive Formations

3

3-1 INTRODUCTION

In this chapter some of the factors governing the flow of fluids from the formation to the well bore are discussed, and an attempt is made to indicate in what ways these factors may affect the production history of the well. The analysis is based on two characteristics of formation behavior, namely, the formation IPR (Sec. 3-2) and the relative-permeability curves (Sec. 1-2). Although these characteristics are in some ways linked (see, for example, Sec. 3-3), nevertheless the variation in pressure over the region of the formation drained by a particular well sometimes necessitates their use as independent parameters in order to explain observed facts concerning the well's performance.

It is important to recognize that for an individual well there are two variables, the effects of which have to be studied separately: these are the gross production rate at a particular stage in the well's history and the cumulative production taken from the well.

3-2 PRODUCTIVITY INDEX AND IPR

The producing pressure p_{wf} at the bottom of the well is known as the *flowing BHP,* and the difference between this and the well's static pressure p_s is the *drawdown.* In symbols,

$$\text{Drawdown} = p_s - p_{wf} \tag{3-1}$$

The ratio of the producing rate of a well to its drawdown at that particular rate is called the *productivity index* (*PI*), denoted by J; if the rate q is given in bbl/day of stock-tank liquid and the drawdown in psi, the PI is defined as

$$J = \frac{q}{p_s - p_{wf}} \quad \text{bbl/(day) (psi)} \tag{3-2}$$

Unless otherwise specified, the PI is based on the *gross* liquid rate (oil rate plus water rate).

The *specific PI,* denoted by J_s, is the number of barrels (gross) of stock-tank liquid produced per day per psi drawdown per foot of net pay thickness, or in symbols,

$$J_s = \frac{J}{h} = \frac{q}{h(p_s - p_{wf})} \quad \text{bbl/(day)(psi)(ft)} \tag{3-3}$$

where h is the net pay thickness in feet.

From Eq. (1-8), it is clear that for radial flow of a single homogeneous liquid of small compressibility contained in a uniform horizontal reservoir

$$J = \frac{q}{p_s - p_{wf}} = \frac{0.007082kh}{B_o\mu \ln (r_e/r_w)} \quad \text{bbl/(day)(psi)} \tag{3-4}$$

and

$$J_s = \frac{J}{h} = \frac{0.007082k}{B_o\mu \ln (r_e/r_w)} \quad \text{bbl/(day)(psi)(ft)} \tag{3-5}$$

Example 3-1 Using the figures of Example 1-1,

$$J = \frac{0.007082}{1.25}\,\frac{50 \times 20}{3 \times 8.06} = 0.234 \text{ bbl/(day)(psi)}$$

$$J_s = \frac{0.234}{20} = 0.0117 \text{ bbl/(day)(psi)(ft)}$$

If the value of a well's PI is taken to be a constant, independent of the well's current production rate, then writing Eq. (3-2) in the form

$$q = J\,\Delta p \tag{3-6}$$

where Δp is the drawdown, makes it evident that the relationship between q and Δp is a straight line through the origin of slope J (Fig. 3-1).

Alternatively Eq. (3-2) may be written

$$p_{wf} = p_s - \frac{q}{J} \tag{3-7}$$

At a particular instant in a pool's life, p_s has a specific value, so if J is also constant, the plot of p_{wf} against q is a straight line (Fig. 3-2). The angle θ that this line makes with the pressure axis is such that

$$\tan \theta = \frac{OB}{OA} = J$$

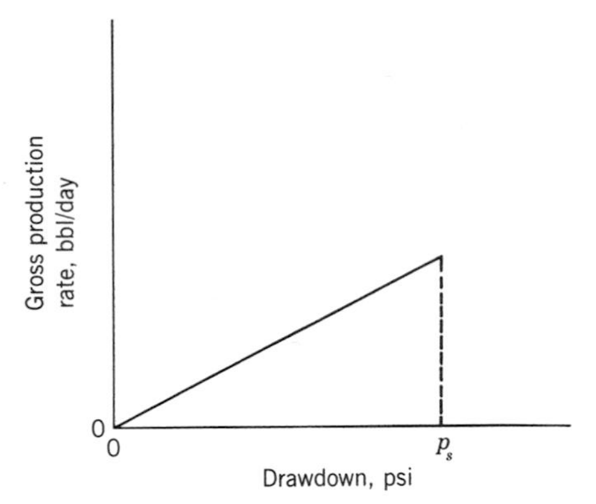

Fig. 3-1 Plot of production rate versus drawdown: constant productivity index.

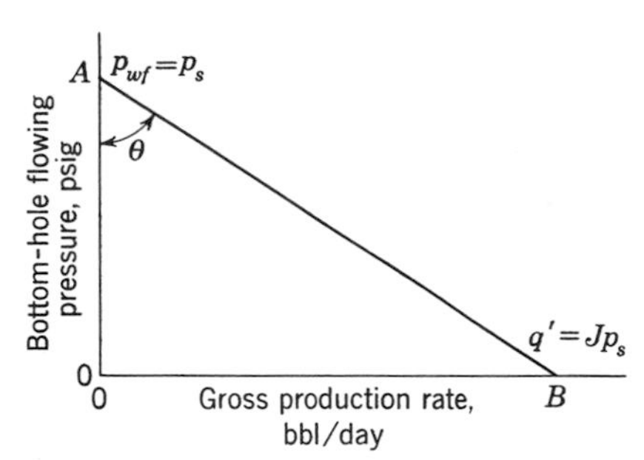

Fig. 3-2 Graphical representation of PI.

The value of q at the point B, namely Jp_s, is called the well's *potential* and will be denoted by the symbol q' throughout this book. It should be stressed, however, that Fig. 3-2 refers to *formation* behavior, that is, to the reaction of the formation to a pressure drawdown at the well bore, so that what is here referred to as the well's potential is in reality the formation's potential: it is the maximum rate at which the formation can deliver liquid into the well bore, and it occurs when the flowing BHP is zero (that is, atmospheric; in those wells to which vacuum is being applied, the well's potential would be defined at a flowing BHP of zero absolute). Just how great a percentage of a well's potential rate can actually be produced into the surface tanks is another question, which will be discussed in subsequent chapters.

In preparing Figs. 3-1 and 3-2 it was assumed that the PI was independent of the production rate, which resulted in the formation's producing possibilities all lying on a straight line; such a result would be in agreement with the radial-flow equation, Eq. (3-4). More generally (Sec. 3-3), the line may be expected to be curved. Retaining the definition of Eq. (3-2) for the PI, the direction of curvature is usually such that the value of J decreases with increasing values of Δp, or of q (Figs. 3-3 and 3-4). In order to cover such cases Gilbert (Ref. 1) used the term *inflow performance relationship* (IPR) to describe the curve of the flowing BHP plotted against the gross production rate (that is, Fig. 3-4) for any particular well.

To illustrate the importance of knowing the IPR of a well, two examples will be discussed.

Consider first the case of a well owned by a private individual (perhaps the reader). Suppose the local regulatory body—conservation board or other government agency—has granted an allowable production of 50 bbl/day clean oil to the well (to simplify the argument, it will be assumed that the well produces with zero water cut). In order to make the allowable, a pump has been installed in the well; for the first few years of its life the well has pro-

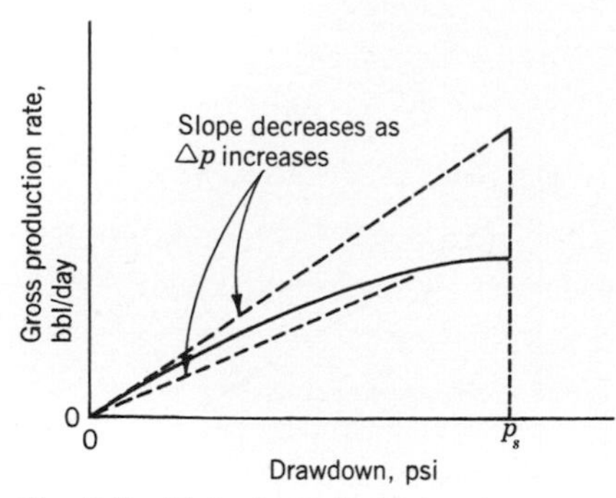

Fig. 3-3 Plot of production rate versus drawdown: general case.

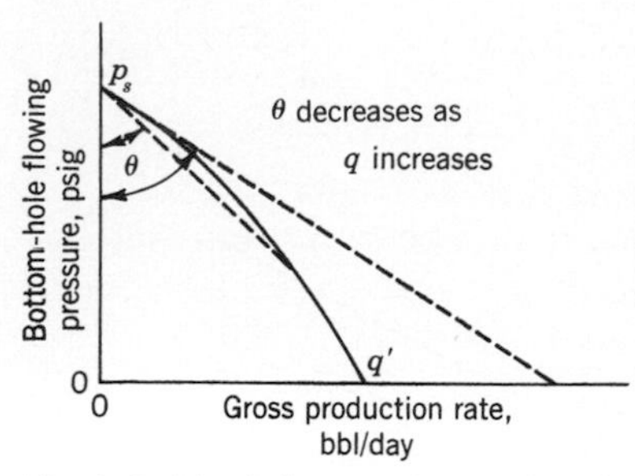

Fig. 3-4 The inflow-performance relationship.

duced 50 bbl/day without difficulty. However, recently production has been less than the allowable. One of two things has happened: either the formation is no longer capable of producing 50 bbl/day, or there is some mechanical defect in the well's equipment resulting in a low lifting efficiency (from the bottom of the well to the surface). It is a fairly costly operation to pull a pump and replace it, and certainly it is not one that would be undertaken without some guarantee that as a result of the work and the expense the well would once again produce 50 bbl/day.

So the first step in a case like this is to determine the well's IPR, if it can be done relatively easily and cheaply. The result might be either as shown in Fig. 3-5 or as shown in Fig. 3-6.

If the IPR were as illustrated in Fig. 3-5, the well's owner could be certain that no amount of pump changing would result in a production rate of 50 bbl/day and would either have to become reconciled to a below-allowable

Fig. 3-5 IPR showing formation incapable of desired production rate.

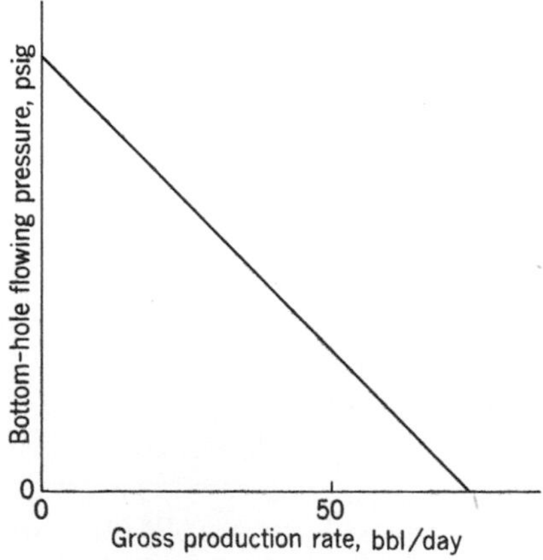

Fig. 3-6 IPR showing formation capable of desired production rate.

rate or else undertake a formation-stimulation work-over such as a fracturing or an acidizing job. If, on the other hand, the IPR were as illustrated in Fig. 3-6, the owner would be reasonably sure that a mechanical work-over of the equipment in the well would restore production to its allowable rate.

As a second example of the importance of knowing the IPR, suppose that a company has been carrying out a formation-stimulation program on some of its wells and that to gauge the success of this program, "before" and "after" production-rate figures are used. Let the results on two wells (both cutting zero water) be as follows:

	Steady Production Rate, bbl/day	
Well	**Before Treatment**	**1 week after Treatment**
A	60, flowing	100, flowing
B	35, pumping	36, pumping

The treatment would probably be accounted successful on well A and unsuccessful on well B. But while this may in fact be true, insufficient evidence has been presented to warrant such a conclusion; the before and after IPRs of the wells might be as illustrated in Figs. 3-7 and 3-8.

The treatment has had no effect at all on the IPR of well A; that is, the formation inflow performance has not been improved in any way, so the treatment was completely unsuccessful. The production increase from 60 to 100 bbl/day was fortuitous and might have been caused by the treatment dislodging some tubing obstruction, by a different-sized tubing having been run into the hole after the job, or by a different choke having been inserted in the flow line at the surface, to name a few of the possibilities.

On the other hand, the treatment on well B has increased the formation's potential considerably and was an undoubted success. Why then were the before and after rates almost identical? Once again, there are several possible

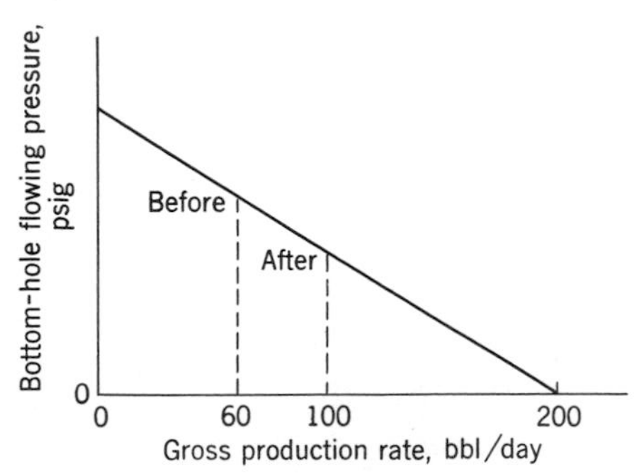

Fig. 3-7 Formation stimulation a failure despite increased production rate.

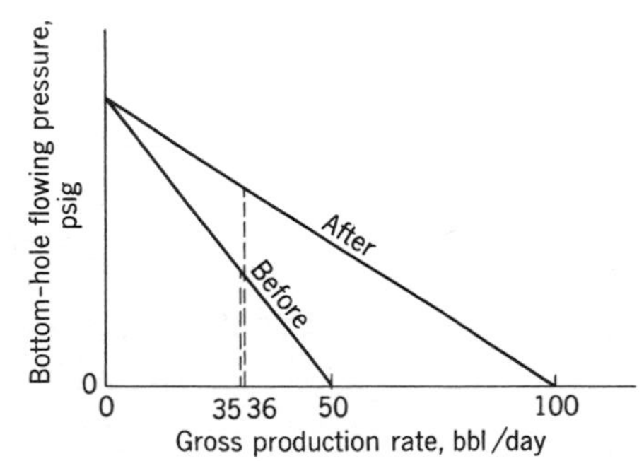

Fig. 3-8 Formation stimulation a success despite unaltered production rate.

reasons: the pump might not have been properly seated after the treatment; the pump might have been damaged in some way when it was pulled for the treatment to be undertaken; the producing GOR of the formation might have been increased by the treatment, resulting in reduced pump efficiencies; or the truth of the matter may lie with one or more of various other possible explanations.

To complicate matters further, the shutdown of a well for treatment is often used as a good opportunity to change the pump or other equipment in the hole and generally to "spring-clean" the well so that any subsequent production rate increase—or decrease—might be a direct result of this housekeeping and have nothing whatever to do with the formation-stimulation treatment.

It will subsequently be seen (Secs. 5-5 and 5-6) that knowledge of IPR behavior is an essential requirement for accurate forecasting of well and field producing potentialities.

3-3 FACTORS INFLUENCING SHAPE OF IPR

The discussion that follows will concentrate on effects resulting from the presence of free gas in the formation and, consequently, will lead to some conclusions relating to the dependence of producing GLRs on drawdown. In this section, production will be considered to be water-free (the problem of water-cut behavior will be addressed in Sec. 3-6).

Single Zone of Constant Permeability

It is evident from the form of the radial-flow equation that the greater part of the pressure drop (from static pressure to flowing BHP) in a producing formation occurs in the neighborhood of the well bore. For example, if the figures of Example 1-1 are used, the formation pressures at various distances from the well bore are as shown in Table 3-1. In this case more than 50 per-

TABLE 3-1 Formation Pressures at Various Distances from Well Bore (Figures of Example 1-1)

Distance from Centerline of Well Bore, ft	Formation Pressure, psig
$^7/_{24}$ (sand face)	500
1	576
2	619
5	676
20	762
100	862
932	1000

cent of the pressure drop is occurring within 20 ft of the well bore and the oil within this 20-ft radius comprises only 0.046 percent of the oil contained within the drainage radius of the well.

Suppose the flowing BHP at the well is below the bubble point of the oil. As a body of oil moves in toward the well, the pressure on it drops steadily, allowing gas to come out of solution. The free-gas saturation in the vicinity of the oil body steadily increases, and so the relative permeability to gas steadily increases at the expense of the relative permeability to oil (Ref. 2) (see Fig. 1-7). The greater the drawdown, that is, the lower the sand-face pressure at the well, the more marked this effect will be, so that it would be reasonable to expect the PI (which depends on the effective oil permeability) to decrease and the producing GOR (which depends on the effective gas permeability) to increase as the drawdown is increased. Such an argument leads to the conclusion that a curved IPR, as shown in Fig. 3-4, is to be expected whenever the flowing BHP is below the bubble-point pressure (Ref. 3).

The drawdown may also have a considerable effect on the producing GOR, not only because of the increased effective permeability to gas with decreasing BHP, but also because all the oil in the vicinity of the well will contribute free gas in addition to the free gas entering the environs of the well from further back in the formation. The greater the drawdown, the greater the contribution of free gas from the oil close to the well bore, and so the greater the producing GOR.

In certain circumstances the producing GOR may first decrease and then increase with increasing production rate so that some nonzero rate of production will result in the minimum producing GOR (Ref. 4). It is believed, however, that this is due to formation stratification (see below) and that the picture presented here, which implies a steady increase in the GOR with increasing production rate, is the basic one for a homogeneous formation. It is hardly necessary to add that for flowing pressures close to the bubble point (that is, gas saturations less than the critical in the formation) and also for highly permeable formations (implying high rates at low drawdowns), the effect of rate on productivity index and on GOR may be either nonexistent or very small, provided Darcy flow is maintained. The effects of turbulent flow around the well bore are discussed in Sec. 3-5 below.

Last, as long as the value of the flowing BHP remains above the saturation pressure, no free gas will be evolved in the formation and the PI will remain constant; that is, the portion of the IPR applicable to values of the flowing BHP higher than the saturation pressure will be a straight line.

Stratified Formation

Practically every producing formation is stratified to some extent; that is to say, it contains layers of differing permeability. To illustrate the type of effects that such stratification may have upon the shape of the IPR and upon the dependence of GOR on production rate, consider an example in which there are three different zones having permeabilities of 10, 100, and 1 md, respectively. It will be assumed that there is no vertical communication between the zones, except through the well bore itself (Fig. 3-9). Production from this formation will evidently be drawn chiefly from the 100-md zone, with the result that the static pressure in this zone will drop below those in the other two, the 1-md zone exhibiting the highest static pressure.

Effect on IPR

Suppose that a stage has been reached in which the pressure in the 100-md zone is 1000 psig, that in the 10-md zone is 1200 psig, and that in the 1-md zone is 1500 psig. The well is now tested at various production rates to establish the IPR. If the individual IPRs of the three zones are as illustrated in Fig. 3-10, the composite IPR, which will be the sum of these three curves, will have the shape shown. It follows as a generalization that many wells will, because of stratification and subsequent differential depletion of the zones on production, exhibit a composite IPR curve of the type illustrated in Fig. 3-11; that is to say, an improving PI with increasing production rate at the lower rates, but a deteriorating PI at the higher rates.

Effect on GOR

It was mentioned in Sec. 1-2 that at any given oil saturation, the ratio of the effective permeability to gas to the effective permeability to oil is higher, the greater the degree of cementation and consolidation, that is, to all intents and purposes, the lower the permeability. If sufficient production has been taken from a stratified formation to ensure a marked degree of differential depletion, and if a series of production tests is made on a well completed in this formation, the rates varying from virtually zero to near potential, then at the low rates the flowing BHP will be high and only the higher-pressured layers will contribute to the production. These layers will be those with the greater degree of consolidation and cementation, that is, with the higher values of the

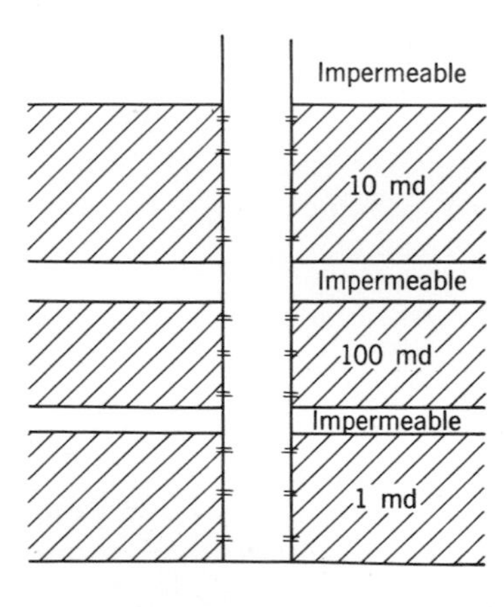

Fig. 3-9 Idealized stratified formation.

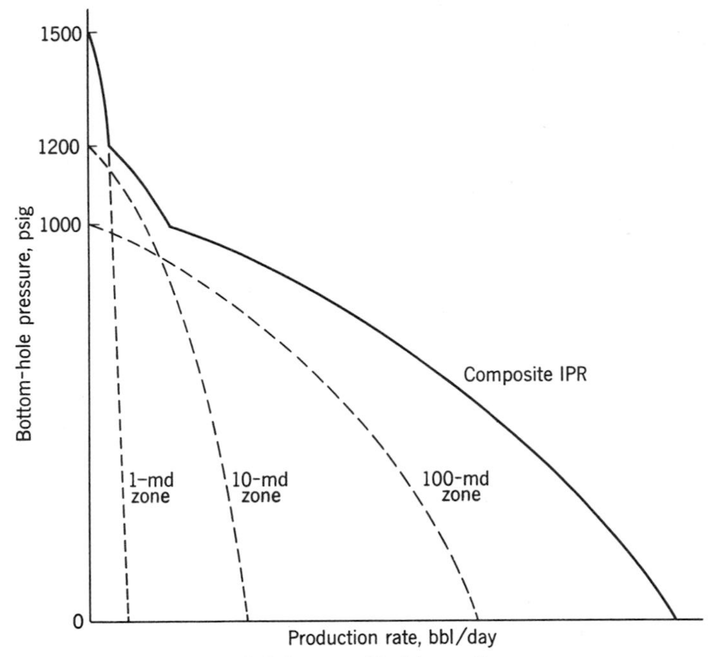

Fig. 3-10 Composite IPR for stratified formation.

gas/oil permeability ratio. In other words, the producing layers at the low rates of flow are those which produce with a high GOR.

As the well's rate of production is gradually increased, the less consolidated layers will begin to produce one by one—at progressively lower GORs—and so the overall ratio of the production will fall as the rate is increased. If, however, the most highly depleted layers themselves produce at high ratios owing to high free gas saturations, the overall GOR will eventually start to rise as the rate is increased and this climb will be continued, after the most permeable zone has come onto production, by virtue of the "vicinity of the well bore" effect, discussed above.

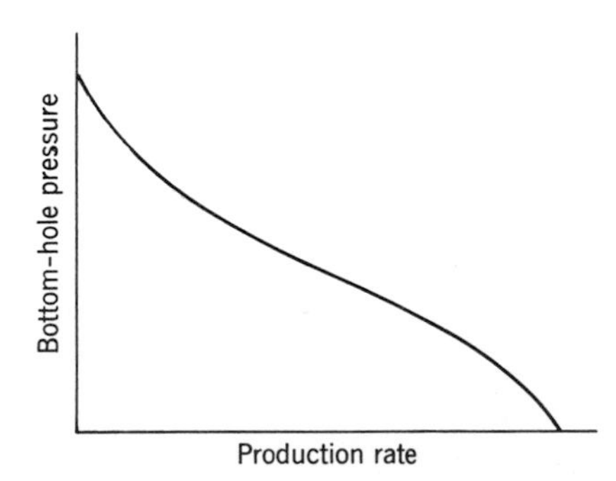

Fig. 3-11 Typical IPR curve.

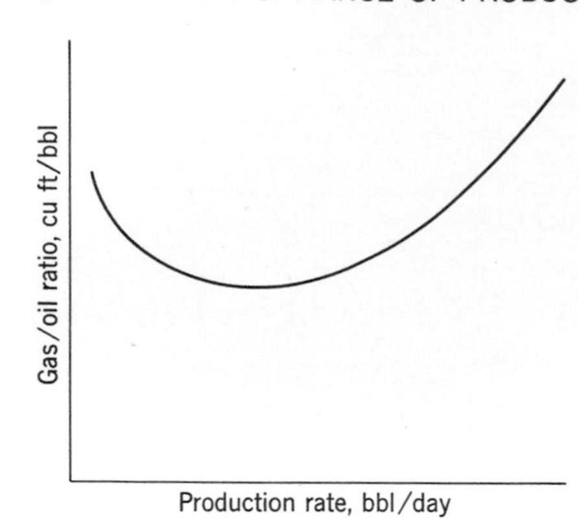

Fig. 3-12 GOR versus production rate: the minimum shown may be due to formation stratification.

Thus it is to be expected that a well producing from a stratified formation will exhibit a minimum in its GOR as the rate of production is increased (Ref. 4) (Fig. 3-12).

3-4 PREDICTING THE IPR: DEPLETION-DRIVE FIELD

There are two separate problems to be solved. The first is the shape of the rate versus pressure curve at a particular moment in time, that is, at a particular stage of depletion. The second is the manner in which the IPR falls away as production continues.

Vogel's Method

Vogel (Ref. 5) has developed an empirical equation for the shape of the IPR of a well producing from a depletion-drive reservoir in which the average reservoir pressure is less than the bubble-point pressure. This equation is

$$\frac{q}{q'} = 1 - 0.2\left(\frac{p_{wf}}{\bar{p}}\right) - 0.8\left(\frac{p_{wf}}{\bar{p}}\right)^2 \tag{3-8}$$

where $\bar{p}$ is the average reservoir pressure.

If the IPR were a straight line running to the maximum production rate q', Eq. (3-2), taken together with the definition of q', shows that q/q' would be equal to $[1 - (p_{wf}/\bar{p})]$. Hence the difference between the value of q derived from Eq. (3-8) and the "straight-line" value of q is

$$q'\left[1 - 0.2\left(\frac{p_{wf}}{\bar{p}}\right) - 0.8\left(\frac{p_{wf}}{\bar{p}}\right)^2 - 1 + \frac{p_{wf}}{\bar{p}}\right]$$

$$= 0.8\,q'\left(\frac{p_{wf}}{\bar{p}}\right)\left(1 - \frac{p_{wf}}{\bar{p}}\right)$$

which is never negative. Indeed, the difference is zero at the endpoints defined by $p_{wf} = 0$ and $p_{wf} = \bar{p}$, and is positive for all intermediate values of p_{wf}, reaching a maximum value when $p_{wf} = \bar{p}/2$.

It follows that Vogel's equation does indeed define an IPR having the general shape of Fig. 3-4.

In his paper, Standing (Ref. 6) rewrote Eq. (3-8) as

$$\frac{q}{q'} = \left(1 - \frac{p_{wf}}{\bar{p}}\right)\left(1 + 0.8\,\frac{p_{wf}}{\bar{p}}\right)$$

From Eq. (3-2), this gives

$$J = \frac{q'}{\bar{p}}\left(1 + 0.8\,\frac{p_{wf}}{\bar{p}}\right) \tag{3-9}$$

neglecting any difference between $\bar{p}$ and p_s.

Let J^* be the initial value of J, that is, the value of the PI for small drawdowns (Fig. 3-13). Allowing p_{wf} to tend to $\bar{p}$, Eq. (3-9) shows that

$$J^* = \frac{1.8q'}{\bar{p}} \tag{3-10}$$

Equations (3-9) and (3-10) give

$$J^* = 1.8J \Big/ \left(1 + 0.8\,\frac{p_{wf}}{\bar{p}}\right) \tag{3-11}$$

which permits J^* to be calculated from a measured value of J. Alternatively, J^* may be calculated from the radial-flow equation. If Eq. (1-10) is used, for example,

$$J^* = \frac{0.007082\, k_o h}{B_o \mu_o \left[\ln\left(\frac{r_e}{r_w}\right) - \frac{3}{4}\right]}$$

where k_o, B_o, μ_o are taken at the average pressure (and fluid saturation) in the drainage volume.

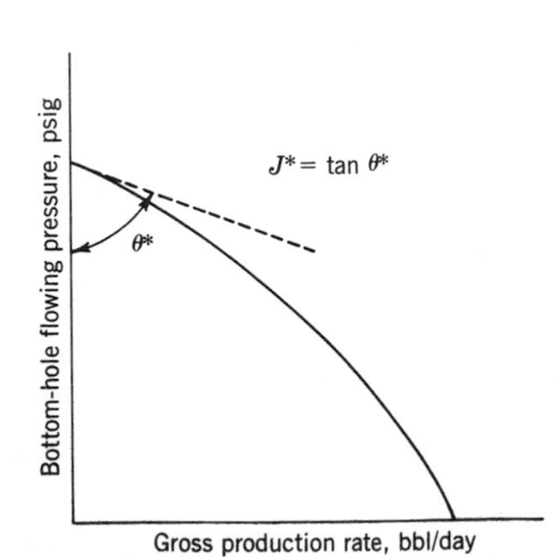

Fig. 3-13 Initial value of PI.

A future value of J^* (J_f^*) can be calculated from the present value (J_p^*) by using Eq. (1-10) for J_p^* and J_f^* and dividing. The result is

$$J_f^* = J_p^* \left(\frac{k_{ro}}{B_o\mu_o}\right)_f \bigg/ \left(\frac{k_{ro}}{B_o\mu_o}\right)_p \tag{3-12}$$

Finally, if q' is eliminated from Eqs. (3-8) and (3-10), the future IPR curve may be plotted from the equation

$$q_o = \frac{J_f^* \bar{p}_f}{1.8}\left[1 - 0.2\,\frac{p_{wf}}{\bar{p}_f} - 0.8\left(\frac{p_{wf}}{\bar{p}_f}\right)^2\right] \tag{3-13}$$

The following example is given by Standing (Ref. 6):

Example 3-2 A well draining 40 acres is flowing at 400 bbl/day with a flowing BHP of 1815 psig. The average pool pressure is 2250 psig. Determine the IPR for the well at the time when the average pool pressure will be 1800 psig, given the following additional information:

	Present	Future
Av pressure, psig	2250	1800
Oil viscosity at $\bar{p}$, cP	3.11	3.59
B_o at $\bar{p}$	1.173	1.150

The value of the average oil saturation in the reservoir at 2250 psig and at 1800 psig can be determined from a reservoir engineering calculation (for example, by Tarner's method). Values in this example turn out to be

S_o at 2250 psig, 0.768
S_o at 1800 psig, 0.741

The value of k_{ro} at each of these oil saturations may then be determined by using a Corey-type formula, for example

$$k_{ro} = \left(\frac{S_o - S_{oc}}{1 - S_w - S_{oc}}\right)^4$$

where S_{oc} is the irreducible oil saturation. In this example the following values were found:

k_{ro} at 2250 psig, 0.815
k_{ro} at 1800 psig, 0.685

The value of J at the production rate of 400 bbl/day is 400/(2250 − 1815) or 0.92 bbl/(day)(psi) so that from Eq. (3-11)

$$J_p^* = 1.8 \times 0.92 \bigg/ \left(1 + 0.8 \times \frac{1815}{2250}\right) = 1.01 \text{ bbl/(day)(psi)}$$

From Eq. (3-12)

$$J_f^* = 1.01 \left(\frac{0.685}{3.59 \times 1.150}\right) \bigg/ \left(\frac{0.815}{3.11 \times 1.173}\right) = 0.750 \text{ bbl/(day)(psi)}$$

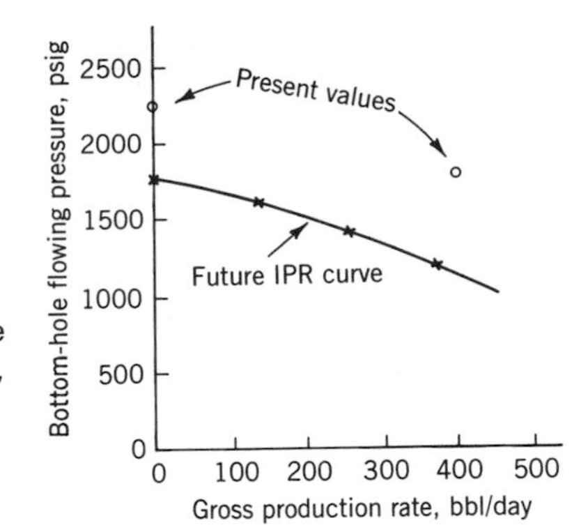

Fig. 3-14 Example 3-2: Future IPR curve. (*From Standing, Ref. 6. Courtesy AIME.*)

Substitution of this value in Eq. (3-13) gives the future IPR curve plotted in Fig. (3-14) (Ref. 6).

Log-Log Method

A second method for defining the shape of the IPR curve is to modify Eq. (3-6) to read

$$q = J(\Delta p)^n \tag{3-14}$$

in which the exponent n is not necessarily equal to unity (Ref. 7). Under this assumption, the potential of the well is Jp_s^n, so that a straight-line IPR would lead to a production rate of $Jp_s^n[1 - (p_{wf}/p_s)]$ at a flowing BHP of p_{wf}. It follows that the ratio of the production rate given by Eq. (3-14) to the production rate at the same drawdown obtained by using a straight-line IPR is

$$\frac{J(p_s - p_{wf})^n}{Jp_s^n(p_s - p_{wf})/p_s}$$

or

$$\left(\frac{p_s - p_{wf}}{p_s}\right)^{n-1}$$

This ratio is always ≤ 1 if $n > 1$, is unity if $n = 1$, and is always ≥ 1 if $n < 1$.

Note from Eq. (3-14) that

$$\frac{dq}{dp_{wf}} = -Jn(p_s - p_{wf})^{n-1}$$

so that for $n < 1$, dq/dp_{wf} tends to infinity as p_{wf} tends to p_s: that is, the curve $q = J(\Delta p)^n$ is horizontal (on the $p-q$ plot) at $p_{wf} = p_s$, $q = 0$.

Moreover, the difference between the curve $q = J(\Delta p)^n$ and the straight line at a particular value of p_{wf} is a maximum when

$$p_{wf} = p_s(1 - 1/n^{1/(n-1)})$$

For example, when $n = 1/2$ this maximum difference occurs at $p_{wf} = 3/4p_s$; when $n = 2/3$ it occurs at $p_{wf} = 19/27p_s$; and when $n = 2$, at $p_{wf} = 0.5p_s$.

Typical curves for the $q = J(\Delta p)^n$ formula on the regular $p-q$ plot are shown in Fig. 3-15, and the case $n < 1$ may be contrasted with the more symmetrical Vogel curve.

Under the mathematical formulation of Eq. (3-14), the symbol J no longer retains a physical meaning that can be related back to the equations of reservoir mechanics, as was the case with Vogel's method. Instead, recourse must be made to field data in an attempt to build a correlation between J and n for the wells in a particular reservoir. If it is possible to develop such a correlation, then a grid can be established on a log-log plot of q against Δp (Fig. 3-16); note here that the various values of n give the slopes of the lines, while the J's are the q values corresponding to $\Delta p = 1$.

In order to use this grid for extrapolation into the future, the following procedure is convenient:

Whenever a flowing well measurement of rate and drawdown is made, the result is plotted on the grid (point A). This point defines the IPR line for the well at that time (shown as a broken line). Some convenient reference drawdown is chosen, say 100 psi, and the producing rate q_{100} that would correspond to that drawdown is plotted (point B on Fig. 3-16; Fig. 3-17). Such a plot is kept up for each well.

To obtain a future IPR for a well, extrapolate—as best as possible—the points of Fig. 3-17, and so estimate the production rate at the reference drawdown at the required future time (or well cumulative). Plot this value back onto the grid (point C on Fig. 3-16), and draw in the line through C that fits the grid (shown dotted on Fig. 3-16). This line defines J and n and hence the IPR for the well at the desired future time.

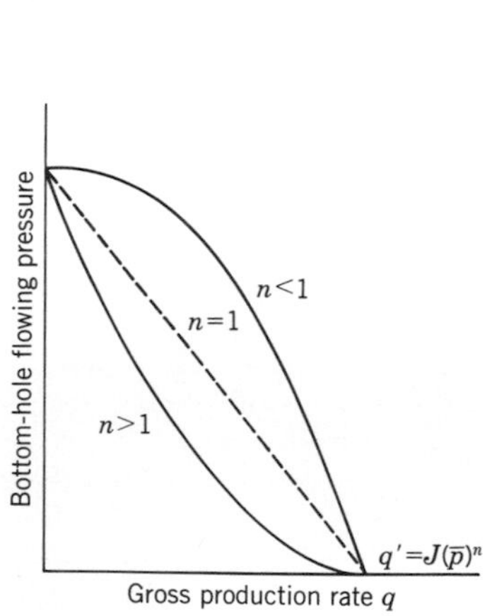

Fig. 3-15 Typical rate versus pressure curves derived from Eq. (3-14).

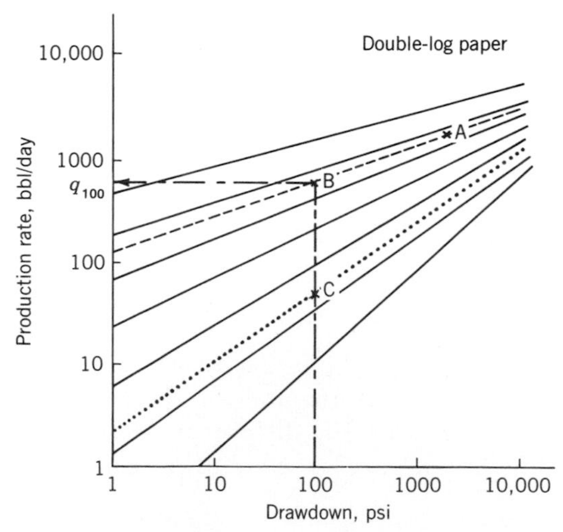

Fig. 3-16 Production rate versus drawndown grid.

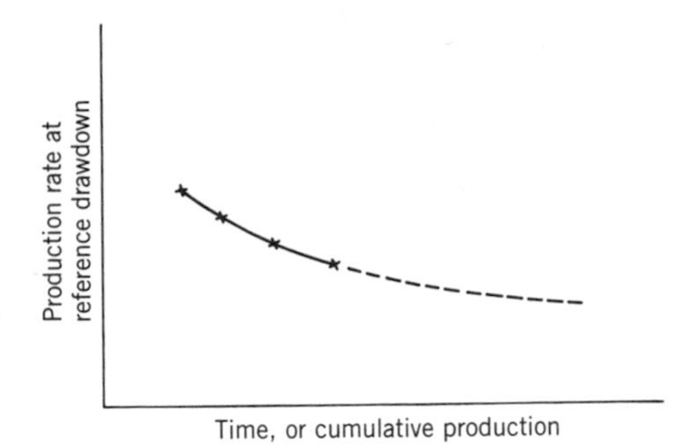

Fig. 3-17 Extrapolation using "production rate at reference drawdown."

The log-log method is evidently relatively complex compared with that developed by Vogel and Standing.

Fetkovich's Method

Fetkovich (Ref. 8) has developed a method which, in many ways, combines the approach of Vogel with that of the log-log assumption.

The method takes as its starting point the equation of Evinger and Muskat (Ref. 9) for two-phase flow in the situation in which a single well of radius r_w is draining a horizontal, homogeneous reservoir of radius r_e. This equation is

$$q_o = \frac{0.007082kh}{\ln\left(\frac{r_e}{r_w}\right)} \int_{p_{wf}}^{p_e} f(p)dp \tag{3-15}$$

where $f(p)$ is a function of pressure. Various expressions may be used for $f(p)$ to cover different cases, for example, single-phase steady-state flow with a constant pressure p_e at the outer boundary, single-phase pseudo-steady-state flow with no flow across the outer boundary, or two-phase flow with various boundary assumptions (Ref. 7).

The simplest two-phase case is that of constant pressure p_e at the outer boundary, with p_e less than the bubble-point pressure so that two-phase flow occurs throughout. Under these circumstances $f(p)$ takes on the value $k_{ro}/\mu_o B_o$, where k_{ro} is the relative permeability to oil at the saturation conditions in the formation corresponding to the pressure p.

Fetkovich makes the key assumption that, to a good degree of approximation, the expression $k_{ro}/\mu_o B_o$ is a linear function of p, the straight line passing, to all intents and purposes, through the origin.

If p_i is the initial formation pressure—in the case under consideration p_i is sufficiently close in practical terms to p_e that these two symbols may be used interchangeably—then the straight-line assumption leads to

$$\frac{k_{ro}}{\mu_o B_o} = \left(\frac{k_{ro}}{\mu_o B_o}\right)_i \frac{p}{p_i} \tag{3-16}$$

Substitution of Eq. (3-16) into Eq. (3-15) gives

$$q_o = \frac{0.007082kh}{\ln\left(\dfrac{r_e}{r_w}\right)}\left(\frac{k_{ro}}{\mu_o B_o}\right)_i \frac{1}{2p_i}(p_i^2 - p_{wf}^2)$$

or

$$q_o = J'_{oi}(p_i^2 - p_{wf}^2) \tag{3-17}$$

where

$$J'_{oi} = \frac{0.007082kh}{\ln\left(\dfrac{r_e}{r_w}\right)}\left(\frac{k_{ro}}{\mu_o B_o}\right)_i \frac{1}{2p_i} \tag{3-18}$$

It is interesting to compare Eq. (3-17) with the standard equation defining PI, namely Eq. (3-6), and to note that Fetkovich's form is that applicable to the flow of clean, dry gas (the powers of 2 in this latter case are required to account for the compressibility of gas).

Equation (3-17) leads to an IPR plot of the "expected" type, illustrated in Fig. 3-4. This can be seen most readily by differentiating Eq. (3-17) with respect to p_{wf}, obtaining

$$\frac{dq_o}{dp_{wf}} = -2J'_{oi}\,p_{wf}$$

which implies that the rate of change of q_o with p_{wf} is negative and that the change in q_o accompanying a particular increment in p_{wf} is lower at the lower values of the inflow pressure.

Example 3-3 Plot the IPR curve for a well in which p_i is 2000 psia and Eq. (3-17) holds, with J'_{oi} equal to 5×10^{-4} bbl/(day)(psia)^2.

The calculation of the oil production rate at various flowing BHPs is shown in Table 3-2 and the results are plotted in Fig. 3-18.

The next challenge is to modify Eq. (3-17) to take into account the fact that in practice p_e is not a constant but decreases as cumulative production increases. The assumption made is that J'_{oi} will decrease in proportion to the decrease in average reservoir (drainage area) pressure. Thus, when the static

TABLE 3-2 Example 3-3: Determination of Oil Rates at Various Assumed Values of p_{wf}, from Eq. (3-17)

p_{wf} (psia)	p_{wf}^2, MM psia^2	$p_i^2 - p_{wf}^2$, MM psia^2	q_o, bbl/day
1500	2.25	1.75	875
1200	1.44	2.56	1280
1000	1.00	3.00	1500
800	0.64	3.36	1680
600	0.36	3.64	1820
400	0.16	3.84	1920
200	0.04	3.96	1980
0	0	4.00	2000

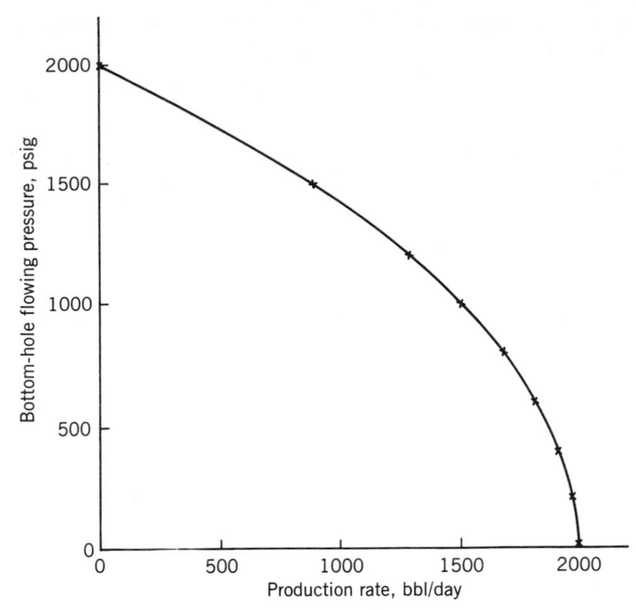

Fig. 3-18 Example 3-3: IPR curve from Fetkovich's method.

pressure is p_s ($< p_i$), the PI equation is

$$q_o = J'_{oi} \frac{p_s}{p_i} (p_s^2 - p_{wf}^2) \tag{3-19}$$

or, alternatively

$$q_o = J'_o (p_s^2 - p_{wf}^2) \tag{3-20}$$

where

$$J'_o = J'_{oi} \frac{p_s}{p_i} \tag{3-21}$$

These equations may be used to extrapolate into the future.

Example 3-4 Use Eqs. (3-20) and (3-21) and the data of Example 3-3 to predict the IPRs of the well at static pressures of 1500 psia and 1000 psia.

From Eq. (3-21), the value of J'_o at 1500 psia is $5 \times 10^{-4} \times 1500/2000$

or

$$3.75 \times 10^{-4} \text{ bbl/(day)(psia)}^2$$

and the value at 1000 psia is

$$2.5 \times 10^{-4} \text{ bbl/(day)(psia)}^2$$

When the appropriate values of J'_o and p_s are inserted into Eq. (3-20), calculations similar to those of Table 3-2 give the results plotted on Fig. 3-19.

The form of Eqs. (3-17) and (3-20), as well as the close analogy with the comparable equation for gas reservoirs, leads naturally to consideration of techniques long used in the testing of gas wells and in the analysis of test results. Discussion of the applicability of isochronal and flow-after-flow tests to oil wells has been presented by Fetkovich (Ref. 8) and, in particular, he has studied the possibility that non-Darcy flow may be a major factor in many oil field situations, such flow being, perhaps, covered analytically by modifying Eq. (3-20) to read

$$q_o = J'_o \, (p_s^2 - p_{wf}^2)^n \tag{3-22}$$

where the value of the exponent n may differ from unity (almost always less than, if indeed not equal to, unity). For further detail on the analysis and results based on Eq. (3-22) the reader is referred to the original work of Fetkovich (Ref. 8), or to the detailed summary and discussion undertaken by Brown (Ref. 7).

In concluding this section on the shape of the IPR and the problem of how the IPR might decline as the formation pressure declines (that is, as fluid saturations change), it should be underlined that prediction in this area is still an uncertain art. It may be that a simpleminded method is (at present) no less reliable than the rather complex procedures suggested by the log-log method or by Vogel, Standing, or Fetkovich. On the other hand, a simpleminded method gives no possibility for logical refinement and the work outlined above is of major importance for the future development of greater accuracy and more sensitive predictability.

One of the simplest of the simpleminded methods that can be used once

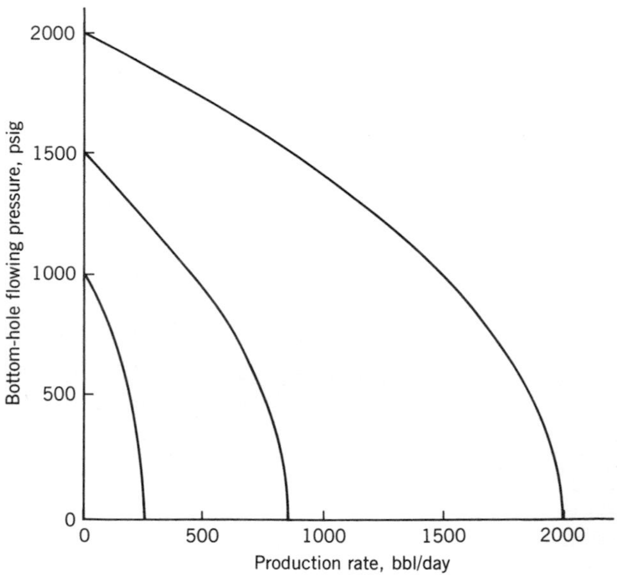

Fig. 3-19 Example 3-4: Future IPR curves from Fetkovich's method.

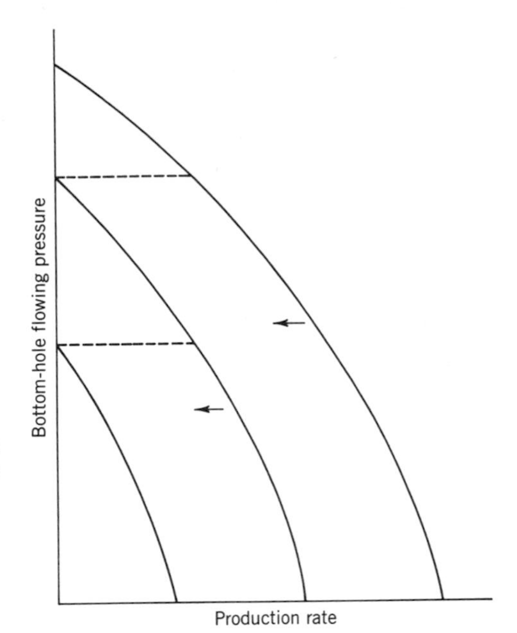

Fig. 3-20 Future IPR curves obtained by shifting present-day curve to the left.

today's IPR is known is to move today's curve progressively to the left, eliminating that part of the curve applicable to pressures in excess of the assumed future average pool pressure (Fig. 3-20).

3-5 EFFECT OF TURBULENT FLOW

It may be argued that the basic assumption made by Fetkovich (Ref. 8), namely that $k_{ro}/\mu_o B_o$ is, approximately, a linear function of p, the straight line passing close to the origin, is indeed a "compromise" assumption relating to wells producing with a high production rate in which turbulent-flow pressure losses in the formation are of the same order of magnitude (at least) as the Darcy flow losses.

Jones and Blount (Ref. 10) have addressed the problem of turbulent-flow losses for both gas and oil well production. They show that for radial flow in a homogeneous horizontal reservoir the pressure drawdown is of the form

$$p_s - p_{wf} = Cq + Dq^2 \tag{3-23}$$

where C is the standard laminar (or Darcy) flow coefficient and D is a turbulence coefficient. Dividing through by q gives

$$\frac{p_s - p_{wf}}{q} = C + Dq \tag{3-24}$$

which indicates that the reciprocal of the measured PI, when plotted against production rate, might be expected to give a straight line. The slope of such a line would be a measure of the degree of turbulence. If this proves not to be small, consideration should be given to remedial work, such as additional perforations over the completed interval or an extension in the length of that interval.

3-6 EFFECT OF DRAWDOWN ON WOR

To this point no consideration has been given in this chapter to the relationship between drawdown and WOR. What follows here is a discussion, based on Gilbert's presentation (Ref. 1), of possible effects stemming from the watering out of one or more productive stringers open to the well bore. If water is moving from the water source to the well via stringers in the formation, it is possible to determine whether, at the well bore, the pressure in the water is greater than or less than the pressure in the oil sands (that is, whether it is *high-pressure* or *low-pressure* water) from an analysis of the gross IPR and three or four water-cut values taken at different gross rates. The method of approach may be illustrated by means of an example.

Example 3-5 A series of tests is made on a certain well with the following results:

Gross rate, bbl/day	Water cut, water/gross %	Flowing BHP, psig
47	85	1300
90	60	920
125	48	630
162	45	310

Determine the static pressure and the productivity index of the oil and water zones, respectively. Based on the results, at what rate could water be expected to flow into the oil sand if the well were left shut in?

Referring to Fig. 3-21, the first step is to plot the gross IPR (line 1). From the gross rate and the measured water cuts the water and oil IPRs are calculated as follows (lines 2 and 3):

Gross rate, bbl/day	Water cut, %	Water rate, bbl/day	Oil rate, bbl/day	p_{wf}, psig
47	85	40	7	1300
90	60	54	36	920
125	48	60	65	630
162	45	73	89	310

Evidently, from the figure,

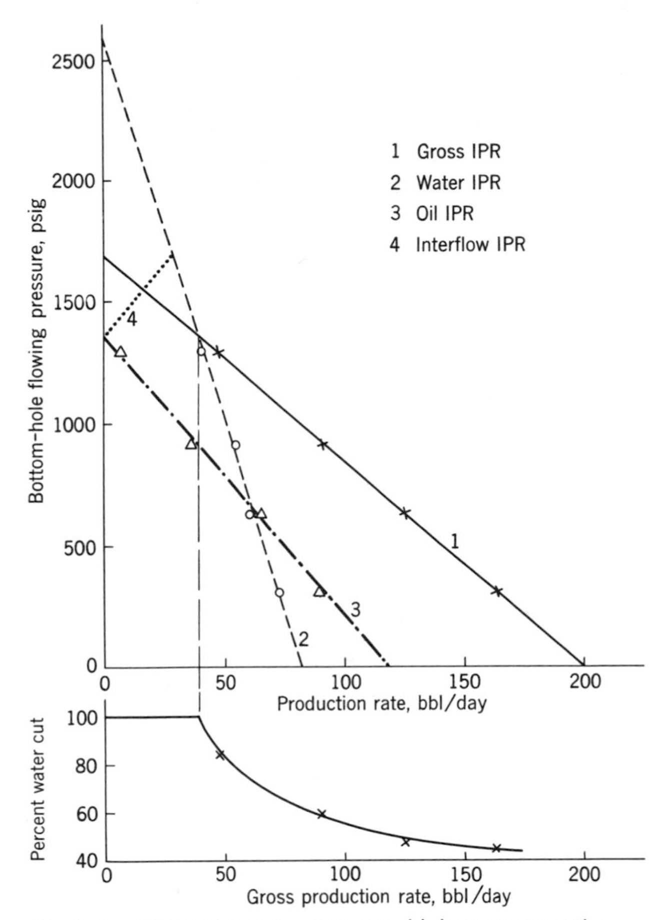

Fig. 3-21 IPR and water-cut curves: high-pressure water.

$$\text{Static pressure of oil zone} = 1350 \text{ psig}$$

$$\text{PI of oil zone} = {}^{120}/_{1350} = 0.089 \text{ bbl/(day)(psi)}$$

$$\text{Static pressure of water zone} = 2600 \text{ psig}$$

$$\text{PI of water zone} = {}^{82}/_{2600} = 0.0315 \text{ bbl/(day)(psi)}$$

When the well is shut in, it might be expected (from the gross IPR) that the BHP would stabilize at about 1700 psig and that water would flow into the oil zone at some 28 bbl/day (line 4 of Fig. 3-21 is the interflow IPR).

It is of interest to note the shape of the water cut versus rate curve (also shown on Fig. 3-21), which is typical of high-pressure water; namely, a 100 percent cut (pure water) is obtained at low rates, the oil content gradually increasing with the offtake rate. In Fig. 3-22 the case of low-pressure water is similarly illustrated, and the typical water cut versus rate curve is shown; namely, the cut starts at or near zero and increases with rate.

In both the examples illustrated, it has been assumed for simplicity that

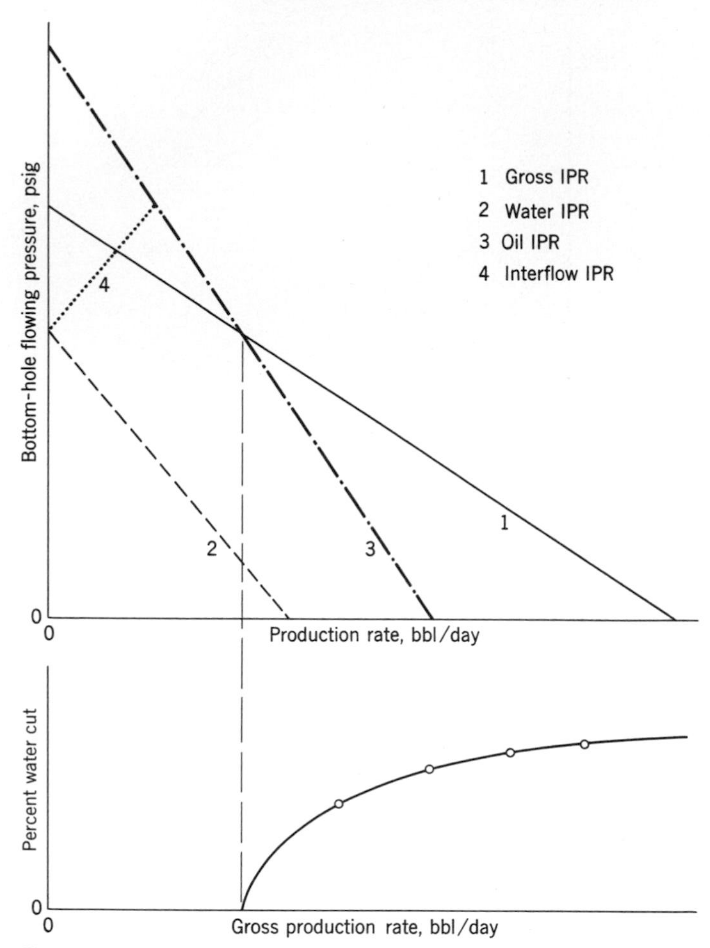

Fig. 3-22 IPR and water-cut curves: low-pressure water.

the IPRs are straight. However, this assumption is by no means essential to the argument, and if sufficient measurements of gross rate, cut, and BHP are available, curved IPRs can be drawn in without difficulty.

This section will be closed with a word of warning by Gilbert (Ref. 1):

> Differential depletion is progressive during sustained flowing periods wherever the ratio of lateral permeabilities to vertical permeabilities is large; and interflow through the bore between producing layers takes place during any subsequent periods of shut-down unless a suitable mud is spotted in the producing interval. Thus any water inflow from a relatively high-pressure source tends to seek out and enter the more depleted oil layers during the shut-down period, and with permanent injury in some fields to effective oil permeabilities.

REFERENCES

1. Gilbert, W. E.: "Flowing and Gas-Lift Well Performance," *API Drill. Prod. Practice,* 1954, p. 126.
2. Mueller, T. D., J. E. Warren, and W. J. West: "Analysis of Reservoir Performance K_g/K_o Curves and a Laboratory K_g/K_o Curve Measured on a Core Sample," *Trans. AIME,* **204:**128 (1955).
3. Handy, L. L.: "Effect of Local High Gas Saturations on Productivity Indices," *API Drill. Prod. Practice,* 1957, p. 111.
4. Sullivan, R. J.: "Gas-Oil Ratio Control in Flowing Wells," *API Drill. Prod. Practice,* 1937, p. 103.
5. Vogel, J. V.: "Inflow Performance Relationships for Solution-Gas Drive Wells," *J. Petrol. Technol.,* **20**(1):83 (1968).
6. Standing, M. B.: "Concerning the Calculation of Inflow Performance of Wells Producing from Solution Gas Drive Reservoirs," *J. Petrol. Technol. Forum,* Sept. 1971, p. 1141.
7. Brown, Kermit E.: *The Technology of Artificial Lift Methods,* vol. 1, Petroleum Publishing Company, Tulsa, Oklahoma, 1977.
8. Fetkovich, M. J.: "The Isochronal Testing of Oil Wells," *SPE Paper No. 4529, 48th Annual Meeting of SPE and AIME,* Las Vegas, Nevada, Sept.-Oct. 1973.
9. Evinger, H. H., and M. Muskat: "Calculation of Theoretical Productivity Factor," *Trans. AIME,* **146:**126 (1942).
10. Jones, Lloyd G., and E. M. Blount: "Use of Short Term Multiple Rate Flow Tests to Predict Performance of Wells Having Turbulence," *SPE Paper No. 6133, 51st Annual Meeting of SPE and AIME,* New Orleans, Louisiana, Oct. 1976.

Vertical Lift Performance

4

4-1 INTRODUCTION

In order to analyze the performance of a conventionally completed flowing well, it is necessary to recognize that there are three distinct phases, which have to be studied separately and then finally linked together before an overall picture of a flowing well's behavior can be obtained. These phases are the inflow performance, the vertical lift performance, and the choke (or bean) performance (Fig. 4-1).

The *inflow performance,* that is, the flow of oil, water, and gas from the formation into the bottom of the well, is typified, as far as gross liquid production is concerned, by the PI of the well or, more generally, by the IPR (Chap. 3).

The *vertical lift performance* involves a study of the pressure losses in vertical pipes carrying two-phase mixtures (gas and liquid). Several methods of approach to this problem are discussed in this chapter, particularly that of Poettmann and Carpenter (Ref. 1), extended by Baxendell (Ref. 2) and by Baxendell and Thomas (Ref. 3), that of Ros (Ref. 4), and that of Gilbert (Ref. 5).

The pressure loss accompanying the flow of oil, water, and gas through a flow-line restriction (choke or bean) at the surface is known as the *choke performance* (Sec. 5-2).

As an introduction to two-phase vertical lift performance, a survey is presented in Sec. 4-2 of the principal types of flow regime that occur in the tubing, that is, of the possible geometrical configurations of the gas and liquid phases in the flow string. In connection with this survey and with the role

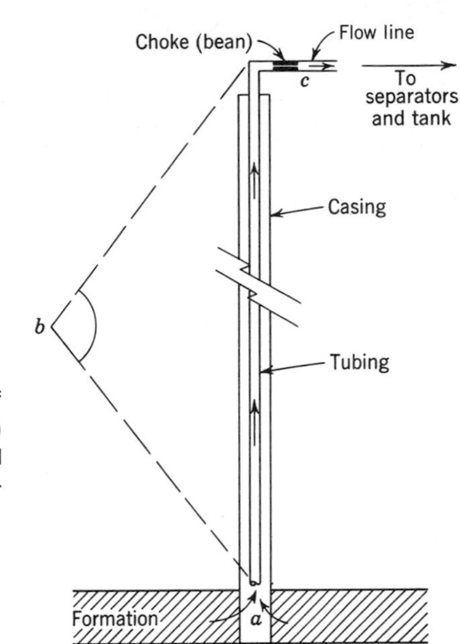

Fig. 4-1 The three phases of flowing well performance. (*a*) Inflow performance; (*b*) vertical lift performance; (*c*) choke performance.

played by the free gas, it may be helpful to consider the action in the stem of a coffee percolator. As the temperature of the water rises, small gas bubbles initially travel up the stem without lifting the water in any way (bubble flow and complete gas slippage). Later, when the water becomes hotter, relatively large gas bubbles enter the stem and lift the water above so that percolating action begins to take place (plug or slug flow). A few minutes spent in watching this type of action may clarify many of the points raised in this chapter.

In much of the work that follows, and specifically in the examples, two simplifying assumptions are made. Both are for the sake of clarity and neither is necessary or limiting. The first is that the tubing string is landed opposite the producing formation so that the inflow pressure at the foot of the tubing is equal to the flowing BHP. The modifications that are required if the tubing shoe is some appreciable distance above the perforations are discussed in Sec. 4-11.

The second assumption that is made is that the IPR is a straight line. The question of defining the curvature of the IPR is discussed in the preceding chapter, and in practice curved IPRs should be used whenever it is deemed appropriate to do so.

4-2 FLOW REGIMES IN VERTICAL TWO-PHASE FLOW

As the pressure on a crude oil containing gas in solution is steadily reduced, free gas is evolved; as a consequence, the liquid volume decreases. This phenomenon affects the relative volumes of free gas and oil present at each point in the tubing of a flowing well. For instance, if the flowing BHP in a particular well is above the bubble point of the crude being produced, liquid only is present in the lower part of the tubing. As the liquid moves up the tubing, the pressure drops and gas bubbles begin to form. Such a flow regime—bubbles of gas dispersed in a continuous liquid medium—is known as *bubble flow.*

As the fluid moves further up the tubing, the gas bubbles grow and become more numerous. The larger bubbles slip upward at a higher velocity than the smaller ones, because the volume of a bubble, and hence the buoyancy effect, depends on the cube of the radius whereas the frictional drag on the surface of the bubble varies only as the square. Thus the larger bubbles grow by entrainment of the smaller bubbles they overtake. A stage is reached in which these large bubbles extend across almost the entire diameter of the tubing, so that the flow regime has become one in which slugs of oil containing small gas bubbles are separated from each other by gas pockets that occupy the entire tubing cross section except for a film of oil moving relatively slowly along the wall of the tubing. This condition is known as *slug* (or *plug*) *flow.*

Still higher in the tubing, that is, at lower pressures, the gas pockets may have grown and expanded to such an extent that they are able to break through the more viscous oil slugs, with the result that the gas forms a continuous phase near the center of the tubing, carrying droplets of oil up with it. Along the walls of the tubing there is an upward-moving oil film. This is *annular flow.* Continued decrease in pressure with resultant increase in gas volume results in a thinner and thinner oil film, until finally the film all but disappears and the flow regime has become *mist flow,* a continuous gas phase in which oil droplets are carried along with the gas.

Not all these flow regimes will occur simultaneously in a single tubing string—the pressure drop that would be required across the tubing would be far greater than is encountered in practice. But frequently two, or possibly three, together with their zones of overlap may be present; this is a factor to be remembered when vertical-flow pressure losses are under discussion.

In addition to the flow regimes themselves, the viscosities of the oil and gas, the variations of these viscosities with temperature and pressure, the PVT characteristics of the reservoir fluids, the flowing BHP, and the tubing-head pressure (THP) all directly affect the pressure gradient at a particular point of the tubing. The most that can be hoped for in the determination of these pressure gradients is that the more important variables can be isolated. Considerable progress has been made in this study and many sets of generalized curves have been published[1] (Ref. 7). In this book one of the first sets of curves—that due to Gilbert (Ref. 5)—is used throughout in the illustrative examples. The justification for this decision is that the methods outlined are general and will still be applicable, perhaps with minor modifications, when a more detailed and complete set, or the accompanying computer program, is

[1] The reader is referred to the excellent bibliography compiled by Brown (Ref. 6).

used. Furthermore, it is to be expected that any qualitative conclusions that can be drawn from the use of Gilbert's curves will remain valid.

To conclude this section, reference must be made to the work of Versluys (Ref. 8), who as early as 1929 attempted a full mathematical treatment of the theory of flowing oil wells. This work has been applied only to a limited extent because of practical difficulties encountered in the evaluation of certain empirical factors that Versluys found it necessary to introduce.

4-3 VERTICAL LIFT PERFORMANCE: POETTMANN AND CARPENTER

For a given value of the flowing BHP, the formation will produce oil, water, and gas into the well at definite rates. The question now to be answered is whether the pressure differential across the length of the tubing, from the assigned value of the flowing pressure at the bottom to some pressure at the top which cannot be less than atmospheric (the THP), is sufficient to permit the fluids entering the well from the formation to flow up the tubing at the necessary rates.

The problem was approached semitheoretically by Poettmann and Carpenter (Ref. 1), who based their analysis on the energy equation. The assumptions made are that the difference in the kinetic energy of the flowing fluid in its initial and final states of flow (bottom and top of the tubing) is negligible; that the external work done by the fluid flow is negligible; and that the energy loss W_f resulting from such irreversible phenomena as slippage and frictional effects against the tubing wall are expressible in the form

$$W_f = 4fv^2 \frac{\Delta h}{2gD} \tag{4-1}$$

where v is the average mixture velocity over the tubing interval of length Δh and D is the inside diameter of the tubing. The factor f must be determined empirically.

Using these assumptions, Poettmann and Carpenter were able to reduce the energy equation to the form

$$144 \frac{\Delta p}{\Delta h} = \bar{\rho} + \frac{\bar{K}}{\bar{\rho}} \tag{4-2}$$

where Δp = pressure drop over the vertical interval Δh ft, psi

$\bar{\rho}$ = average density of the fluid in this interval, lb/cu ft

$$\bar{K} = \frac{fq^2M^2}{(7.413 \times 10^{10} D^5)} \tag{4-3}$$

q = liquid production rate (stock-tank oil and water), bbl/day

M = total mass of gas and liquid associated with 1 bbl of stock-tank liquid. lb

D = inside diameter of the tubing, ft
f = energy-loss factor defined by Eq. (4-1)

For an illustration of the method of using Eqs. (4-2) and (4-3), consider the case of a well drilled into a formation, the inflow properties of which are known; that is, the formation's IPR, its static pressure, its water-cut behavior, and its GOR behavior are known. Also, it is supposed that certain properties of the oil and gas such as the oil formation volume factor, the gas solubility, the gas and oil densities, and the gas formation volume factor have been determined or can be estimated.

If a certain value is assumed for the flowing BHP p_{wf}, the rate at which the formation supplies oil, water, and gas to the bottom of the well can readily be found from the known inflow properties. Now divide the tubing into equal parts, H_1H_2, H_2H_3, H_3H_4, and so on, each of length Δh (Fig. 4-2), and let the pressures in the flowing column in the tubing be $p_2, p_3, p_4, \ldots$ at the points $H_2, H_3, H_4, \ldots$. Since the pressure at H_1 is p_{wf}, it follows that, for the interval H_1H_2, the pressure may be taken as p_{wf} as a first approximation, and so the various factors needed for use in Eq. (4-2) may be determined, namely, $\bar{\rho}$, q, and M. If it is assumed for the moment that there is some method for finding the value of the energy-loss factor f, Eq. (4-2) may be used to calculate the pressure drop Δp over the interval H_1H_2. In this way a first approximation to p_2, the pressure at H_2, is found. If a better approximation is felt to be desirable, then the average of p_{wf} and this first approximation may be used in place of p_{wf} in the calculation of $\bar{\rho} + \overline{K/\bar{\rho}}$. This process is repeated for the interval H_2H_3, and so on up the tubing, either until the surface is reached and

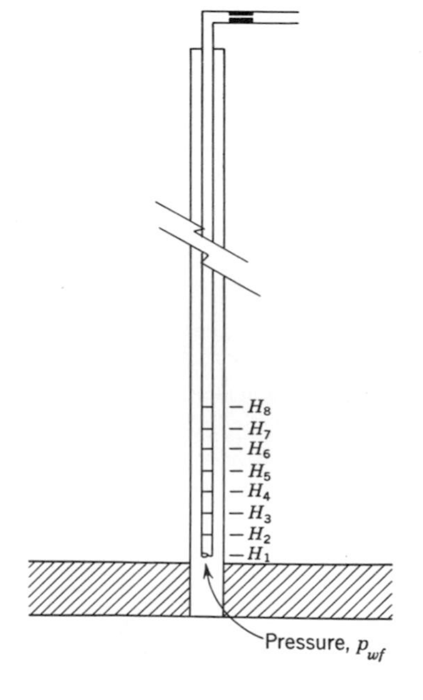

Fig. 4-2 Division of tubing into intervals of equal length for calculation purposes.

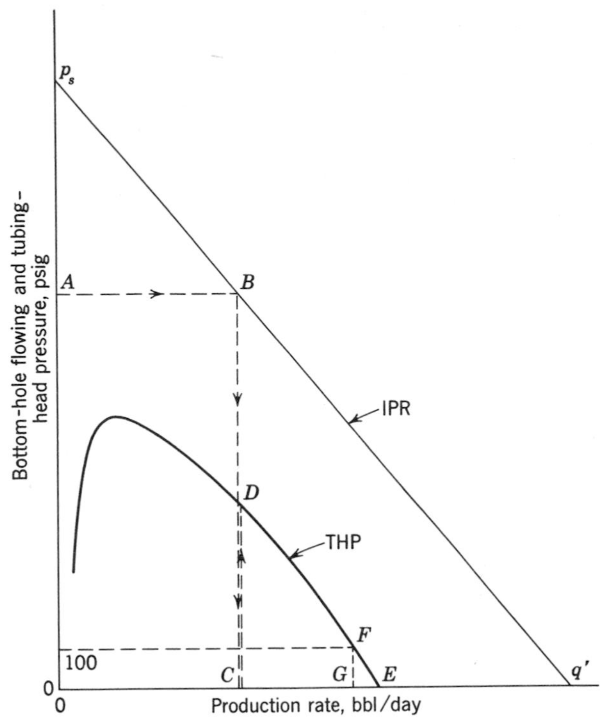

Fig. 4-3 Flowing BHP and THP as functions of production rate.

the pressure at this stage (that is, the THP) is greater than zero or until the calculated pressure at some stage is zero or negative, in which case it may be concluded that the well will not flow with the value of BHP assumed at the start of the calculations.

Taking different values of the flowing BHP as starting points, it is possible to develop a graph of the type shown in Fig. 4-3 for any particular size of tubing. With the value of the flowing BHP defined by point *A* as the starting point, the formation's rate of production is determined by moving horizontally to point *B* on the IPR and then vertically downward to point *C*. With the production rate defined by *C*, the Poettmann-Carpenter equation is used to calculate the value of the THP (point *D*). Evidently *BD* is a measure of the pressure loss in the tubing. The maximum rate at which the well is able to flow is defined by point *E*, which corresponds to a zero THP. Since there will always be a positive pressure loss in the tubing, point *E* must lie to the left of the well's potential q', as shown; in other words, a well can never realize the full formation potential on natural flow. In practice, a well is never produced with a zero THP because the flow line and separator at the surface will always exert some back pressure. If it is decided to flow the well with a THP of

100 psi, then a horizontal line is drawn at a height equivalent to 100 psi and the point *F* at which this intersects the THP curve defines the flowing rate *G*.

A graph of the type shown in Fig. 4-3 can also be used to determine the optimum size of tubing to run in a given well, that is, the size of tubing that will permit the well to flow at its maximum rate at some predetermined THP. Repeating the Poettmann-Carpenter calculations for, say, $2^3/_8$-in. and $3^1/_2$-in. tubing, curves such as those shown in Fig. 4-4 might result. In the example illustrated, $2^3/_8$-in. tubing would be better than $3^1/_2$-in. at a THP of 200 psi, but the reverse would be true at a THP of 100 psi.

The manner in which the energy-loss factor *f* was correlated in terms of the other variables is best described in the words of Poettmann and Carpenter themselves (Ref. 1):

> However, in order to establish the values of *f* as some function of the variables involved in the flow of oil, water, and gas in the tubing, the reverse calculation was made. This involved making use of field data in the form of production data, PVT data, and pressure and temperature traverses. Values of *f* were calculated in this manner from the field data and were found to correlate best as a function of the product of the inside diameter of the tubing and the mass velocity of the fluid in the tubing [see Fig. 4-5]. In terms of units obtained from

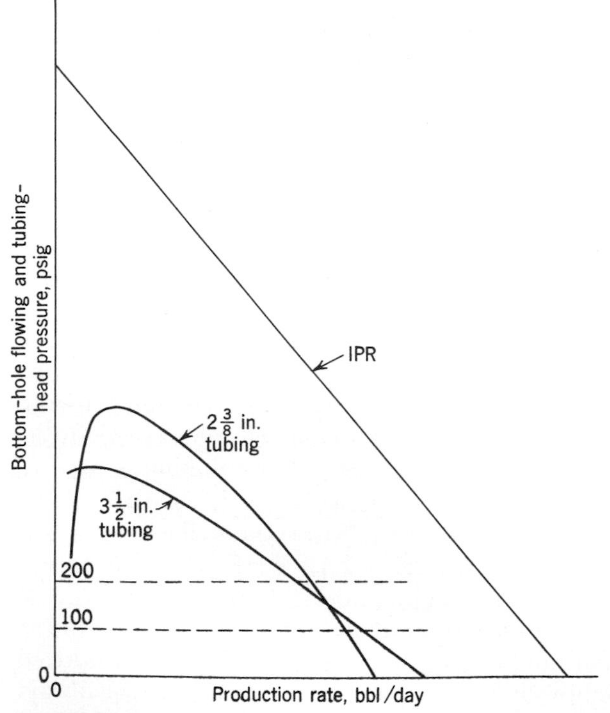

Fig. 4-4 Determination of optimum tubing size.

field data:

$$Dv\rho = (1.4737 \times 10^{-5}) \frac{Mq}{D} \tag{4-4}$$

. . . . . .

The fact that viscosity is not one of the variables involved in the vertical multi-phase flow is both fortunate and to be expected. This is because of the fact that the degree of turbulence is of such a magnitude that, of the total energy loss W_f, that portion resulting from viscous shear is negligible.

It may be mentioned here that, because of the logarithmic plot, the scatter of the points in Fig. 4-5 is in fact greater than would appear at a glance. For instance, when $D\rho v$ is 2.8, the curve gives a value of 1.8 for f, but the smallest value of f calculated from observations is 0.68 and the largest is 3.9. When $D\rho v$ is 25, the curve gives f equal to 0.012; the smallest value from the observations is 0.0042 and the largest is 0.033.

Poettmann and Carpenter's method has been adapted to the problem of annular-flow gradients by Baxendell (Ref. 2) and extended to that of high-rate tubing-flow gradients by Baxendell and Thomas (Ref. 3). This work was based on the results of tests run in La Paz field, Venezuela, from which it was possible to compute values of the factor f for high flow rates. The suggested high-rate correlation is shown in Fig. 4-6, and it is apparent that the proposed modification and extension tie in well with Poettmann and Carpenter's original results. In particular, it should be noted that the curve proposed by Baxendell and Thomas applies to values of the energy-loss factor less than 0.008, whereas Poettmann and Carpenter's curve at these low f values was based on six points only (Fig. 4-5).

It was remarked by Baxendell and Thomas that there was a high degree of consistency in the field results at tubing rates sufficiently high to give a value of qM/D greater than 3.0×10^6 (roughly speaking, production rates greater than 500 bbl/day). In fact, these workers state their belief that in this flow range the accuracy of pressure calculations should be ±5 percent (Ref. 3). Such consistency is in marked contrast to the results obtained by Poettmann and Carpenter at the lower flow rates (Fig. 4-5). As pointed out by Ros (Ref. 4), the reason is probably that at the high velocities accompanying high flow rates, the free gas does not have an opportunity to slip through the oil in the tubing (see Sec. 4-7), so energy losses in the tubing are almost entirely the result of wall friction; consequently, the assumption of a simple expression such as Eq. (4-1) for the energy losses resulting from irreversible effects is reasonably good. But at low liquid velocities in the tubing (that is, at low flow rates) a considerable portion of the energy lost to the system is carried by the gas as it bubbles up through the liquid (gas slippage); hence, the assumption of a single energy-loss factor is inadequate.

Several attempts to overcome this difficulty have been made; one, by Ros (Ref. 4), will be discussed briefly in the next section. A second is that of Tek (Ref. 9), who worked from the idea that the energy-loss factor should be cor-

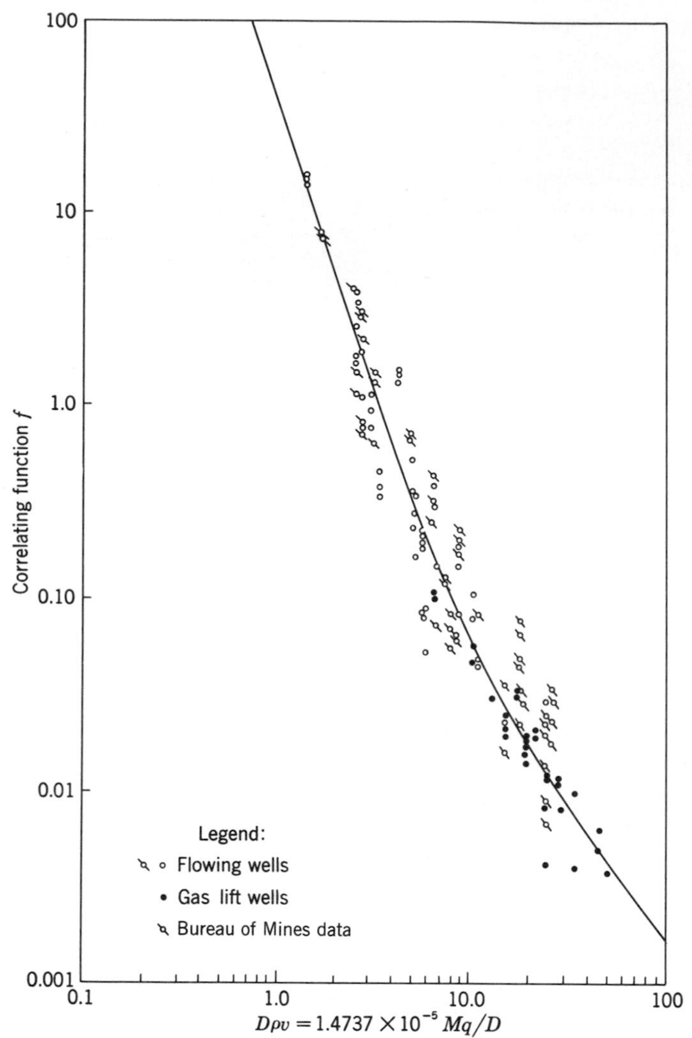

Fig. 4-5 Correlation of field data on flowing and gas-lift wells. (*After Poettmann and Carpenter, Ref. 1. Courtesy* API Drill. Prod. Practice.)

related against a dimensionless quantity rather than against the dimensional product qM/D. Toward this end he introduced, with a marked degree of success, a two-phase Reynolds number and correlated f as a function of this number and of the mass ratio of gas to liquid, based on separator and stock-tank quantities.

The weakness in Poettmann and Carpenter's approach to the problem of the pressure losses accompanying two-phase vertical flow is the inaccuracy in the determination of the correlating function f, but the work of Ros (Ref. 4),

Tek (Ref. 9), and others appears to be overcoming this problem. Gilbert (Ref. 5) has developed an elegant graphical method for finding the vertical-flow pressure losses (Sec. 4-5). The curves required in Gilbert's methods can be computed and drawn from any of the different methods for pressure gradient calculation, and, for the sake of completeness, an illustration using Poettmann and Carpenter's equations follows. The description given is based on a procedure outlined by Baxendell (Ref. 2).

The pressure versus depth curve in a $2^7/_8$-in. tubing is required when the liquid flow rate is 200 bbl/day, with a WOR of 0.2 and a GLR of 0.5 mcf/bbl. If Y lb is the mass of 1 bbl of stock-tank liquid (16.7 percent water and 83.3 percent oil) and d lb/cu ft is the density of the gas at standard conditions, then

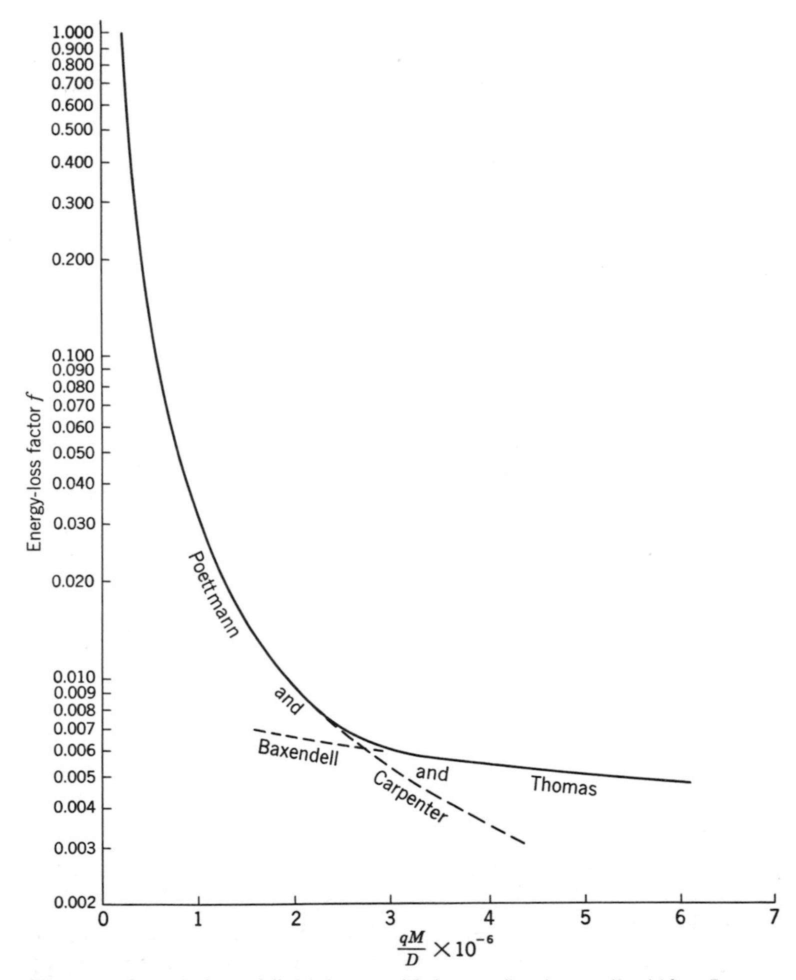

Fig. 4-6 Correlation of field data on high-rate flowing wells. (*After Baxendell and Thomas, Ref. 3. Courtesy AIME.*)

M, the mass of gas and liquid associated with 1 bbl of stock-tank liquid, is given by

$$M = Y + 500d \qquad \text{lb} \tag{4-5}$$

Since q, the liquid production rate, is 200 bbl/day and D, the inside diameter of the tubing in feet, is roughly 2.5/12, the value of $1.4737 \times 10^{-5}\,(Mq/D)$, which is the function plotted along the abscissa of Fig. 4-5, is

$$1.4737 \times 10^{-5}\,\frac{(Y + 500d)200}{2.5/12}$$

The values of Y and d are obtainable from a standard laboratory PVT analysis of the oil under consideration, so the energy-loss factor may be read from Fig. 4-5. Thus the function $\overline{K}$ defined by Eq. (4-3) may be calculated.

In order to obtain a value for $\Delta p/\Delta h$, it is necessary to determine the average density $\bar{\rho}$ in lb/cu ft of the fluid in the interval Δh [Eq. (4-2)]. The density is the mass divided by the volume, and the mass associated with 1 bbl of stock-tank liquid is given by Eq. (4-5). The volume is equal to the volume of oil plus the volume of water plus the volume of free gas. If the average pressure in the interval Δh under consideration is p, the PVT analysis will give the values of the oil formation volume factor B_o, the gas formation volume factor B_g, and the gas solubility R_s in that interval. Since the WOR at the surface is 0.2, each barrel of stock-tank liquid is made up of 0.833 bbl of oil and 0.167 bbl of water. Thus the oil volume per barrel of stock-tank liquid in the interval Δh is $0.833B_o$, and the water volume is 0.167 (neglecting the compressibility of water and the small solubility of gas in water). Moreover, R_s scf of gas goes into solution in 1 stock-tank bbl of oil at the pressure p; hence, of 500 cu ft of free gas per barrel of liquid at surface conditions, $0.833R_s$ cu ft is in solution in the interval Δh. The free-gas volume is therefore $(500 - 0.833R_s)$ scf, or $B_g(500 - 0.833R_s)$ bbl in the tubing at the pressure p.

Thus the total volume at the pressure p of the fluids associated with 1 stock-tank bbl of liquid is

$$0.833B_o + 0.167 + B_g(500 - 0.833R_s) \qquad \text{bbl}$$

or

$$5.614\,[0.833B_o + 0.167 + B_g(500 - 0.833R_s)] \qquad \text{cu ft}$$

When this expression is used in conjunction with Eq. (4-5), it is evident that

$$\bar{\rho} = \frac{Y + 500d}{5.614\,[0.833B_o + 0.167 + B_g(500 - 0.833R_s)]} \qquad \text{lb/cu ft} \tag{4-6}$$

The pressure gradient $\Delta p/\Delta h$ can now be calculated from Eq. (4-2).

If a value is assumed for the THP (for instance, 200 psi) and if pressure increments of, say, 200 psi are used, the gradient associated with each incremental step can be calculated as outlined above. Dividing the pressure increment of 200 psi by this pressure gradient gives the length of tubing over which the increment occurs. In this way a plot of pressure versus depth may be made. This plot may then be extrapolated to zero pressure and the depth scale so adjusted that zero pressure corresponds to zero depth. By repeating this procedure for a range of assumed flow rates, GLRs, and tubing sizes, families of pressure-distribution curves similar to those shown in Figs. 4-11 through 4-20 may be prepared.

4-4 VERTICAL LIFT PERFORMANCE: ROS

Ros (Ref. 4) took as a basis for his study the pressure-balance rather than the energy-balance equation. For single-phase flow, the pressure gradient is equal to the sum of the static, the friction, and the acceleration gradients, so that

$$\frac{dp}{dh} = g\rho + 4f\frac{{}^1\!/_2\rho v^2}{D} \tag{4-7}$$

the acceleration gradient being neglected. In this equation, which is equivalent to that of Poettmann and Carpenter (Eq. 4-2), the symbol ρ stands for the density of the fluid, g is the acceleration due to gravity, and v is the fluid velocity.

For two-phase flow, the static gradient term in Eq. (4-7) should be modified to allow for the possibility of investigating the effects of gas slippage through the liquid, and Ros therefore replaced it by

$$\epsilon_l \rho_l g + \epsilon_g \rho_g g \tag{4-8}$$

where ρ_l and ρ_g are the densities of the liquid and gas, respectively, ϵ_l is the *liquid holdup,* and ϵ_g is the *gas holdup.* The liquid holdup may be defined as the volume of liquid actually present in a certain length of pipe, divided by the volume of that length of pipe; the definition of gas holdup is similar. Evidently,

$$\epsilon_g + \epsilon_l = 1 \tag{4-9}$$

Although this modification introduced by Ros does not alter the meaning attached to the fluid density by Poettmann and Carpenter, it does separate the effects of the gas from those of the liquid and so permit an experimental investigation to be made to determine the factors influencing each of these parts of the composite static-gradient term and also their relationship to the friction term. If Eq. (4-8) is introduced into Eq. (4-7) and use is made of Eq. (4-9), it follows that

$$\frac{1}{\rho_l g}\frac{dp}{dh} = \epsilon_l + (1 - \epsilon_l)\frac{\rho_g}{\rho_l} + \text{friction term} \tag{4-10}$$

Since in most field situations the density of the gas is very small compared with the density of the oil, the term $(1 - \epsilon_l)\rho_g/\rho_l$ may be neglected, and Eq. (4-10) reduces to

$$\frac{1}{\rho_l g}\frac{dp}{dh} = \epsilon_l + \text{friction term} \tag{4-11}$$

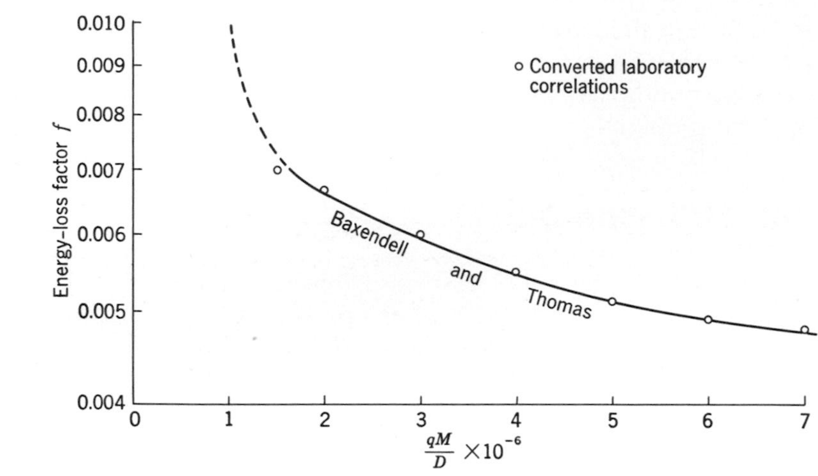

Fig. 4-7 Comparison between laboratory correlations and field data, high-rate flowing wells. (*After Ros, Ref. 4. Courtesy AIME.*)

Ros next carried out a dimensional analysis to isolate those dimensionless groups of major importance in determining the pressure gradients under two-phase vertical flow; he then designed an experimental program to determine the influences of these groups in particular and also those of the groups of secondary importance in regions of special interest. In all, the program gave some 20,000 data points and covered a wide range of pipe sizes, oil and gas rates, and oil viscosities. The resulting empirical correlations for liquid holdup and for wall friction are complex and the computations of pressure gradients based on them are programmed for the computer.

A comparison of results calculated from these correlations with the energy-loss factor obtained by Baxendell and Thomas from field tests (Fig. 4-6) is illustrated in Fig. 4-7 and indicates excellent agreement between laboratory and field data.

4-5 VERTICAL LIFT PERFORMANCE: GILBERT

Gilbert's approach (Ref. 5) to the vertical two-phase flow problem was empirical: based on measured values of tubing-flow pressure losses, families of curves were derived that can be used for extrapolation and interpolation purposes. Gilbert's work will be summarized in this section, but for a more complete account, and for the historical development, the reader is referred to the original work by Shaw (Ref. 10), Babson (Ref. 11), and Gilbert (Ref. 5).

Suppose that the following measurements have been taken in a large number of flowing wells:

Tubing depth, ft
Flowing BHP (that is, tubing-intake pressure), psi

THP, psi
Gross liquid rate, bbl/day
GLR, mcf/bbl
Tubing size

Assuming that the flowing BHP depends (so far as the vertical flow up the tubing is concerned) on the other five variables only, the first step in an attempted correlation is to pick out all those wells producing through the same size of tubing at some fixed GLR and gross liquid rate (in practice, of course, all those wells having GLRs and liquid rates lying within small ranges would be taken together).

If the flowing BHPs are plotted as a function of depth for this group of wells, a result of the type illustrated in Fig. 4-8 is obtained. Each of the curves *a*, *b*, *c*, and *d* corresponds to a different THP, the THPs being the points of intersection of these curves with the zero-depth line (points *A*, *B*, *C*, and *D*). Each of these curves represents the pressure distribution along the tubing for a well flowing at a fixed rate and GLR.

Now suppose a well is producing with a THP *A* and that a valve is connected at the depth *OF*. At a certain instant of time this valve is opened in such a way that the pressure in the tubing at point *F* is unchanged and, simultaneously, the tubing-head valve is closed. Then the pressure distribution in the tubing below point *F* will not be altered, and the well will continue to flow at the same rate. If point *F* is now looked upon as the tubing head, the well is flowing with a THP *E*, which, in Fig. 4-8, is the same as *B*. That is, if curve *a*

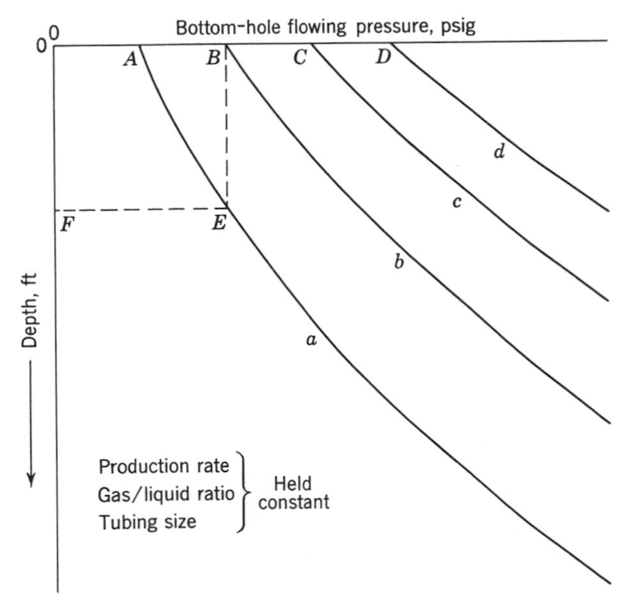

Fig. 4-8 Flowing BHP as function of THP and tubing length: constant GLR, production rate, and tubing size.

below point E is moved upward a distance BE, it must coincide with the curve b. In other words the curves a, b, c, and d are in reality all parts of just one curve, and they can be made to coincide by moving them vertically. In Fig. 4-9 this one curve is shown, with the THPs A, B, C, and D, corresponding to those of Fig. 4-8, marked. The curve c, for instance, of Fig. 4-8 is the curve of Fig. 4-9 with the point X taken as the zero of depth.

To use the curve of Fig. 4-9 to determine the flowing BHP from the THP, given the footage of tubing in the hole, the depth on the arbitrary scale corresponding to the known THP is noted. The *equivalent length* of the tubing is then determined by adding the actual length of tubing to this "THP depth," and the flowing BHP corresponding to this equivalent length of tubing is read from the curve. The curve of Fig. 4-9 is the *pressure-distribution curve* for a certain size of tubing through which liquid is flowing at a fixed rate and gas at a fixed-output GLR.

Before presenting Gilbert's families of curves, it should be noted that there remains some doubt about the correct shape of the curve at low pressures, and indications are that reverse curvature may be present; that is, the pressure-distribution curves may be concave downward instead of concave upward in the low-pressure range (Fig. 4-10). This type of reversal in curvature is common in the curves calculated by using Poettmann and Carpenter's method (Sec. 4-3), but there has always been the question whether the apparent anomaly might be due to a deficiency in the method. To check this point,

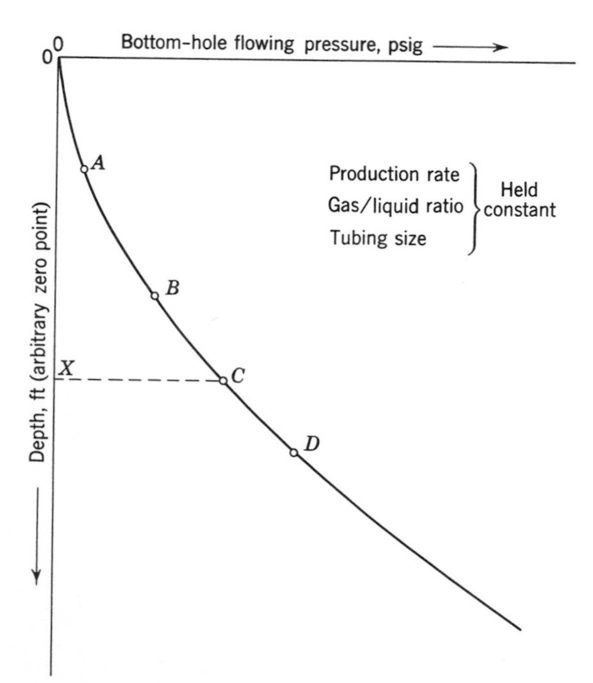

Fig. 4-9 Pressure-distribution curve; vertical two-phase flow.

Baxendell and Thomas (Ref. 3) carried out a carefully controlled well test, and their results confirm the presence of reverse curvature. Figure 4-10 illustrates the measured results together with the calculated curve, which is seen to fit the observed data very closely.

Gilbert's families of curves use GLR as a parameter, and there is one family for each tubing size and liquid rate. The curves themselves are shown in Figs. 4-11 through 4-20. In referring to these figures, note that pressures are given in psi, depths in thousands of feet, production rates in bbl/day, GLRs in mcf/bbl, and tubing sizes in inches (outer diameter). It will also be noted that two depth scales are shown on each graph; the reason for this will be discussed in Sec. 4-7. The graphs shown cover the cases of 1.66-in., 1.9-in., $2^3/_8$-in.,

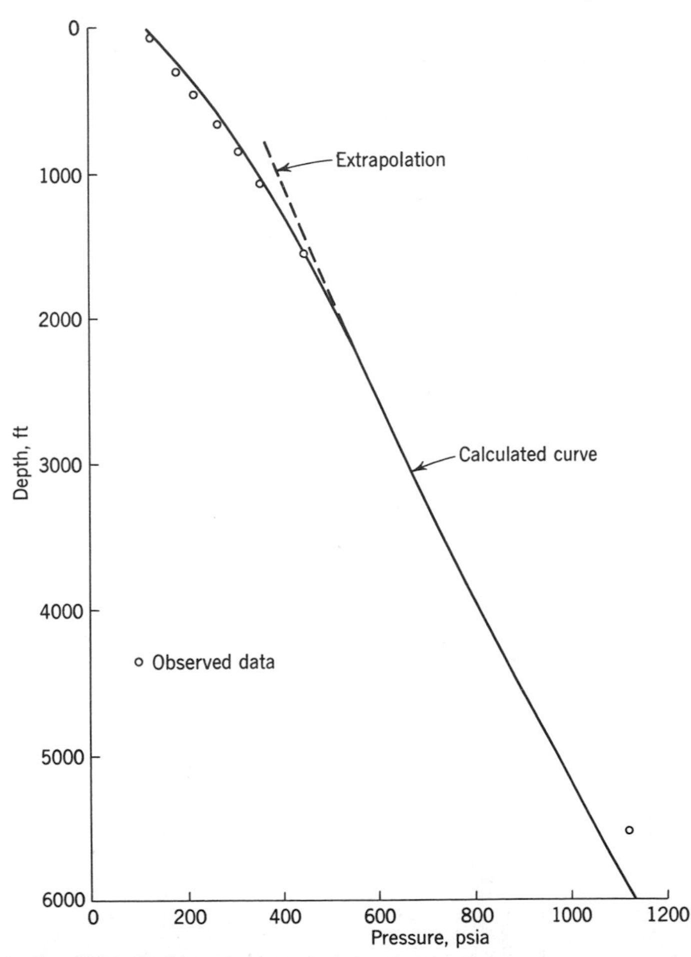

Fig. 4-10 Experimental validation of reverse curvature in low-pressure range of calculated pressure-distribution curves. (*After Baxendell and Thomas, Ref. 3. Courtesy AIME.*)

$2^7/_8$-in., and $3^1/_2$-in. (OD) tubing; gross production rates of 50, 100, 200, 400, and 600 bbl/day; GLRs of up to about 7 mcf/bbl (the actual value varies with the tubing size and the production rate); and total pressure drops of up to 3500 psi in the case of $2^7/_8$-in. tubing and 2500 psi for the other tubing sizes.

Example 4-1 Find the flowing pressure at the foot of 13,000 ft of $2^3/_8$-in. tubing if the well is flowing 100 bbl/day at a GLR of 1.0 mcf/bbl with a THP of 200 psi.

Referring to the family of curves for $2^3/_8$-in. tubing at 100 bbl/day (Fig. 4-18) and using that pressure-distribution curve which corresponds to a GLR of

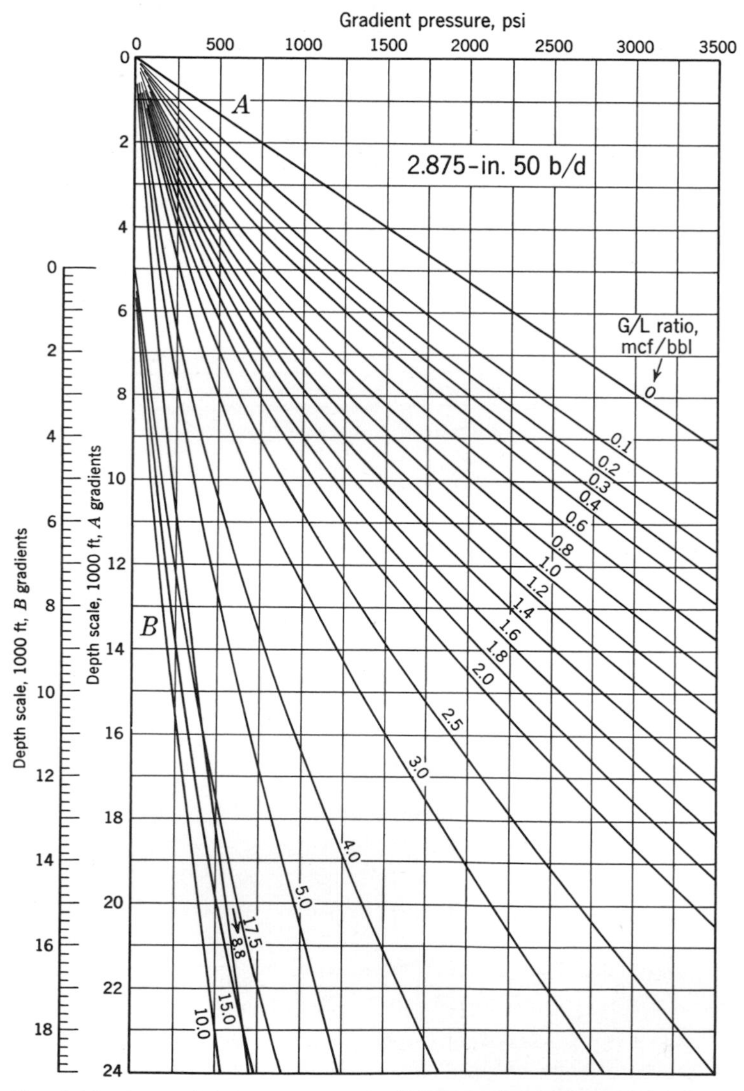

Fig. 4-11 Approximate pressure-distribution curves for 2⅞-in. tubing at 50 bbl/day. (*From Gilbert, Ref. 5. Courtesy* API Drill. Prod. Practice.)

1.0 mcf/bbl, it is seen that the equivalent depth of 200-psi THP (point *A* of Fig. 4-9) is 2600 ft. The length of tubing is 13,000 ft, so the tubing shoe is 13,000 ft below this equivalent depth of 2600 ft. Thus, the equivalent depth of the tubing shoe is 15,600 ft, and the pressure-distribution curve gives the value of the pressure at this depth as 1900 psi.

Example 4-2 What is the THP of a well, completed with 8000 ft of 2³/₈-in. tubing, that is flowing at 600 bbl/day and a GLR of 0.4 mcf/bbl if the pressure at the bottom of the tubing is 2200 psi?

Referring to the appropriate pressure-distribution curve (Fig. 4-19)

Equivalent depth of pressure at tubing shoe = 12,000 ft
Actual depth of tubing shoe = 8000 ft
Equivalent depth of THP = 4000 ft
THP = 530 psi

Example 4-3 It is hoped to flow a well having a PI of 0.4 bbl/(day)(psi) and a static pressure of 1500 psi at a rate of 400 bbl/day through 4000 ft of 2⁷/₈-in. tubing. The GLR is 0.2 mcf/bbl. Will the well flow at the desired rate?

The flowing BHP when the production rate is 400 bbl/day may be derived from Eq. (3-2), namely,

$$J = \frac{q}{p_s - p_{wf}}$$

or, substituting the given values,

$$0.4 = \frac{400}{1500 - p_{wf}}$$

so $p_{wf} = 500$ psi.

From the relevant pressure-distribution curve (Fig. 4-14), the equivalent depth of the tubing (that is, of the pressure at the tubing shoe) is seen to be 3200 ft and the actual length of the tubing is 4000 ft. It thus appears that the well could not flow at the desired rate.

In general, there are two possible approaches to the solution of a flowing well problem.

Method 1 is to calculate the pressure at the foot of the tubing at various assumed production rates, both from the formation IPR, which is assumed to be known or determinable, and from an assumed value of the THP by means of the pressure-distribution curves. Evidently the point at which the values of the flowing BHP calculated in the two ways are in agreement gives the production rate of the well at the assumed THP and also the pressure at the foot of the tubing (Fig. 4-21).

Method 2 is to calculate the flowing BHP from the inflow characteristics of the formation and then determine the pressure loss in the tubing at various assumed rates. Thus a curve of THP against production rate is obtained (Fig. 4-22), as was explained in Sec. 4-3. The production rate *D* corresponding to any assumed value *A* for the THP can thus readily be found, as can the corresponding flowing BHP *C*.

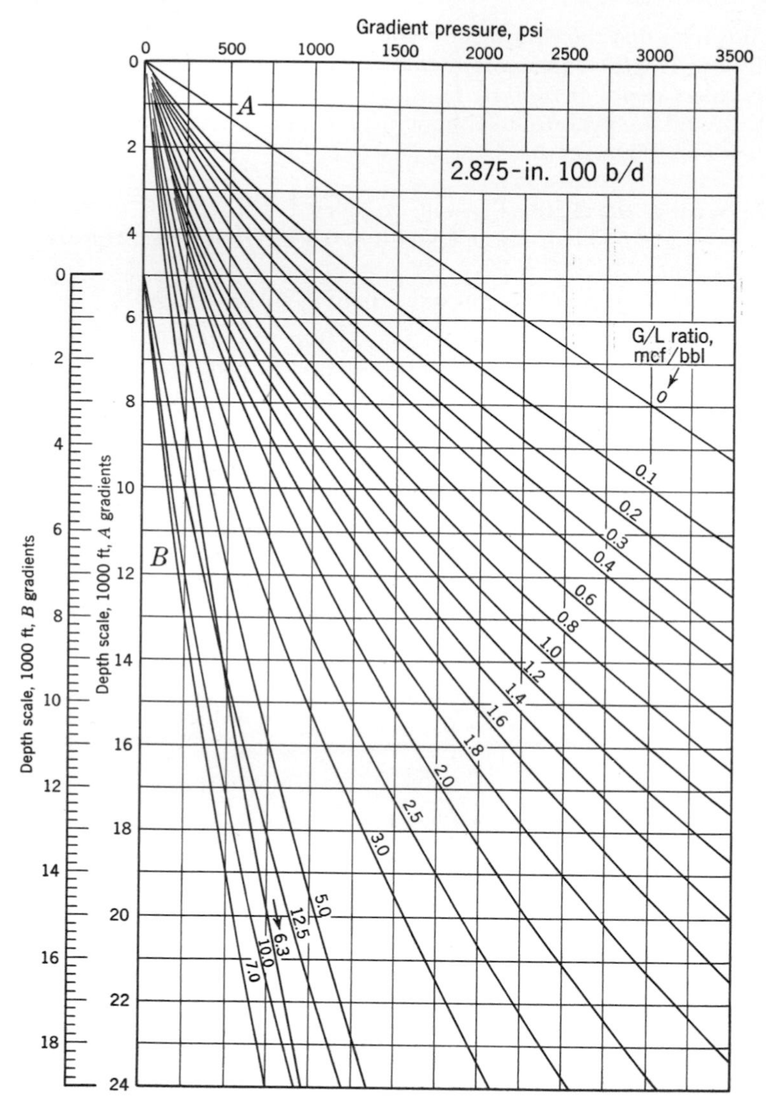

Fig. 4-12 Approximate pressure-distribution curves for 2⅞-in. tubing at 100 bbl/day. (*From Gilbert, Ref. 5. Courtesy* API Drill. Prod. Practice.)

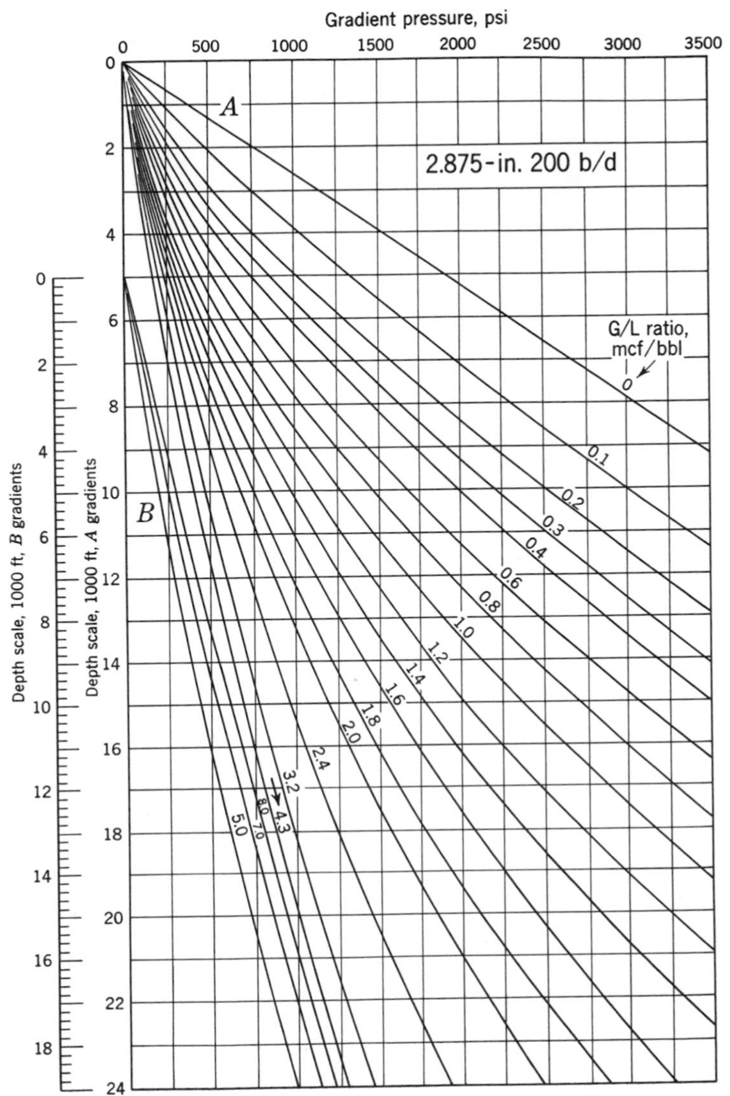

Fig. 4-13 Approximate pressure-distribution curves for 2⅞-in. tubing at 200 bbl/day. (*From Gilbert, Ref. 5. Courtesy* API Drill. Prod. Practice.)

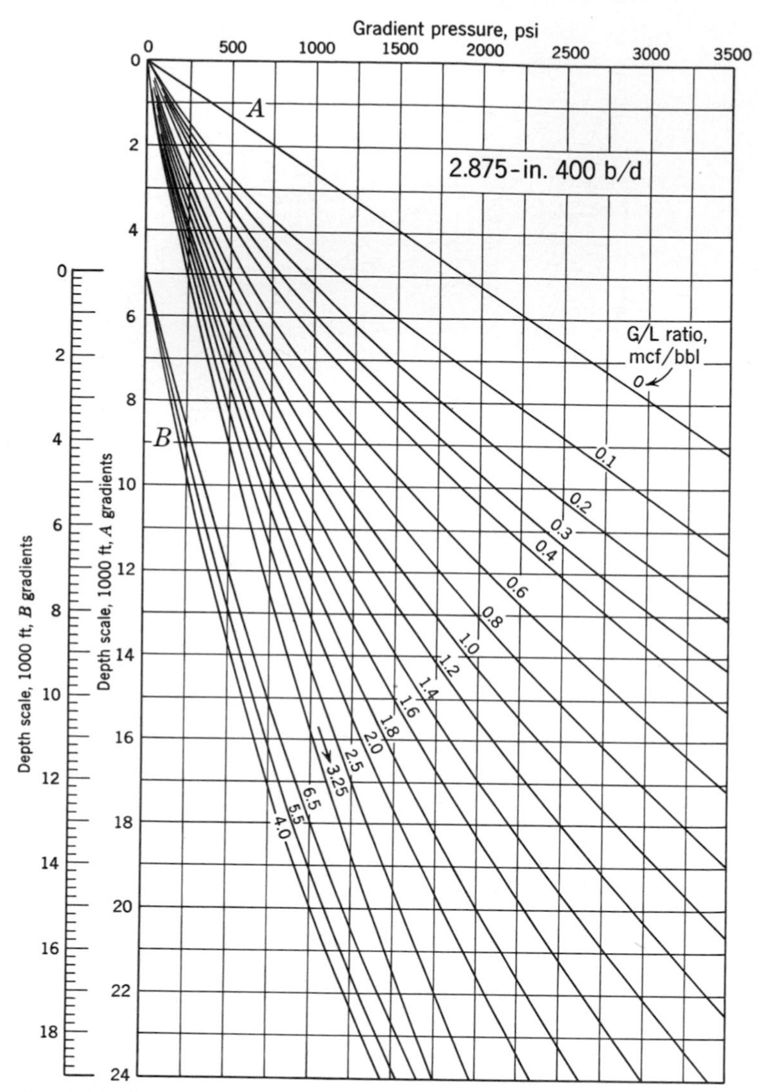

Fig. 4-14 Approximate pressure-distribution curves for 2⅞-in. tubing at 400 bbl/day. (*From Gilbert, Ref. 5. Courtesy* API Drill. Prod. Practice.)

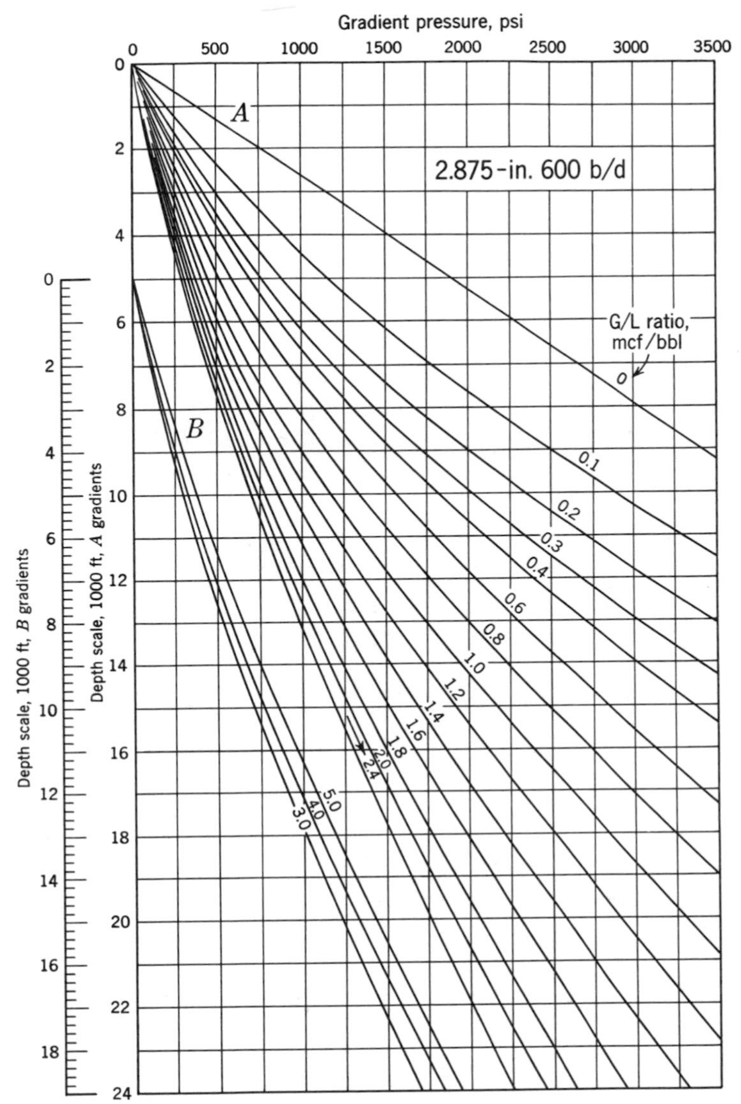

Fig. 4-15 Approximate pressure-distribution curves for 2⅞-in. tubing at 600 bbl/day. (*From Gilbert, Ref. 5. Courtesy* API Drill. Prod. Practice.)

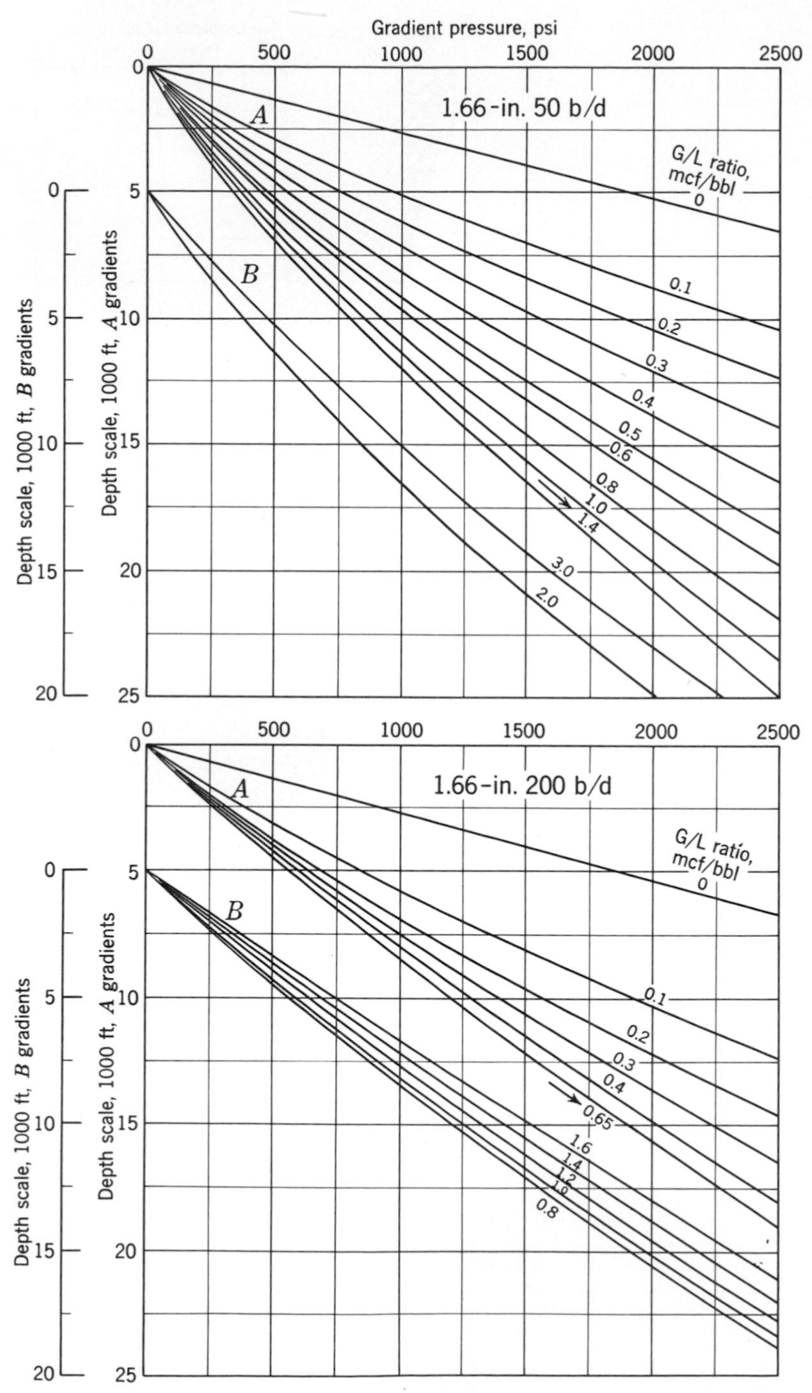

Fig. 4-16 Approximate pressure-distribution curves for 1.66-in. tubing at 50, 100, 200, and 400 bbl/day. (*From Gilbert, Ref. 5, Courtesy* API Drill. Prod. Practice.)

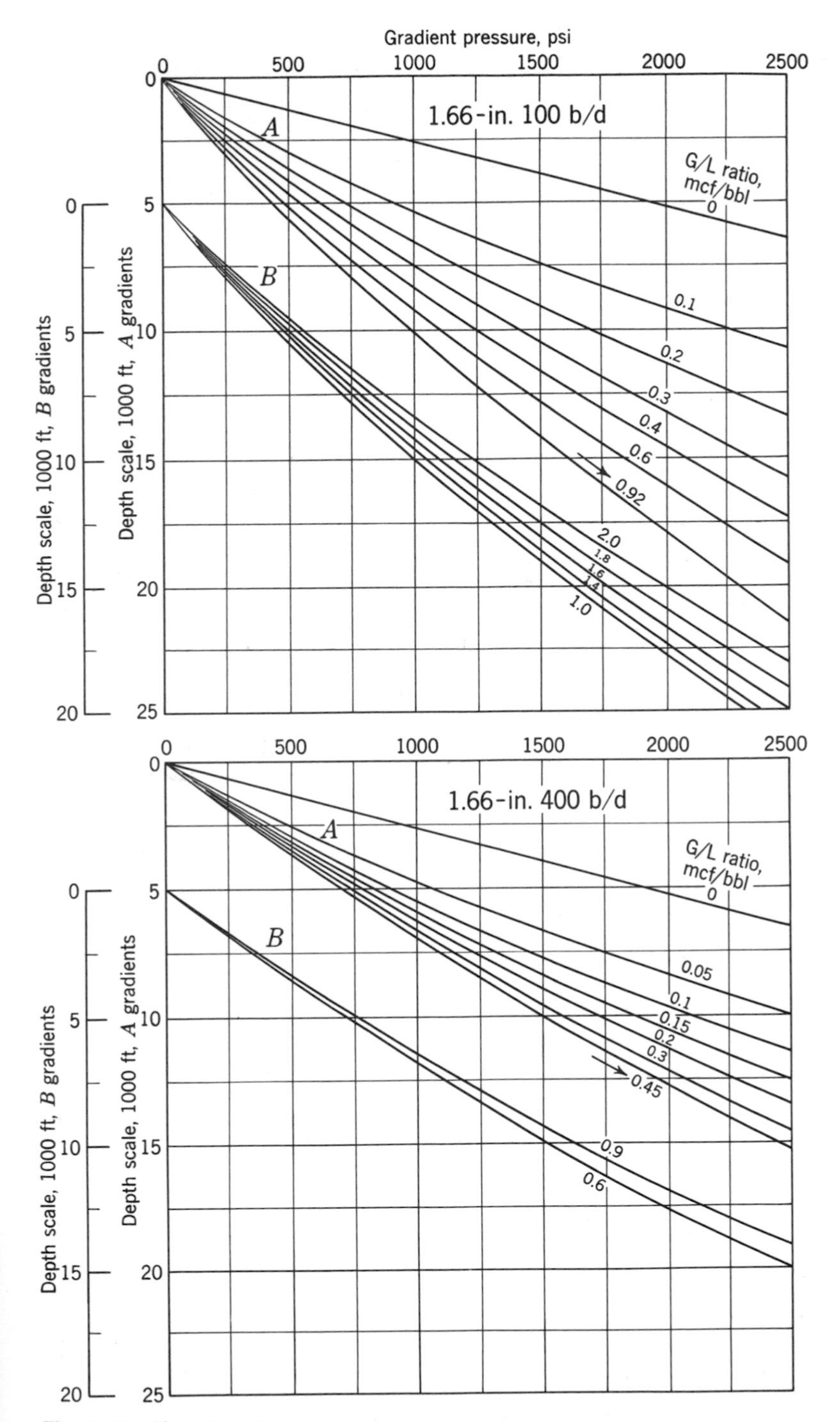

Fig. 4-16 *(Continued)*

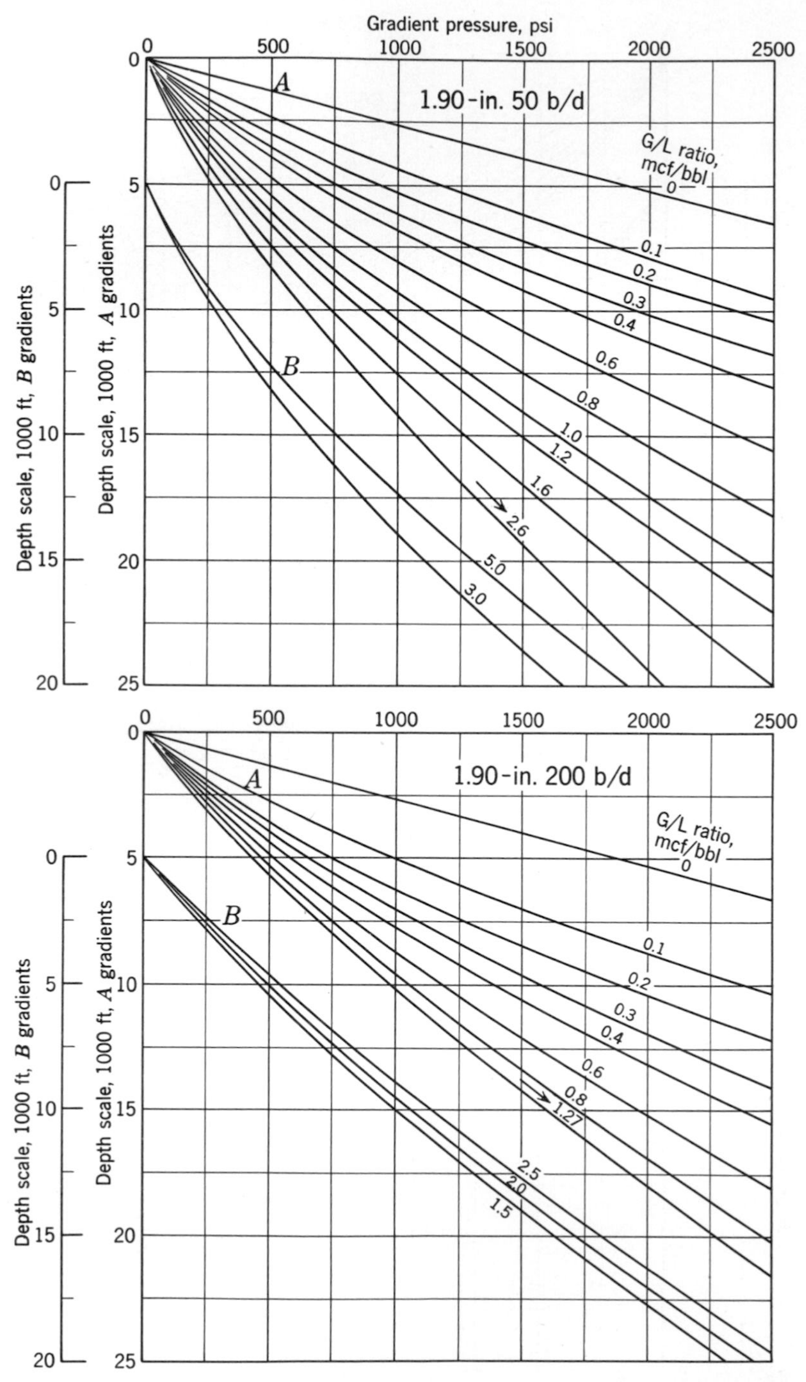

Fig. 4-17 Approximate pressure-distribution curves for 1.9-in. tubing at 50, 100, 200, and 400 bbl/day. (*Gilbert, Ref. 5. Courtesy* API Drill. Prod. Practice.)

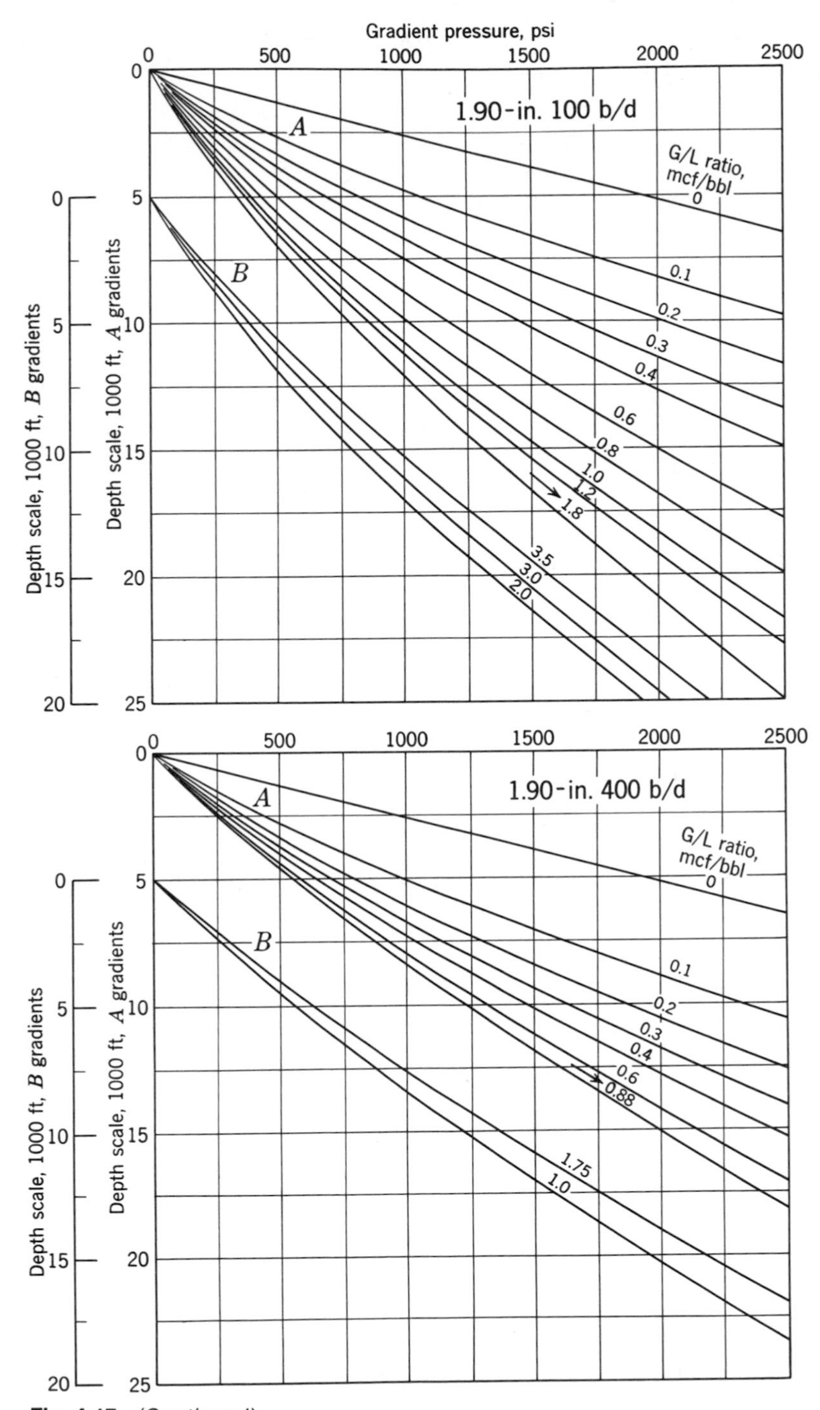

Fig. 4-17 (Continued)

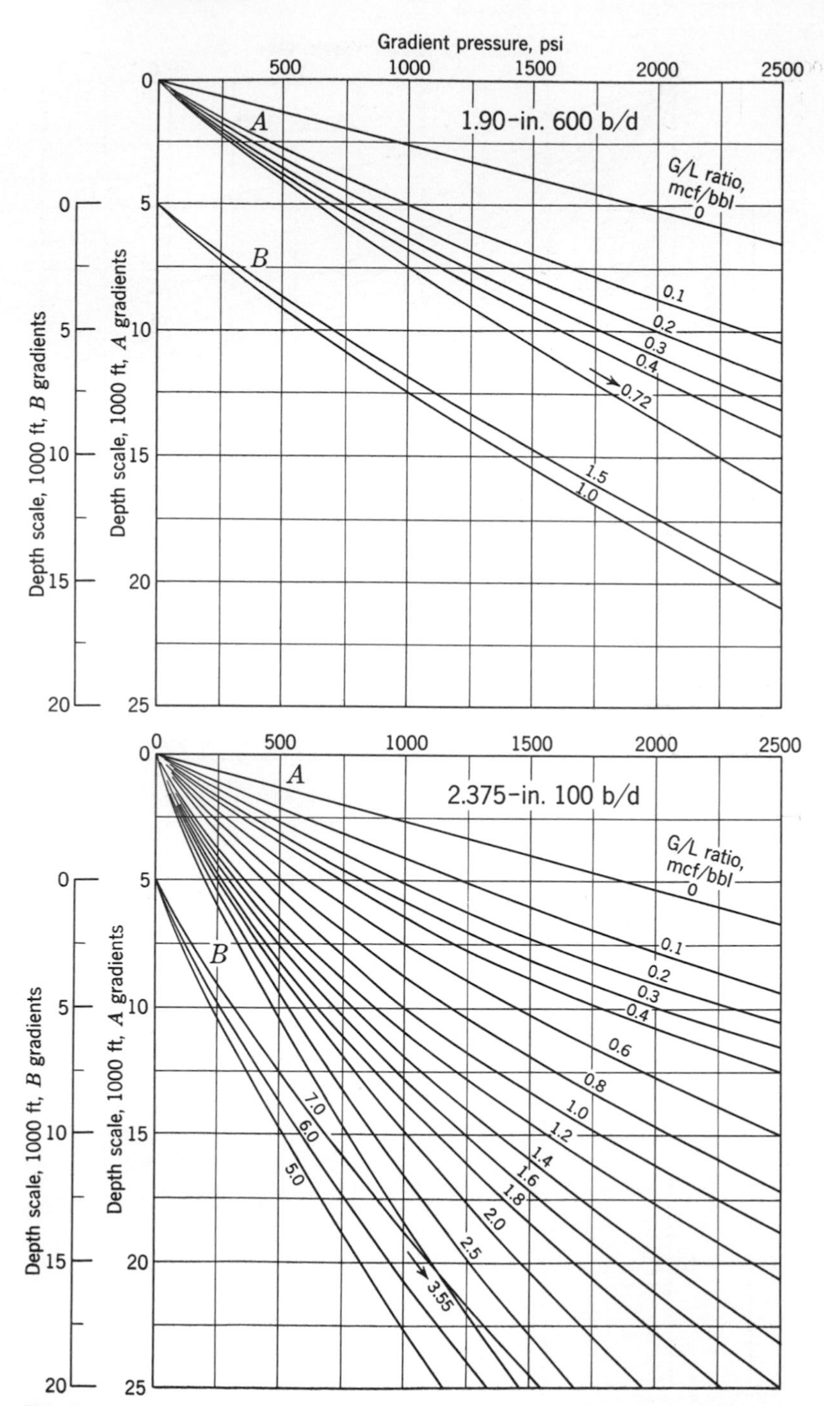

Fig. 4-18 Approximate pressure-distribution curves for 1.9-in. tubing at 600 bbl/day and for $2^3/_8$-in. tubing at 50, 100, and 200 bbl/day. (*From Gilbert, Ref. 5. Courtesy* API Drill. Prod. Practice.)

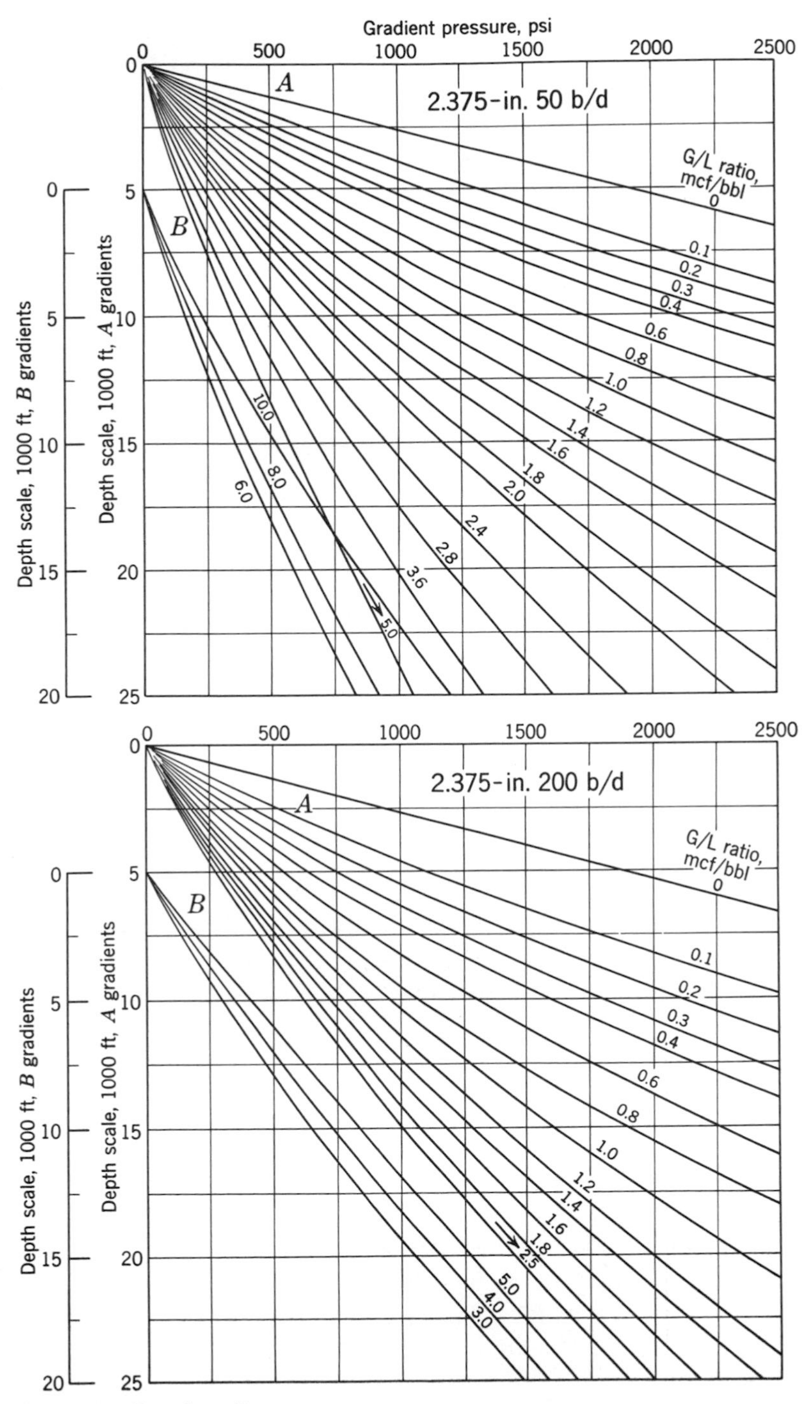

Fig. 4-18 *(Continued)*

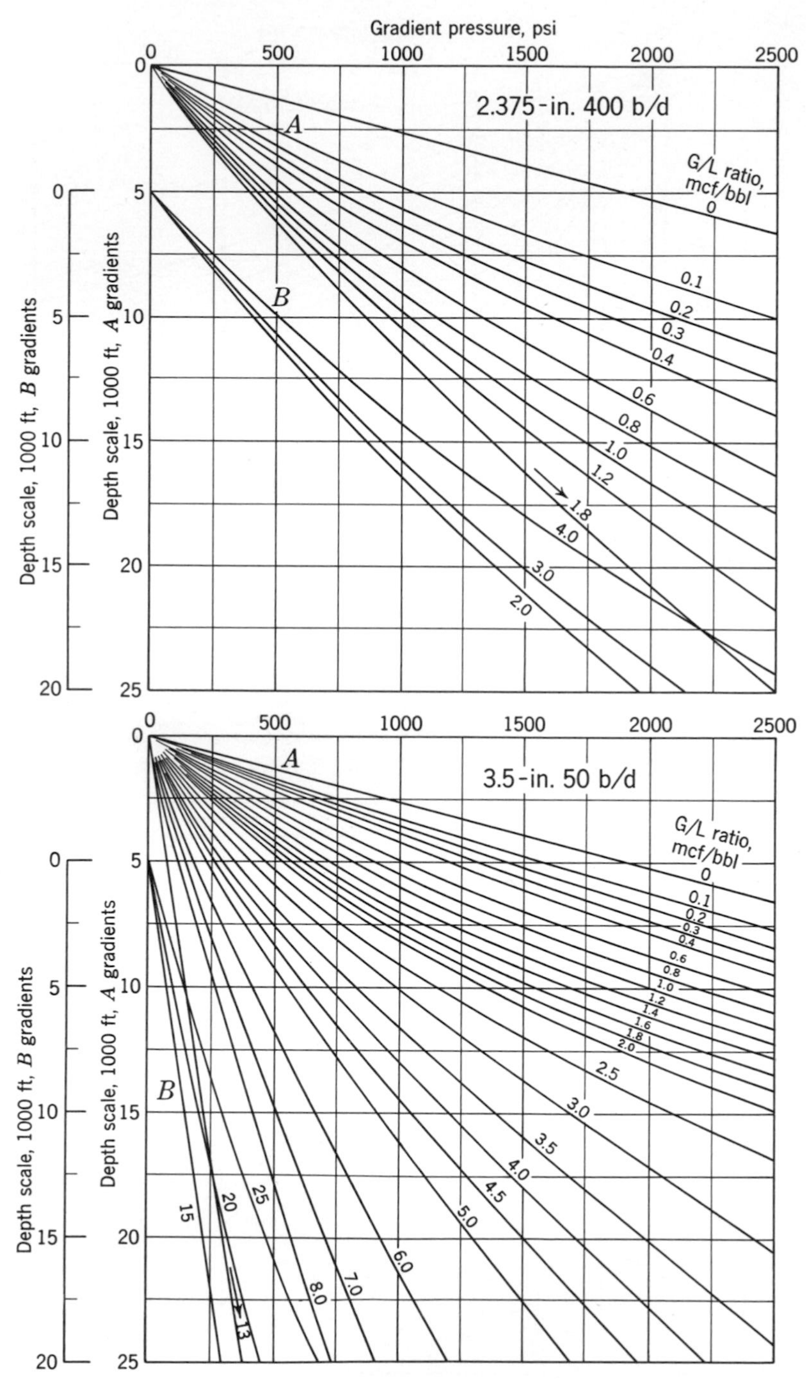

Fig. 4-19 Approximate pressure-distribution curves for 2⅜-in. tubing at 400 and 600 bbl/day and for 3½-in. tubing at 50 and 100 bbl/day. (*From Gilbert, Ref. 5. Courtesy* API Drill. Prod. Practice.)

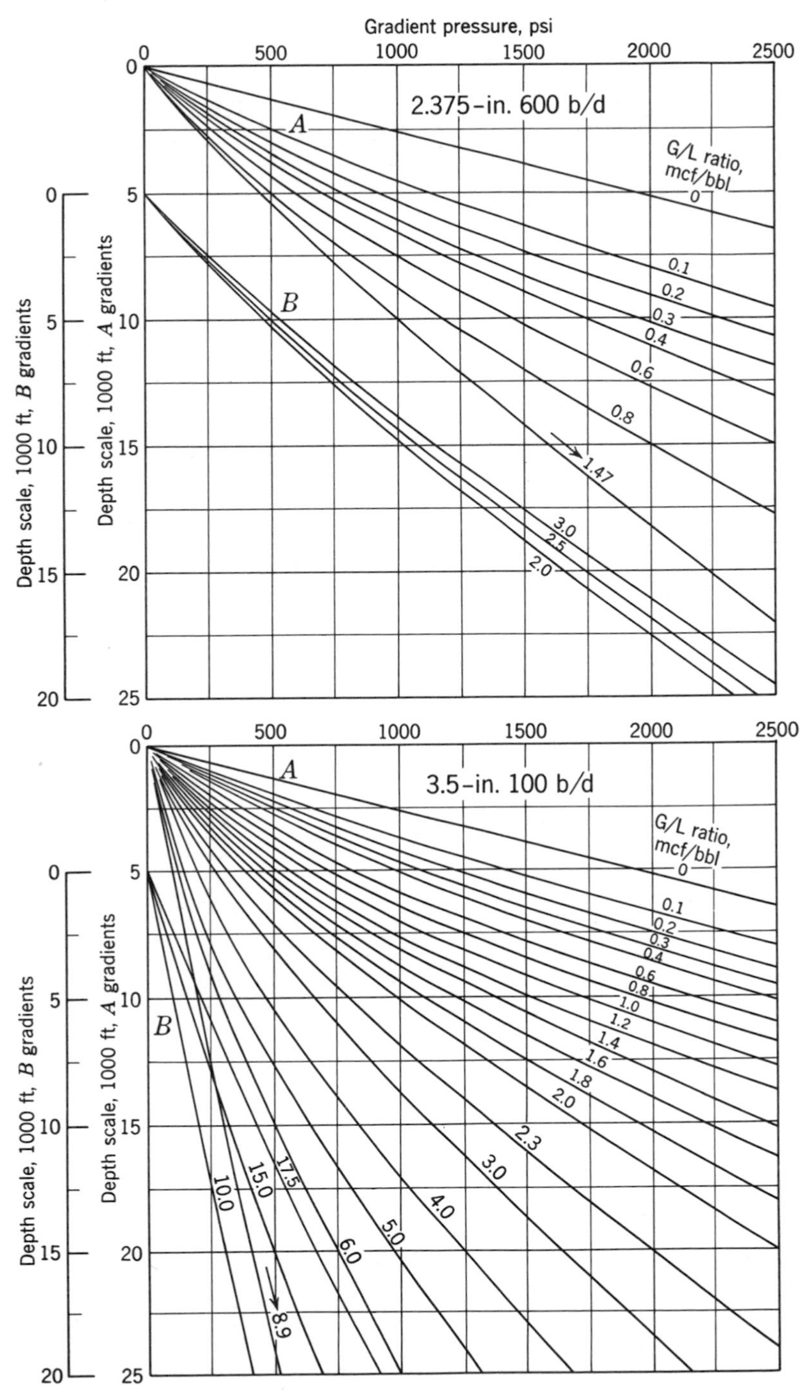

Fig. 4-19 *(Continued)*

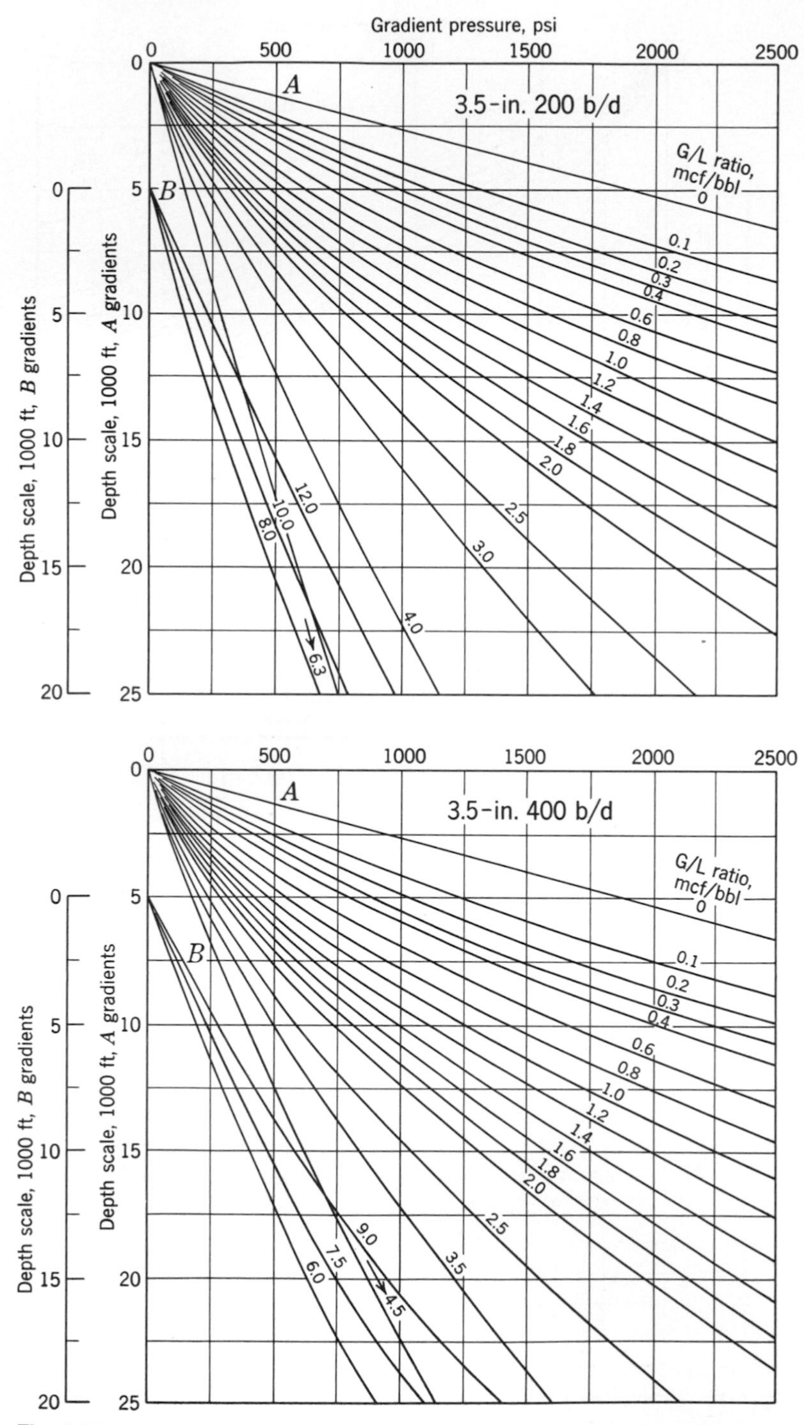

Fig. 4-20 Approximate pressure-distribution curves for 3½-in. tubing at 200, 400, and 600 bbl/day. (*From Gilbert, Ref. 5. Courtesy* API Drill. Prod. Practice.)

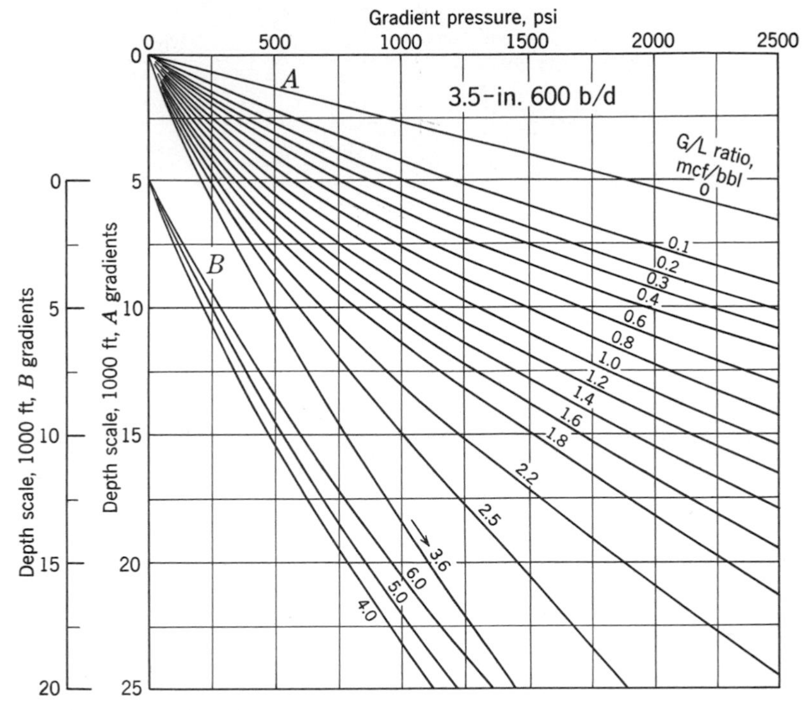

Fig. 4-20 *(Continued)*

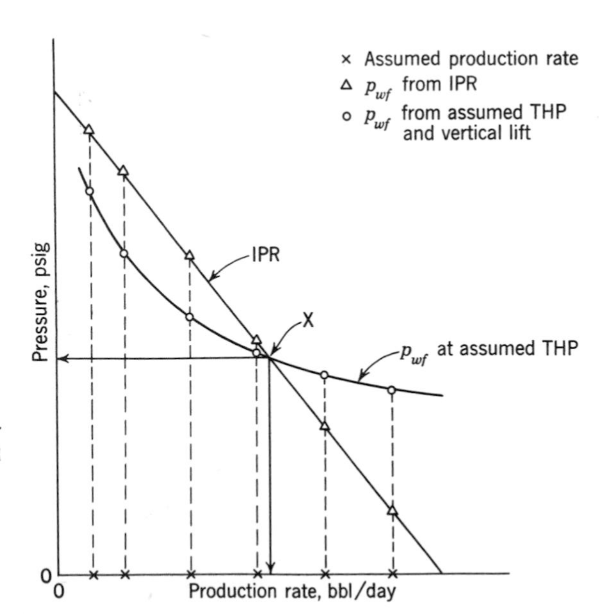

Fig. 4-21 Method 1: Determination of flowing BHP from IPR and assumed THP.

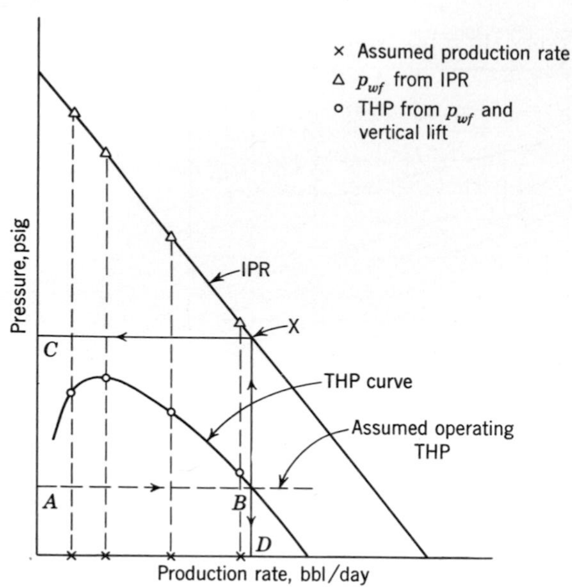

Fig. 4-22 Method 2: Determination of THP from IPR.

Example 4-4 A well producing from a pay zone between 5000 and 5052 ft is completed with $2^7/_8$-in. tubing hung at 5000 ft. The well has a static BHP of 2000 psi and a PI of 0.3 bbl/(day)(psi) and produces with a GOR of 300 cu ft/bbl and a water cut of 10 percent. At what rate will the well flow with a THP of 100 psi?

Calculation of GLR

Let q = gross production rate, bbl/day
q_o = oil production rate, bbl/day
q_w = water production rate, bbl/day

Then
$$\frac{q_w}{q} = 0.1$$

and
$$\frac{\text{gas}}{q_o} = 300$$

$$\text{GLR} = \frac{\text{gas}}{q} = \frac{300q_o}{q} = \frac{300(q - q_w)}{q}$$
$$= 300\left(1 - \frac{q_w}{q}\right) = 300(1 - 0.1)$$
$$= 270 \text{ cu ft/bbl}$$

IPR

The static pressure p_s is 2000 psi and the well's potential q' is 0.3 × 2000, or 600, bbl/day. The IPR can be drawn as shown in Fig. 4-23.

There are now two calculation procedures possible; in some problems one may prove preferable to the other. In this example, for the sake of illustration, both methods will be presented.

Calculation Using Method 1

This involves the calculation of p_{wf} at various values of q, assuming a THP of 100 psi. The steps are shown in Table 4-1. Values of p_{wf} are plotted in Fig. 4-23 (circles), and these points are joined by a smooth curve.

Evidently,

$$q = 300 \text{ bbl/day}$$

so

$$q_w = 30 \text{ bbl/day (10 percent water cut)}$$
$$q_o = 270 \text{ bbl/day}$$
$$p_{wf} = 1000 \text{ psi}$$

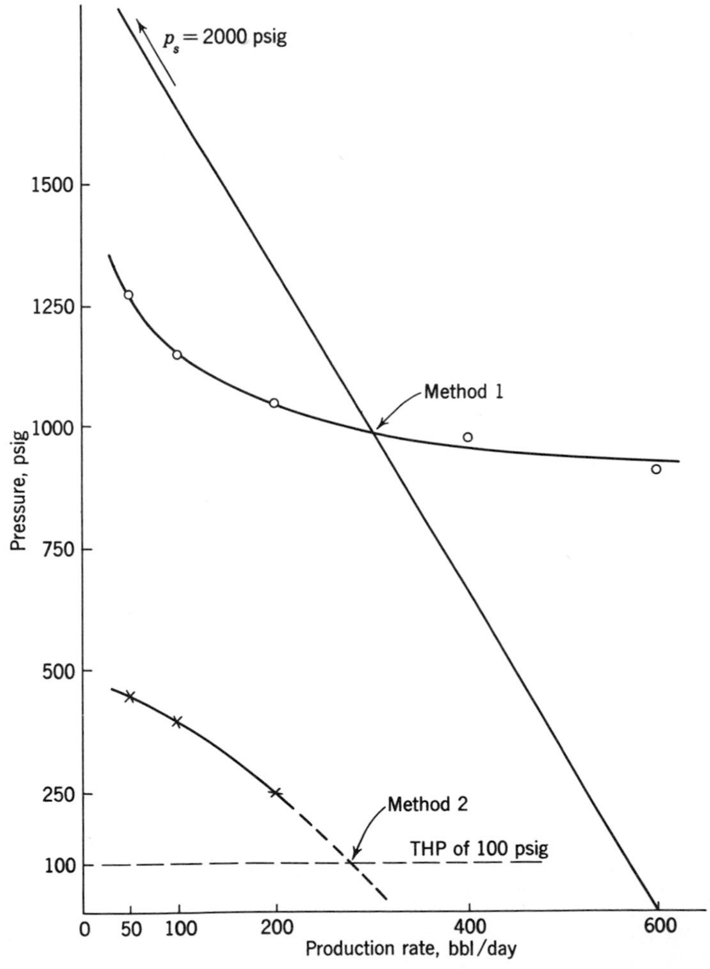

Fig. 4-23 Example 4-4: Graphical determination of anticipated flowing production rate.

TABLE 4-1 EXAMPLE 4-4: Determination of p_{wf} at Various Rates, Using THP of 100 psig

q, bbl/day[a]	Equiv. Depth of THP of 100 psi, ft	Equiv. Depth of Well, ft	p_{wf}, psi
50[b]	500	5500	1275
100[c]	700	5700	1150
200[d]	800	5800	1050
400[e]	800	5800	975
600[f]	800	5800	910

[a] The flow rates in the first column are assumed, and the calculations are carried out as in Example 4-1 with a GLR of 0.27 mcf/bbl.
[b] Fig. 4-11.
[c] Fig. 4-12.
[d] Fig. 4-13.
[e] Fig. 4-14.
[f] Fig. 4-15.

Calculation Using Method 2

This involves the calculation of the THP at various production rates, using the value of p_{wf} from the formation inflow performance, and the steps are shown in Table 4-2. Values of the THP are plotted in Fig. 4-23 (crosses), and these points are joined by a smooth curve. This curve, as drawn, intersects the 100-psi THP line at a production rate of 280 bbl/day. From the IPR, the value of p_{wf} is 1070 psi at this production rate.

In this example, because of the difficulty in drawing in the method 2 curve of THP with sufficient accuracy (only three points on the curve can be found), the

TABLE 4-2 EXAMPLE 4-4: Determination of THP at Various Rates, from Formation and Tubing Performance[a]

q, bbl/day	p_{wf}, psi	Equiv. Depth of p_{wf}, ft	Equiv. Depth of THP, ft	THP, psi
50[b]	1833	7300	2300	450
100[c]	1667	7500	2500	400
200[d]	1333	6700	1700	250
400[e]	667	4200		
600[f]	0			

[a] The figures in the second column may be derived from the IPR of Fig. 4-23 or may be calculated from the formula for the PI, namely,

$$J = \frac{q}{p_s - p_{wf}}$$

The remainder of the calculations are carried out as shown in Example 4-2, using a GLR of 0.27 mcf/bbl.
[b] Fig. 4-11.
[c] Fig. 4-12.
[d] Fig. 4-13.
[e] Fig. 4-14.
[f] Fig. 4-15.

figure of 300 bbl/day gross production rate given by method 1 is probably the more accurate.

4-6 THE PRESSURE-RATE-DEPTH GRID

From knowledge of the IPR of a well it is possible, by using the pressure-distribution curves, to plot the pressure versus rate relationship at several depths. For example, if the top of the producing formation (and the foot of the tubing) is 7448 ft below the tubing head, pressure versus rate curves could be constructed at, say, 1000-ft intervals—that is, at depths of 6448 ft, 5448 ft, and so on—either until the tubing head is reached, or until it is no longer possible to construct such a curve because all the pressure points are negative. Such a pressure-rate-depth grid is illustrated in Fig. 4-24.

It will be found that a grid of this kind has several uses, not the least of which is in the determination of the optimum depth at which to locate equipment. Particular examples will be given when discussing the question of gas-lift valve placement (Secs. 7-4 and 7-6), and in the consideration of the op-

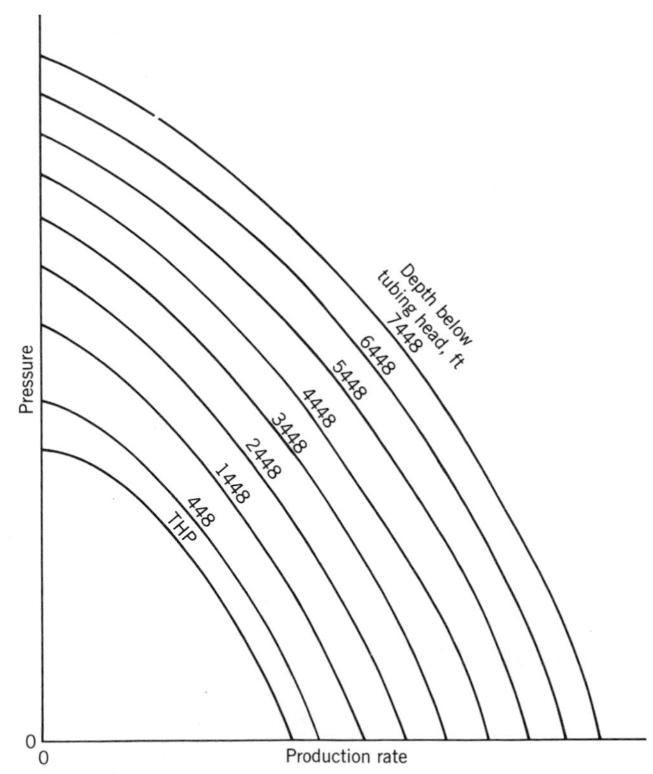

Fig. 4-24 Pressure-rate-depth grid.

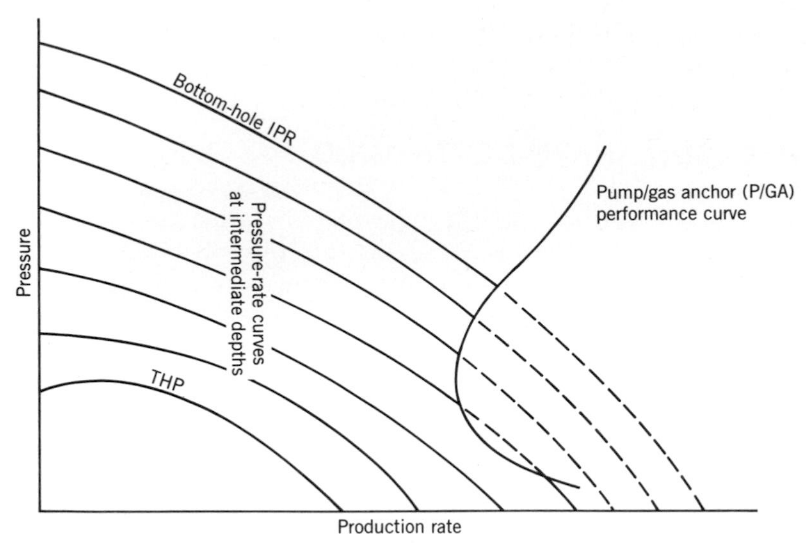

Fig. 4-25 Use of pressure-rate-depth grid in placement of down-hole production equipment at optimum depth.

timum setting depth for a pump/gas anchor (PG/A) combination (Sec. 11-5). In general terms it may be said that the method involves superposing on the grid of Fig. 4-24 the curves defining the operation of the lift equipment. Points of intersection give the depth of placement of the equipment as well as the production rate to be expected. Problems associated with the stability of the various intersection points must be analyzed, as must the question of whether the particular lift device under consideration can indeed handle the production rates suggested.

It should be emphasized that this method is based on natural flow up to the depth of the equipment to be installed; above that depth, and to the surface, natural flow is replaced by some artificial lift process. This implies that much of the grid is of theoretical interest only in some particular instance. Figure 4-25 illustrates this in the case of the suggested P/GA combination. The argument presented in Chap. 11 suggests that the pressure versus rate curve for this combination cuts across the grid in the manner shown, so that the (flowing) curves to the right of the P/GA curve are not in fact applicable, while points of intersection define possible producing situations (Sec. 11-5).

4-7 SLIPPAGE AND FLOW RESISTANCE: OPTIMUM GLR

Consider an experiment in which liquid and free gas are pumped up from the bottom to the top of a constant-diameter string fixed in a vertical position.[2] A

[2] This example is adapted from Gilbert (Ref. 5, p. 132).

pressure regulator holds a constant back pressure on the upper end of the tubing string, and a record is kept of the pressure at the lower end required to force liquid and free gas at various rates through the tubing. Suppose first that the GLR is held constant while the liquid rate is varied. When the liquid rate is extremely low (say, 0.01 bbl/day), the free gas rate is also small. For example, if the GLR is 1.2 mcf/bbl, then at a liquid rate of 0.01 bbl/day the free-gas rate from the top of the tubing is

$$\frac{0.01 \times 1200 \times 1728}{1440 \times 60} = 0.24 \text{ cu in./sec}$$

Evidently the situation in the tubing is that of an almost stationary liquid column through which gas is slowly bubbling. Thus the pressure at the lower end is equal to the pressure due to the liquid column plus the pressure due to the flow resistance, the second term being very small at the flow rate of 0.01 bbl/day.

If now the liquid throughput rate is increased but the GLR held constant, the liquid velocity will increase and there will be less time for gas slippage through the liquid. In terms of the flow regimes described in Sec. 4-2, slug, annular, and mist flow will take the place of bubble flow and the gas will help to lift the liquid out of the tubing. It is found that the decreased slippage of gas and the consequent increase in its ability to lift the liquid results in decreased pressure at the foot of the tubing despite the larger total fluid volume being driven through the tubing per unit time. There is a limit, however, to this phenomenon as the liquid rate increases, and increasing pressures will eventually accompany the higher fluid velocities associated with increased throughput rates.

The above discussion shows that for any given tubing size and depth, there is an optimum liquid rate for production at a fixed GLR, that is, a rate that results in the minimum pressure loss in the tubing. At rates less than the optimum the pressure loss increases as the rate decreases because of *gas slippage* and the attendant loss in the ability of the gas to lift the liquid. At rates higher than the optimum, the increasing fluid volumes (liquid plus free gas) being forced through the tubing per unit time result in higher velocities and consequently increased *resistance losses.* At very low rates the pressure drop approaches the static liquid-column pressure, so the curve of pressure loss against liquid production rate must tend to the static pressure loss as the liquid production rate tends to zero. The general shape of the curves of pressure loss against liquid production rate (GLR held constant) is shown in Fig. 4-26.

Tests may also be carried out to determine the effect of increasing the GLR when the liquid throughput rate is held constant. At low GLRs the flow regime is that of bubble flow with small gas bubbles dispersed in a continuous column so that the lifting effect of the gas is small, and the pressure drop from the bottom to the top of the tubing is equal to the sum of the liquid weight and the liquid flow resistance. In this case the pressure loss due to the flow resistance does not tend to zero as the GLR tends to zero, so the curve of pressure loss against GLR tends to a pressure-loss value higher than the static

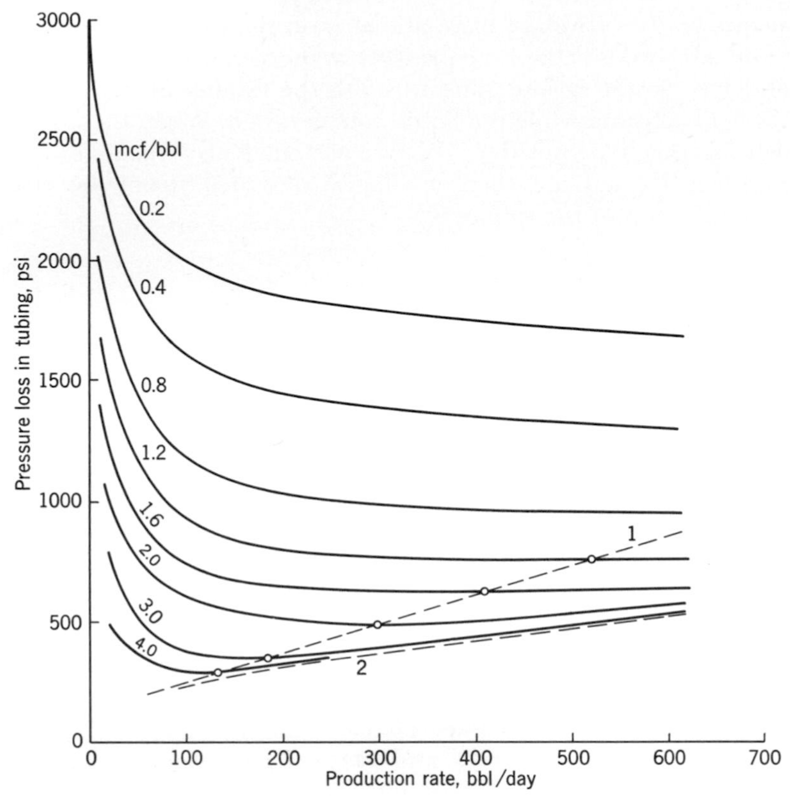

Fig. 4-26 Pressure loss as a function of production rate at various GLRs. (*After Gilbert, Ref. 5. Courtesy* API Drill. Prod. Practice.)

as the GLR tends to zero. Evidently this end value of the pressure loss increases with the liquid flow rate. Increases in the GLR will cause transitions to occur from bubble flow to slug, annular, and eventually mist flow; in other words, the assistance given by the gas in lifting the oil will steadily increase, and the pressure loss over the tubing will decrease. But, once again, if the GLRs become too large, the high velocities in the tubing result in high resistance losses and increasing pressure losses. The general shape of the curve of pressure loss against GLR (liquid production rate held constant) is shown in Fig. 4-27.

Figures 4-26 and 4-27 may be combined in a single three-dimensional picture (Fig. 4-28). In Gilbert's own words (Ref. 5):

> Several general characteristics of two-phase vertical flow may be observed from this picture. In particular it may be noted that:
>
> 1. For any constant gas-liquid ratio there is a rate of flow which requires minimum intake pressure. Also, this rate of flow for minimum pressure and

the minimum pressure itself both increase as the gas-liquid ratio is decreased as indicated by curve 1. (These observations are of interest in connection with flowing wells because of the tendency of flowing wells to have a more or less constant gas-liquid ratio at any one time).

2. For any constant rate of flow, there is a gas-liquid ratio which provides minimum intake pressure. This minimum intake pressure is directly related to the rate of flow, while the gas-liquid ratio for minimum intake pressure is inversely related to the rate of flow as indicated by curve 2. (These observations are of interest in connection with gas-lift which permits control of gas-liquid ratios).

The GLR that provides the minimum intake pressure at any particular rate of flow is termed the *optimum GLR,* and the value of this optimum as determined by Gilbert is shown by the arrow opposite the GLR figure on each of the sets of curves in Figs. 4-11 through 4-20. This optimum GLR is the reason for the two depth scales shown in these figures (Sec. 4-5). Since pressure gradients in the tubing increase with the GLR above the optimum ratio, curves corresponding to above-optimum GLRs would cross those corresponding to below-optimum GLRs if they were all started at a common origin. This would lead to a very confused graph, and so in Figs. 4-11 through 4-20 the depth scale *B* for the above-optimum GLRs has been set 5000 ft lower than the depth scale *A* for the below-optimum GLRs.

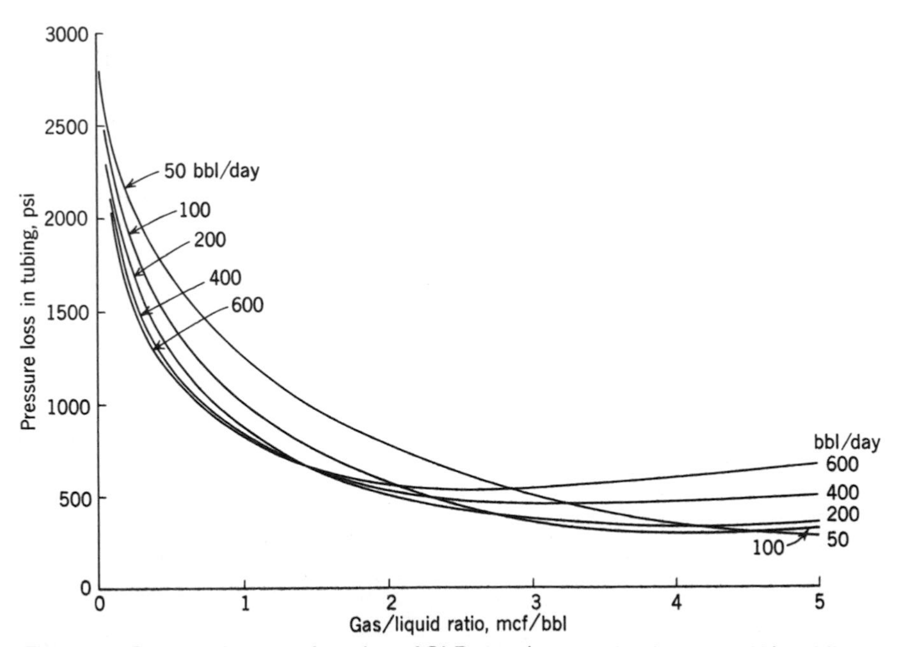

Fig. 4-27 Pressure loss as a function of GLR at various production rates. (*After Gilbert, Ref. 5. Courtesy* API Drill. Prod. Practice.)

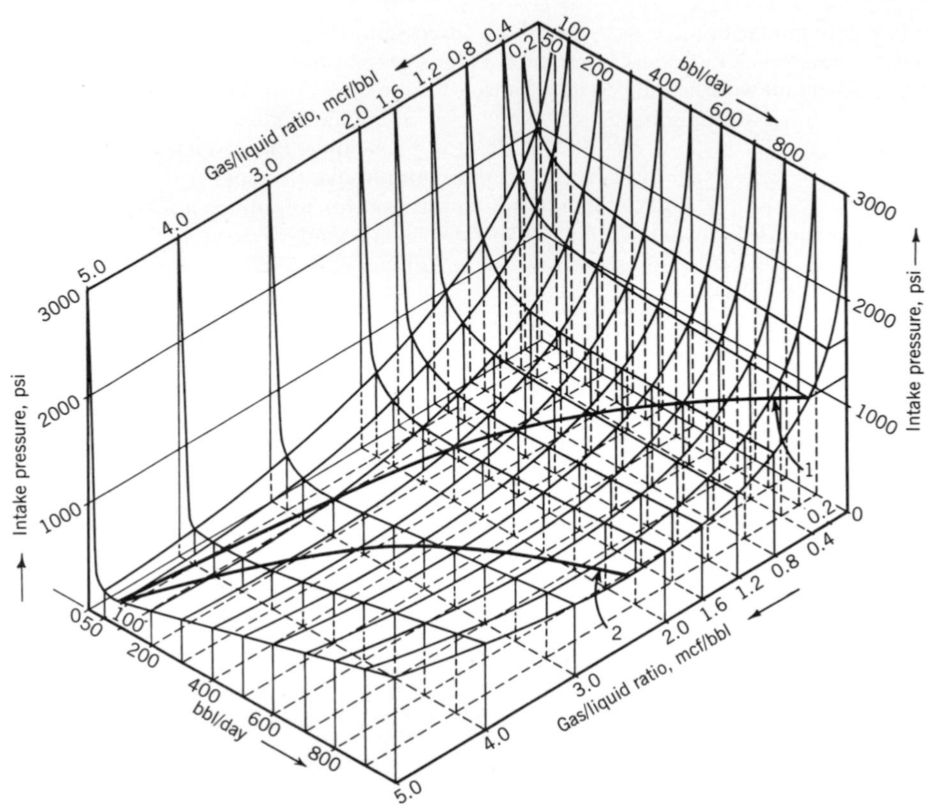

Fig. 4-28 Pressure loss as a function of GLR and production rate. (*From Gilbert, Ref. 5. Courtesy* API Drill. Prod. Practice.)

4-8 EFFECT OF TUBING SIZE: ANNULAR FLOW

To demonstrate the effect of tubing size on flow performance, Table 4-3 has been prepared from Figs. 4-11 through 4-20. The results have been plotted on Figs. 4-29 through 4-32, the first two of which illustrate the effect of liquid flow rate on pressure loss for various tubing sizes and the second two, the effect of tubing diameter on pressure loss for various liquid flow rates. From Figs. 4-29 and 4-30, it is evident that at low flow rates and low GLRs the smaller tubing sizes are the more efficient. This is due to the fact that slug and annular flow may occur in the smaller-diameter tubings while, at the same GLR, bubble flow may still be dominant in the larger-diameter tubings. Thus the slippage loss is higher in the larger strings. As the flow rate is increased at a constant GLR, there is less time for gas slippage to occur; slippage losses decrease, and the pressure loss in the tubing initially decreases. However, as the flow rate is further increased, resistance losses begin to build up and the pressure loss in the tubing rises. Evidently these resistance losses are higher in smaller tubing, so that the minimum-pressure-loss rate will be smaller, the

TABLE 4-3 Effects of Rate and Tubing Size on Two-Phase Vertical-Flow Pressure Losses: Pressure Losses in 10,000-ft Tubing String (Zero THP)

Tubing Size, in.	Liquid Flow Rate, bbl/day 50	100	200	400
	GLR of 1.0 mcf/bbl			
1.66	840	990	1250	1670(E)*
1.9	950	900	1020	1210
2⅜	1250	1000	960	1020
2⅞	1800	1450	1250	1160
3½	2000	1700	1390	1250
	GLR of 0.4 mcf/bbl			
1.66	1300	1250	1300	1510
1.9	1680	1430	1390	1460
2⅜	2080	1800	1580	1600
2⅞	2500	2150	1970	1890
3½	2800(E)*	2400	2180	2000

* (E) = estimated pressure.

smaller the tubing diameter. As the GLR increases, the larger tubing sizes gradually take over from the smaller as the more efficient (compare Figs. 4-29 and 4-30) because the high-total-volume throughput results in large resistance losses in the small strings.

Figures 4-31 and 4-32 show clearly that in the smaller tubing sizes there is less pressure loss at the lower than at the higher flow rates. However, for the range of conditions covered, this situation has completely reversed itself in both of the larger sizes of tubing investigated ($2^7/_8$- and $3^1/_2$-in.). The illustrations also show the sharp dependence of pressure loss on tubing diameter at the low flow rates and the fact that, at the higher flow rates, tubing diameter,

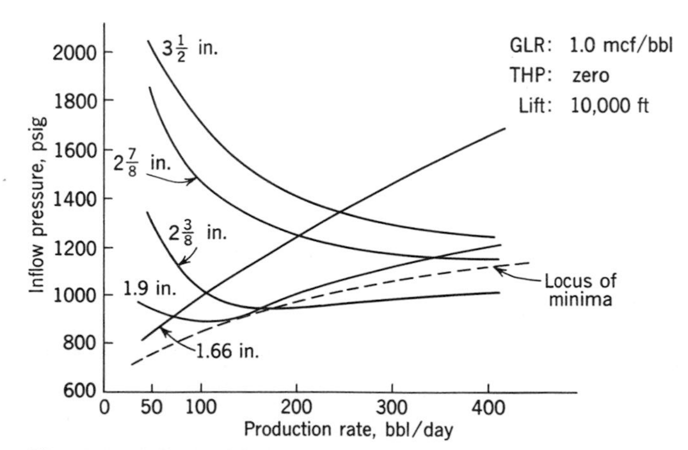

Fig. 4-29 Effect of flow rate on vertical-flow pressure losses: various tubing sizes.

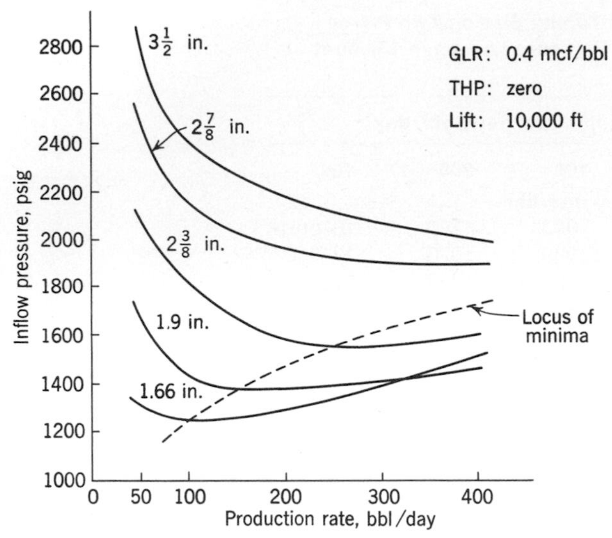

Fig. 4-30 Effect of flow rate on vertical-flow pressure losses: various tubing sizes.

provided it is greater than a certain minimum, may be changed (that is, strings of different size may be used) without too severe an influence upon the pressure loss sustained in the vertical two-phase flow. These remarks, in conjunction with the results illustrated in Figs. 4-29 through 4-32, make it obvious that, in general, the tubing size run in a particular well must be tailored to that well's producing potentialities, that is, to its IPR and its GLR, as well as to the flow rate it is hoped to maintain.

Although annular flow will not be treated in any detail in this book and annular pressure-distribution curves are not generally available, it may be remarked that gas slippage tends to be decreased in the annular ring (as com-

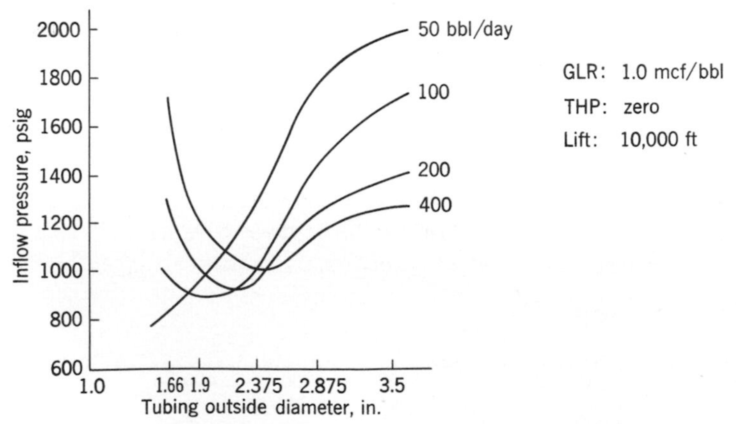

Fig. 4-31 Effect of tubing size on vertical-flow pressure losses: various flow rates.

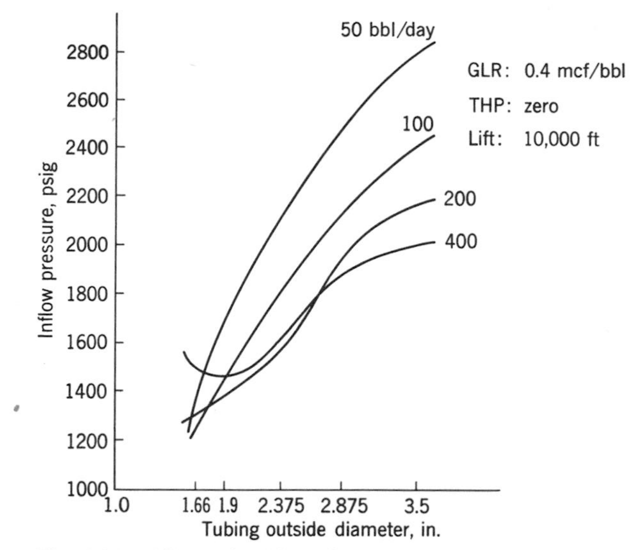

Fig. 4-32 Effect of tubing size on vertical-flow pressure losses: various flow rates.

pared with a circular tubing of the same cross-sectional area) because of the decrease in the distance between wall faces. This is particularly true in the case of a tubing string centered in the casing. Thus it may be that, under certain circumstances, the annulus between the casing and a centralized tubing is a more efficient eductor tube for the oil and gas than the tubing itself.

4-9 PRESSURE-DISTRIBUTION CURVES: COMPARATIVE SUMMARY

One major point of difference between the published families of curves involves the optimum GLR. For example, curves published by Garrett Oil Tools (Ref. 12), such as those shown in Fig. 4-33, and those included by Brown (Ref. 6) such as those of Fig. 4-34, are drawn in such a way that no optimum GLR is apparent, but instead in any family the curves for higher GLRs converge into curves for successively lower ratios. Gilbert's curves, on the other hand, do not show this convergence; rather, each curve is a separate entity all the way to the surface. The Garrett curves were derived by the Poettmann and Carpenter method, and although such calculations do lead to curves that exhibit an optimum GLR beyond which the pressure gradients start to increase once again, this was generally found by Garrett to be extremely high and so was not shown.[3]

A second point of difference, this time between the Gilbert and Garrett

[3] This point was brought out in a letter to the author from R. V. McAfee of Garrett Oil Tools, Division of U.S. Industries, Inc., Houston, Tex., June 7, 1960.

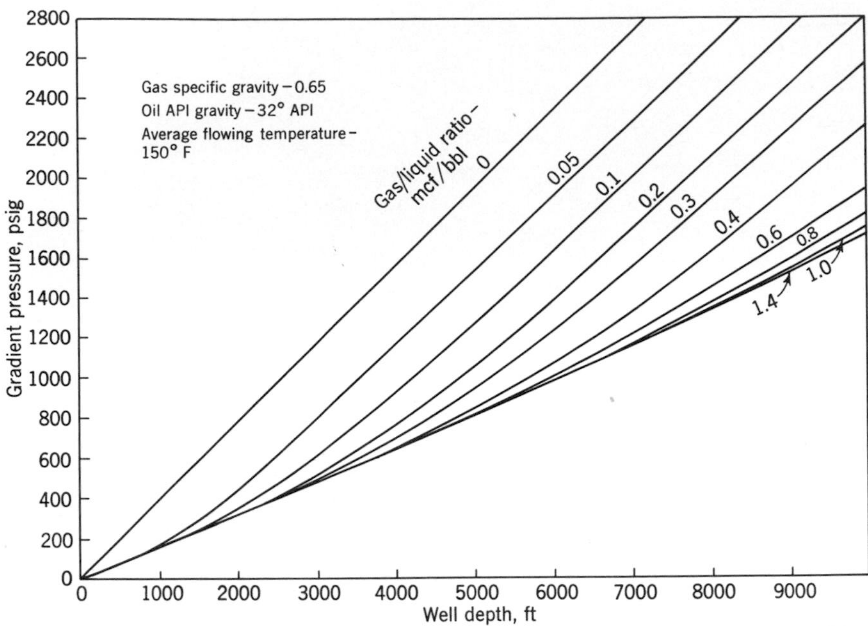

Fig. 4-33 Approximate pressure-distribution curves for 2⅜-in. tubing, 800 bbl/day (oil). (*After Garrett Oil Tools,* Handbook of Gas Lift, *Ref. 12. Courtesy U.S. Industries Inc.*)

curves on the one hand and the curves presented by Brown on the other, is that the latter generally show a greater slope at the lower pressures but a lower slope at the upper end of the pressure scale. This result will be discussed in the next section, where it will be shown that the slope exhibited by Brown's curves is explicable if consideration is given to the fact that free gas will be evolved from solution in the oil as pressures decrease toward the surface.

In generating its sets of curves, Garrett Oil Tools ran many calculations to determine the extent of error incurred by assigning average values to certain variables. For a given liquid rate and tubing size it was found that the only variables of a significant nature were GLR and the supercompressibility of gas; variations in other factors such as liquid surface tension, gas and liquid viscosity and gravity, and flowing temperature had relatively little effect on the pressure-distribution curves.[4] Moreover, although the WOR affected the curves, the extent of the variation was not very great, and Garrett did not find it necessary to expand beyond two sets of curves, one set for 100 percent oil and one set for 100 percent water.

To indicate the degree of difference between various sets of published curves, Table 4-4 has been drawn up. The degree of discrepancy apparent in

[4] Low-gravity, high-pour-point crudes and emulsions (that is, having gravities of 15° API or less and pour points of 100° F or more) show much greater tubing-flow pressure losses than those indicated by either the Garrett or the Gilbert pressure-distribution curves.

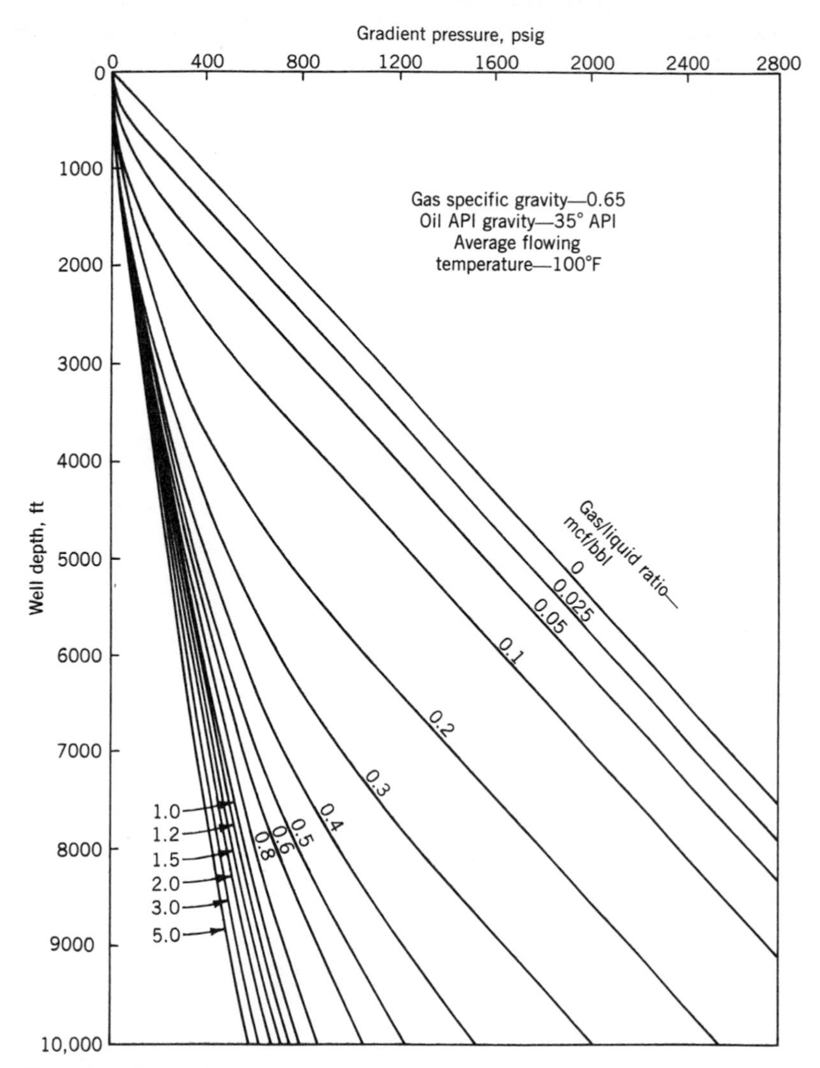

Fig. 4-34 Approximate pressure-distribution curves for 2^3/$_8$-in. tubing, 100 bbl/day (oil). (*From Brown, Ref. 6. Courtesy Petroleum Publishing Company.*)

this table is so great that little reliance can be placed on any set of curves in the first instance.

Some of the uses of the pressure-distribution curves for two-phase flow in the tubing are to determine:

The best tubing size for particular flow or lift conditions
Whether a dead well can be brought back on natural flow or whether some method of artificial lift is required
Gas-lift possibilities
The flowing BHP, and hence the IPR, under certain circumstances

TABLE 4-4 Comparison of Intake Pressures Determined from Gilbert's (Ref. 5), Garrett's (Ref. 12), and Brown's (Ref. 6) Pressure-Distribution Curves (All Pressures Are Tubing-Intake Pressures in psig, Assuming a Zero THP)

				p_{wf}			
					Garrett		
Tubing Size, in.	Liquid Rate, bbl/day	GLR, mcf/bbl	Tubing Depth, ft	Gilbert, Oil	Oil	Water	Brown, Oil
1.66	200	0.6	10,000	1220	2690	2940(E)*	1400
1.90	200	0.1	7,000	1500	2160	2600	1880
	600	1.0	10,000	1400	2000	2200	—
2⅜	100	1.0	10,000	1000	1940	1950	780
	600	0.1	5,000	1120	1150	1520	1360
	600	0.1	8,000	2000	2200	2900(E)	2480
2⅞	200	1.0	10,000	1250	1240	3700(E)	760
	600	0.2	8,000	1660	1640	2050	1840
3½	400	0.1	5,000	1230	840	1350	—
		0.1	10,000	3000(E)	2560	3700(E)	—
		1.0	5,000	480	380	380	—
		1.0	10,000	1260	850	1000	—

* (E) = estimated pressure.

It is evident that the first two uses listed above do not require a high degree of accuracy in the curves themselves. There are only certain definite tubing sizes that can be used, so nothing would be gained by having a set of curves so accurate that they could predict an optimum tubing size of, say, internal diameter 2.081 in. Again, if a dead well can be brought back onto flowing production at only a low rate with a near-zero THP, nothing would be gained by trying to induce flow because the well would soon die again in any event. If the calculations showed the well to be capable of flowing only weakly, artificial lift would be immediately installed. The last two items mentioned above do require a reasonable degree of accuracy in the pressure-distribution curves before any sound engineering decisions can be taken. All in all, although a high degree of accuracy is probably not essential, the degree of discrepancy pointed up by Table 4-4 is undoubtedly too great; it is essential that better curves become available for use by production engineers.

4-10 EQUATION OF PRESSURE-DISTRIBUTION CURVES

In this section an equation is developed that gives the shape of the pressure-distribution curves over a wide range of conditions. The equation is derived from geometrical, rather than from energy, considerations and contains two variables apart from pressure and depth. One of these variables is a constant of integration and its value in any particular case would depend upon the flow rate, tubing size and type, oil viscosity and probably several other factors as well; in practice, and unless and until sound correlations are developed, its value for any particular well must be determined from two pressure-depth readings—typically the THP and a measurement of the flowing pressure at the foot of the tubing. The second variable is related to the PVT characteristics of the crude oil under production, and specifically to the gas solubility.

The advantage of knowing the shape to be expected of a pressure-distribution curve comes when it is necessary to develop a set of curves (similar, for example, to those of Gilbert) for a particular group of wells. That is, sets of curves can be prepared for a specific situation rather than relying on one of the published, generalized sets or on a broadly applicable computer program.

Writing r for the liquid holdup (that is, the volume of liquid per unit volume of tubing) at the depth H (feet below surface) at which the pressure in the tubing is p (psia), the liquid volume per unit length of tubing at that depth is ra, and the corresponding free gas volume is $(1 - r)a$, where a is the cross-sectional area of the tubing.

If the pressure is reduced to $p - \delta p$, then the free-gas volume becomes[5]

$$\frac{p}{p - \delta p}(1 - r)a$$

plus the additional volume of gas released from solution in the liquid. To determine this latter volume it will be assumed that, to a good degree of approximation, the gas (measured in scf/bbl) in solution in crude oil is a straight-line function of the pressure to which the gas-oil mixture is subjected. That is, the number of standard cubic feet of gas in solution in a volume ra of crude at pressure p is $mpra$, where m is a constant (defined by the PVT properties of the oil and gas). Hence the number of standard cubic feet released when the pressure drops to $p - \delta p$ is $mra\ \delta p$. At the pressure $p - \delta p$ this occupies a volume of

$$\frac{14.7\ mra\ \delta p}{p - \delta p} \quad \text{scf}$$

or

$$\frac{14.7\ mra\ \delta p}{5.614\ (p - \delta p)} \quad \text{bbl}$$

Writing $14.7m/5.614$ as M, and neglecting terms in $(\delta p)^2$, this expression reduces to $Mra\ \delta p/p$, so that the total free-gas volume at the pressure $p - \delta p$ is

[5] This analysis does not take account of temperature variations along the length of the tubing. Such variations are usually not great while the well is flowing.

$$\frac{p}{p - \delta p}(1 - r)\,a + \frac{Mra}{p}\,\delta p$$

Neglecting the change in oil volume (as gas is released from solution), the total oil plus free-gas volume at the pressure $(p - \delta p)$ is

$$ra + \frac{p}{p - \delta p}(1 - r)a + \frac{Mra}{p}\,\delta p$$

so that the new value $r - \delta r$ of the liquid holdup is

$$r - \delta r = \frac{ra}{ra + \dfrac{p}{p - \delta p}(1 - r)a + \dfrac{Mra}{p}\,\delta p}$$

which, on simplification, and with neglect of second-order terms, reduces to

$$\frac{\delta r}{\delta p} = \frac{r[1 - r(1 - M)]}{p} \tag{4-12}$$

Integration and rearrangement of Eq. (4-12) gives

$$r = \frac{p}{A + (1 - M)p} \tag{4-13}$$

where A is a constant of integration.

The change in fluid load on a cross section of the tubing on moving down from the depth $H - \delta H$ to the depth H is $a\ \delta p$. Neglecting the weight of the free gas, this change in load is also $\rho\ \delta H\ ra$ where ρ is the pressure gradient (psi/ft) exerted by the liquid. Equating these two expressions:

$$r = \frac{1}{\rho}\frac{dp}{dH} \tag{4-14}$$

From Eqs. (4-13) and (4-14) it follows that

$$\frac{1}{\rho}\frac{dp}{dH} = \frac{p}{A + (1 - M)p}$$

so that

$$\rho\frac{dH}{dp} = \frac{A}{p} + (1 - M)$$

Integrating from the tubing head $(H = O, p = p_{th})$ to the depth H gives

$$\rho H = A \ln\left(\frac{p}{p_{th}}\right) + (1 - M)(p - p_{th}) \tag{4-15}$$

which is the required equation.

Values of M may be expected to range from 0 to 2, the lower limit of zero being attained or nearly attained when producing crudes that are considerably undersaturated, so that relatively little free gas is evolved during movement from the reservoir to the surface. On the other hand, if M equals 2, the slope of the line giving gas solubility in cu ft/bbl as a function of pressure in psia is 0.76 cu ft/(bbl)(psia), which would imply that over 3000 cu ft of gas is in solution in a barrel of oil at a saturation pressure of 4000 psia.

As the value of M increases, Eq. (4-15) defines a curve that is increasingly vertical at the lower pressures, but increasingly flat at the higher pressures. The general shapes of the curves are illustrated in Fig. 4-35, which has been drawn on the basis of a THP of 200 psia and a flowing BHP at 12,000 ft of 3000 psia.

In order to illustrate the calculation using Eq. (4-15), two "anchor points" will be taken from the Gilbert pressure-distribution curve for $2^7/_8$-in. tubing, 50 bbl/day, and a GLR of 3.0 mcf/bbl (Fig. 4-11). Two points on the Gilbert curve are (1) a pressure of 200 psig at an equivalent depth of 4000 ft (the assumed tubing head) and (2) a pressure of 2700 psig at an equivalent depth of 23,200 ft. Thus the depth of the well (length of tubing) is 19,200 ft. One further piece of information is required, namely the value of ρ, which for Gilbert's curves is 0.38 psi/ft.

From the given data the value of A may be determined once a value for M has been assumed. Gilbert did not take into account the possible release of gas in the tubing string, so it is reasonable to take $M = O$ in this example.

Substitution of the anchor-point data in Eq. (4-15) gives

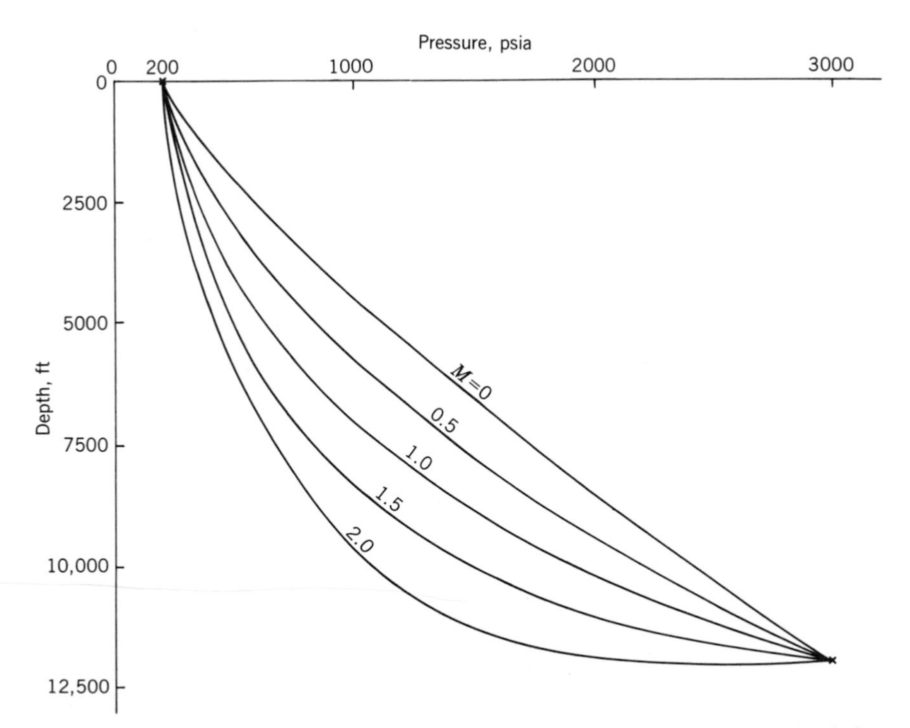

Fig. 4-35 Typical pressure-distribution curves arising from use of Eq. (4-15) for various values of the "free-gas factor" M.

TABLE 4-5 Illustration of Use of Eq. (4-15) Using Data from Gilbert's Curve, Fig. 4-11, GLR 3.0 mcf/bbl

p, psig	p, psia	$\frac{p}{215}$	$\ln\left(\frac{p}{215}\right)$	$1890 \ln\left(\frac{p}{215}\right)$	p-215	0.38H	H, ft below Tubing Head
500	515	2.40	0.876	1660	300	1960	5,150
1000	1015	4.72	1.551	2940	800	3740	9,840
1500	1515	7.04	1.950	3690	1300	4990	13,110
2000	2015	9.35	2.238	4230	1800	6030	15,850
2500	2515	11.68	2.46	4650	2300	6950	18,300

Note: Before plotting, the equivalent depth of the tubing head (4000 ft) must be added to the figures for H.

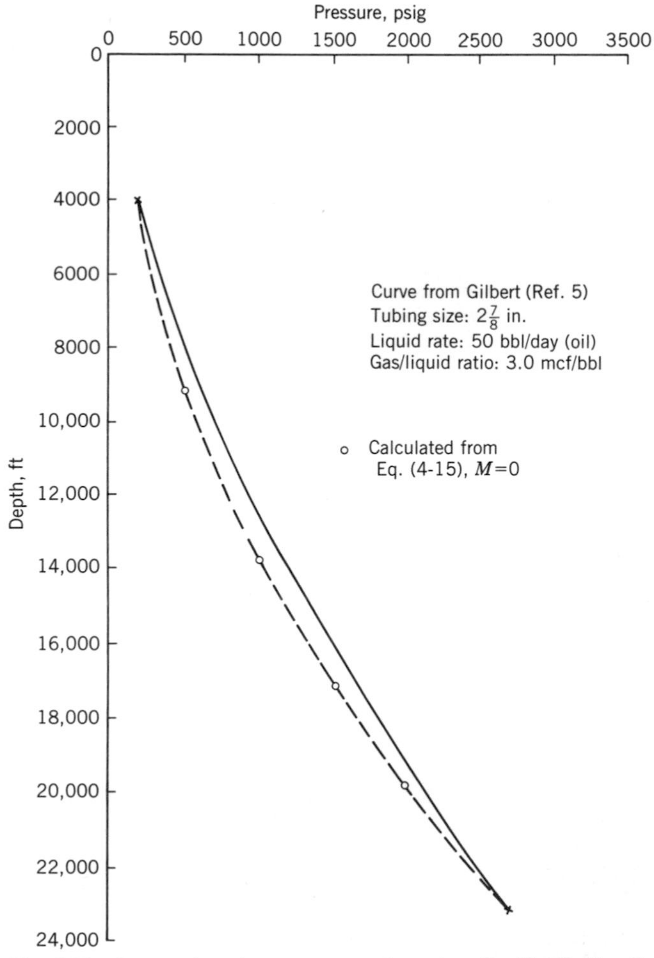

Fig. 4-36 Comparison between curve based on Eq. (4-15), $M = 0$, and one of Gilbert's pressure-distribution curves (Ref. 5).

$$0.38 \times 19{,}200 = A \ln\,(^{2715}/_{215}) + (2715 - 215)$$

from which $A = 1890$.

Hence, pressures at intermediate depths are given by the equation

$$0.38\,H = 1890 \ln \left(\frac{p}{215}\right) + (p - 215)$$

This may be solved by assuming intermediate values for p and then calculating H. The calculations are shown in Table 4-5 and plotted in Fig. 4-36,

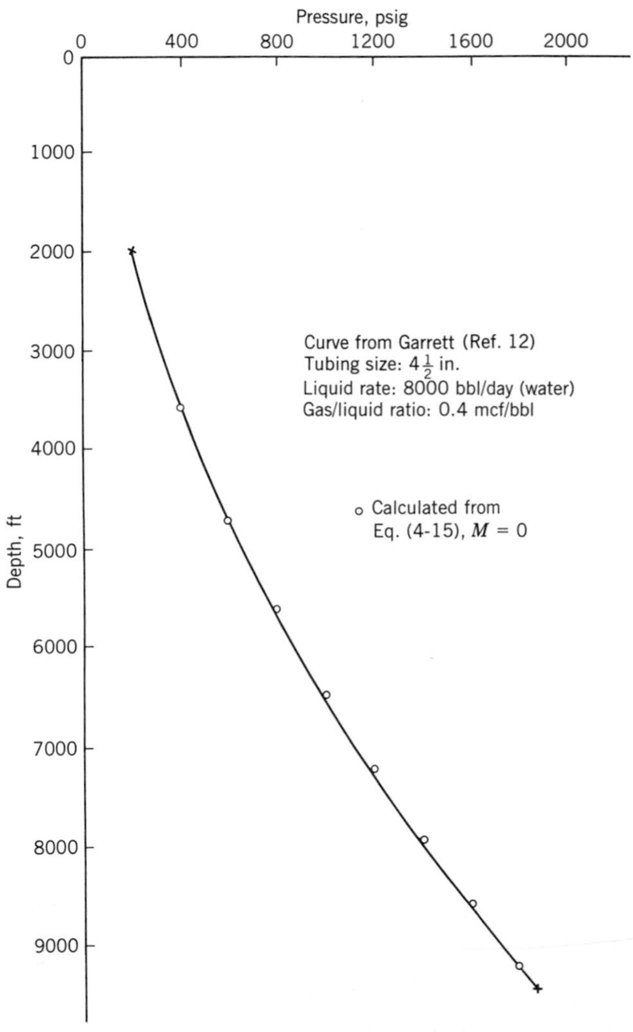

Fig. 4-37 Comparison between curve based on Eq. (4-15), $M = 0$, and one of Garrett's pressure-distribution curves (Ref. 12).

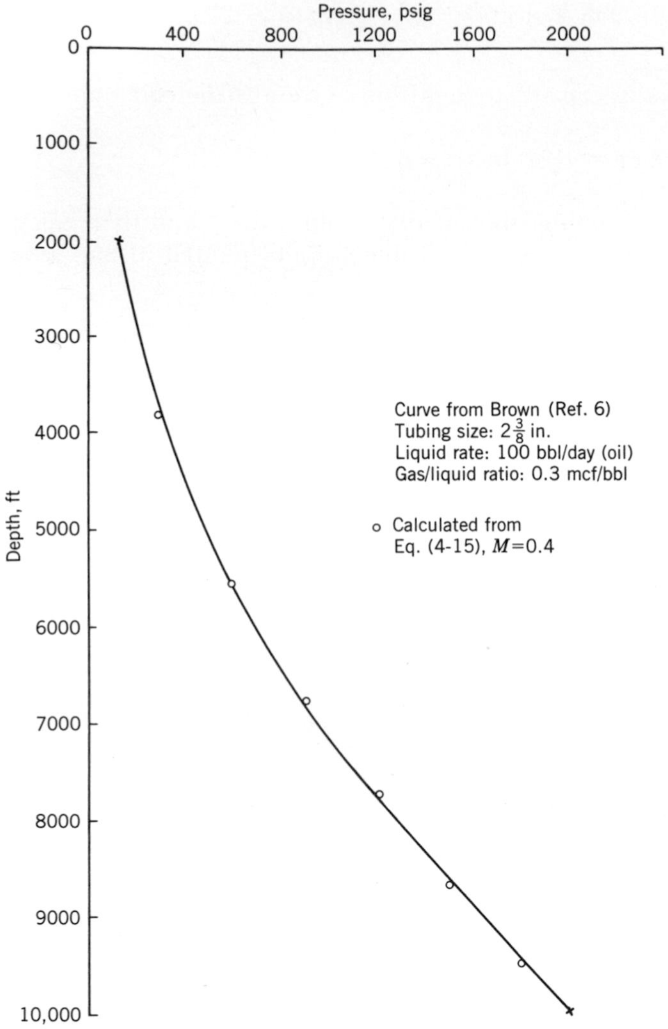

Fig. 4-38 Comparison between curve based on Eq. (4-15), M = 0.4, and one of Brown's pressure-distribution curves (Ref. 6).

which shows the calculated points to be in reasonable agreement with Gilbert's curve.

The validity of an equation such as Eq. (4-15) depends in the final analysis on how accurate it appears to be. Figures 4-37 to 4-41 give various examples. Figure 4-37 is based on a curve chosen at random from Garrett's handbook (Ref. 12) and Fig. 4-38 is from a curve chosen from Brown (Ref. 6), while Figs. 4-39, 4-40, and 4-41 are field examples, the curves being derived from actual pressure-bomb measurements in the tubing while the well in question was

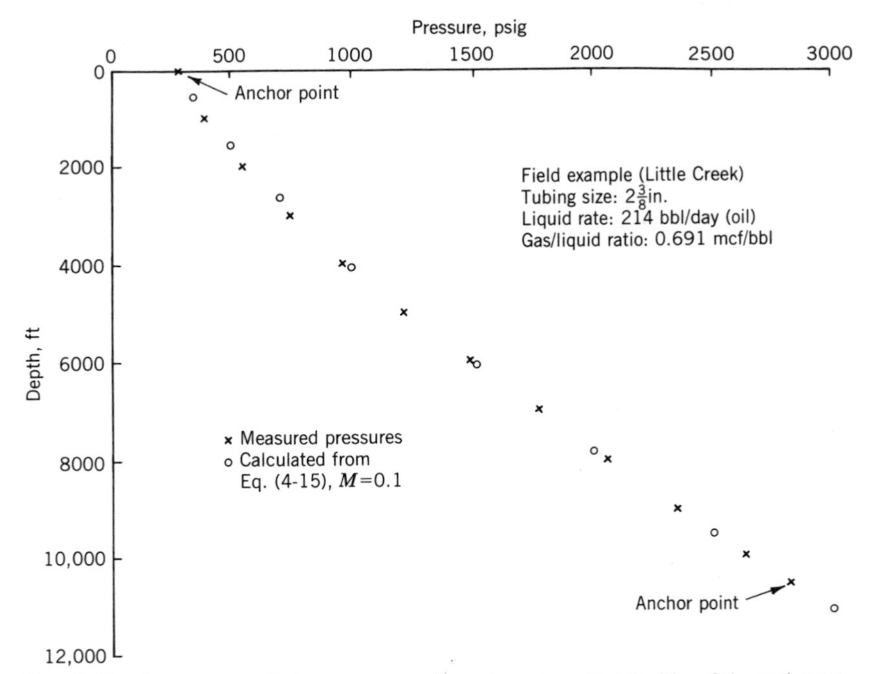

Fig. 4-39 Comparison between curve based on Eq. (4-15), M = 0.1, and pressures measured in a producing well.

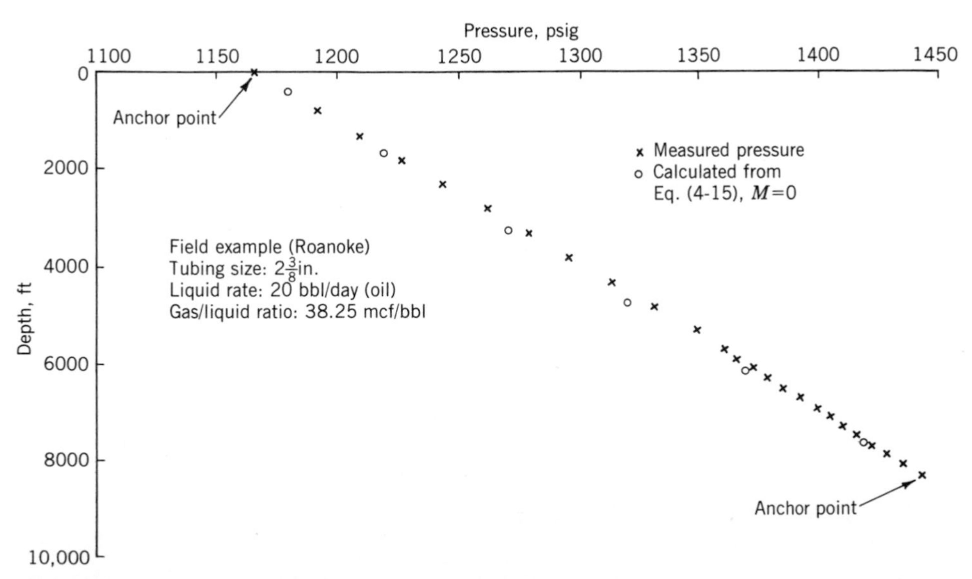

Fig. 4-40 Comparison between curve based on Eq. (4-15), M = 0, and pressures measured in a producing well.

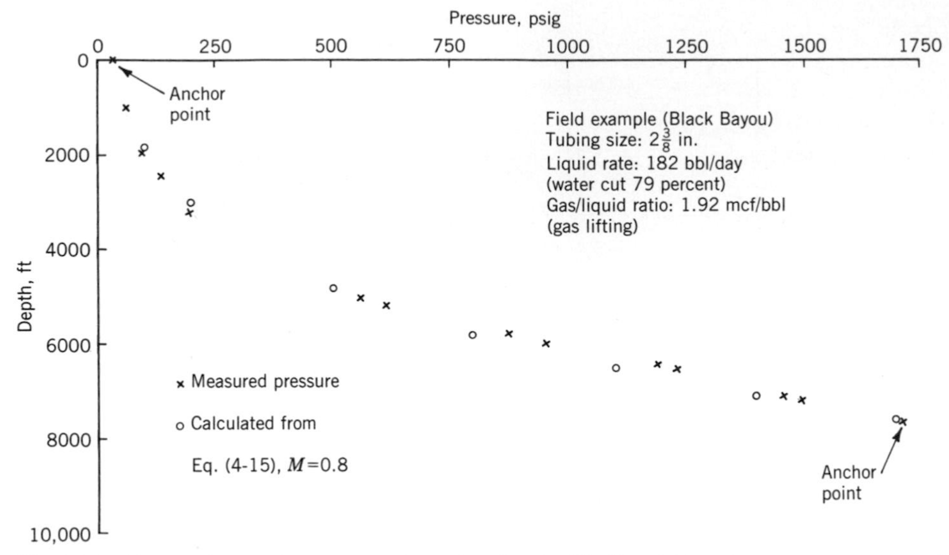

Fig. 4-41 Comparison between curve based on Eq. (4-15), $M = 0.8$, and pressures measured in a producing well.

producing. In each figure the anchor points are marked with crosses and the points calculated from the equation with circles.

4-11 NOTES ON TUBING DEPTH: HOLE DEVIATION

It was mentioned in the introduction to this chapter (Sec. 4-1) that the tacit assumption has been made throughout that the foot of the tubing string is at the level of the top of the producing horizon. If the tubing were hung some distance up the hole, it would be necessary to run the calculation in two stages. From the producing horizon to the tubing shoe, pressure-distribution curves for flow in the casing would be used to determine the pressure at the tubing shoe. This calculation would, in effect, define an "IPR" at the foot of the tubing (Sec. 4-6 and Fig. 4-24), from which level calculations for flow in the tubing would proceed in the standard manner. For an example of a calculation of this type, see Sec. 6-4.

If the well in question were deviated considerably from the vertical, special sets of curves applicable to deviated holes would have to be used (Ref. 6).

4-12 IPR: FLOWING WELLS

In order to define the IPR of a well it is necessary to have at least two flowing BHP versus production-rate points (or one such point together with the well's static BHP). The running of a pressure bomb is the most accurate way of determining the pressure, but this method would be costly if it were applied as an annual or semiannual routine to every flowing well in a field or area, and although a certain percentage (say, 10 to 15 percent) of the wells may be surveyed in this manner for reservoir control purposes, the production engineer must, in general, be content with a simpler, and at the same time somewhat less accurate, method of determining a particular well's PI.

The majority of new flowing wells are completed with a casing-tubing packer in the hole, and in such cases it will be necessary to estimate flowing BHPs from THP readings (see Example 4-1). Such calculations should be carried out periodically and whenever there is a change in the production rate from the well. It should, however, be noted that small, frequently irregular THP changes may occur even at steady production rates, the occurrence of such variations being in no way dependent upon the type of formation under production. These pressure changes are of short cyclical duration and have little effect on the continuity of production except perhaps in very weakly flowing wells. They are caused by the segregation and subsequent accumulation of free gas into slugs in the tubing string. Although only of minor importance so far as a well's production is concerned, they do make dependence on a single reading only of THP an unreliable method for determination of BHPs.

If there is no casing-tubing packer, a more reliable method is one that uses the casing-head pressure (CHP). In such circumstances and in a steadily producing well, flowing at such a rate that the flowing BHP is below the bubble point of the reservoir crude, the casing-tubing annulus will be filled with gas. To demonstrate this result, suppose that the well has been closed in for a few days and that there is some oil in the annulus (as illustrated in Fig. 4-42). If, when the well is opened up on the tubing, the flowing BHP is less than the bubble-point pressure, free gas, in addition to oil, is produced into the well bore from the formation. The majority of the free gas bubbles are entrained with the oil and pass up the tubing, but a certain number segregate by gravity while crossing the annulus from the formation to the tubing shoe, pass up into the annulus, and displace any liquid accumulated there (Fig. 4-43). This process continues until, when equilibrium is established, the oil level in the annulus is at the tubing shoe.

Suppose that it is required to find the IPR of a well for which the conditions indicated above hold, namely, that the annulus is not packed off, that the well is flowing steadily, and that the flowing BHP is below the bubble point (methods for establishing whether or not these conditions prevail for a particular well are discussed below). Then the flowing BHP is equal to the CHP plus the pressure exerted by the column of gas in the annulus, or in symbols,

$$p_{wf} = p_c + \text{pressure exerted by gas column} \qquad (4\text{-}16)$$

The pressure exerted by the gas column will depend upon the density gradient in the gas column, that is, on the gas pressure (and so on p_c), the gas composition (which will determine its compressibility factor), the geothermal temperature gradient, and the length of the column.

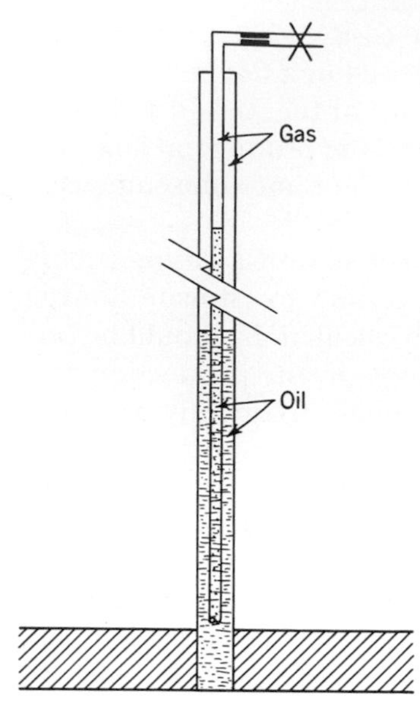

Fig. 4-42 Typical gas-oil contacts in a closed-in well (no packer).

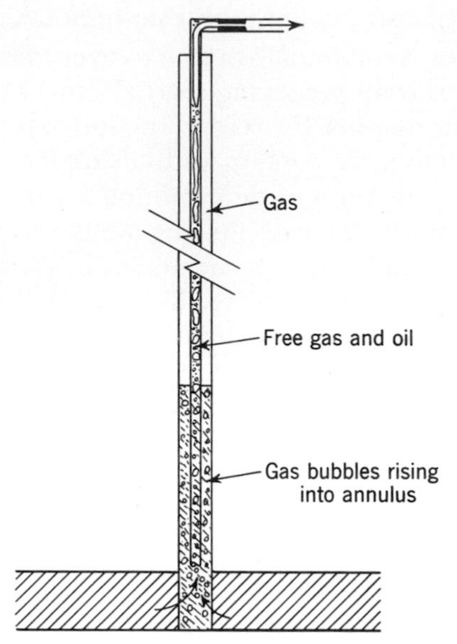

Fig. 4-43 Situation shortly after opening up a flowing well (no packer).

Fortunately, with oil wells (as opposed to gas or condensate wells), the correction to be made to p_c due to the pressure exerted by the gas column is usually relatively small (compared with p_c itself), so that it need not be determined to a high degree of accuracy. Since most oil field gases have the same type of composition (80 percent or more methane content) and since the geothermal gradient does not vary over very wide limits, it would appear that the main factors controlling the pressure exerted by the gas column are the CHP and the length of the column. By making a plot of measured values of the flowing BHP (obtained by using a pressure bomb) divided by the CHP against the tubing depth for a number of California wells, Gilbert (Ref. 5) derived the empirical formula

$$\text{Pressure exerted by gas column} = p_c \frac{D^{1.5}}{100} \tag{4-17}$$

where p_c is the CHP in psia and D is the depth of the tubing in thousands of feet.

Substitution in Eq. (4-16) gives

$$p_{wf} = p_c \left(1 + \frac{D^{1.5}}{100}\right) \tag{4-18}$$

where p_c and p_{wf} are expressed in psia.

Equation (4-17) is only one of many expressions that have been suggested for determining the pressure due to a column of annulus gas [see, for example, Refs. 13 and 14 for more precise formulas, which, however, require more information for their use than that needed for Eq. (4-17)]. It cannot be expected to be entirely accurate, particularly for deep, very high pressure wells; nevertheless, it has been found adequate in many instances. Its use in the determination of a well's inflow performance can best be illustrated by an example.

Example 4-5 A flowing well with 3000 ft of tubing in the hole exhibits a CHP of 550 psig when the production rate is 42 bbl/day and 320 psig when the production rate is 66 bbl/day. What are the PI, static pressure, and potential of this well?

In the first flowing test, the CHP was 550 psig, or 565 psia. From Eq. (4-18), the flowing BHP was

$$565\left(1 + \frac{3^{1.5}}{100}\right) = 595 \text{ psia}$$
$$= 580 \text{ psig}$$

Note here that the correction for the pressure exerted by the gas column is only 580 − 550, or 30, psi; hence, the accuracy of the determination of this correction is not too important. Thus, when q was 42 bbl/day, p_{wf} was 580 psig. This result is plotted at point A in Fig. 4-44.

In the second flowing test, the CHP was 320 psig, or 335 psia. So, from Eq. (4-18), the flowing BHP was

$$335\left(1 + \frac{3^{1.5}}{100}\right) = 353 \text{ psia}$$
$$= 338 \text{ psig}$$

The production rate during the second flowing test was 66 bbl/day, and this result is plotted at B in Fig. 4-44.

If a straight-line IPR is assumed, it is evident from Fig. 4-44 that the static pressure of the well is 1000 psig, that the well's potential is 100 bbl/day, and that the PI is

$$\frac{100}{1000} = 0.1 \text{ bbl/(day)(psi)}$$

This example illustrates a simple method of determining the inflow performance of a flowing well not equipped with a tubing-casing packer: all that has to be done is to gauge the well at two different steady rates and to record the stabilized CHPs at these rates.

Another possibility would be to record the CHP at one gauged rate and to refer to the reservoir engineer's performance graphs to obtain an approximate value for the well's static pressure p_s; once again two points on the IPR would be established.

It must be stressed that these methods involving the CHP can only be used if three conditions hold. First, there must be no annulus packer in the hole. This is readily checked from the well records. Second, the well must be flowing steadily. This can be ascertained most easily by ensuring that the CHP

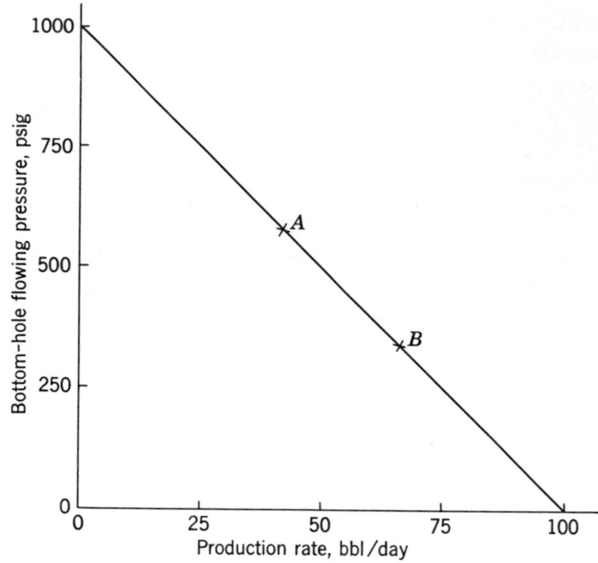

Fig. 4-44 Example 4-5: IPR determined from flowing well data taken at two different rates.

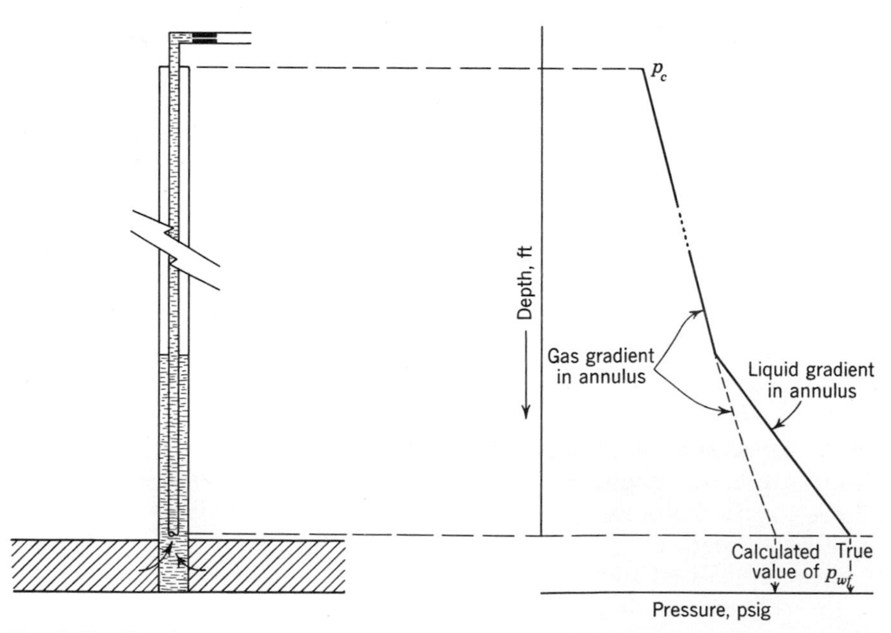

Fig. 4-45 Flowing well conditions when flowing BHP exceeds bubble-point pressure (no packer).

is only fluctuating within certain reasonable limits (not more than a 50-psi range, say). If there is considerable fluctuation in the CHP, the well is *heading* (Sec. 5-8). Third, the flowing BHP must be below the bubble point. If this were not so and if the flowing BHP were calculated from the CHP, with the annulus assumed to be full of gas [for example, from Eq. (4-18)], then the calculated value would be lower than the true value (see Fig. 4-45). As a consequence, the calculated PI would also be low, and the well's potentialities might be seriously underestimated.

4-13 TWO EXAMPLES

In order to illustrate some of the methods discussed in this chapter, two examples are given in conclusion.

Example 4-6

Well depth	5200 ft
7-in. casing	5050 ft
Static pressure at 5000 ft	1850 psig
GLR	0.4 mcf/bbl
$2^3/_8$-in. tubing set at	5000 ft
No casing-tubing packer	

The well is currently flowing at 250 bbl/day with a CHP of 1245 psig, but the tubing is corroded and must be pulled and replaced. In addition to $2^3/_8$-in., 1.9-in. and $3^1/_2$-in. tubing strings are available. Which size of tubing should be run if it is desired to flow the well at the maximum possible rate with a THP of 170 psig?

The values of D and p_{wf} in Eq. (4-18) are, in this example, 5 and 1260, respectively, so

$$p_{wf} = 1260\left(1 + \frac{5^{1.5}}{100}\right) \text{ psia}$$

$$= 1387 \text{ psig}$$

and the PI is given by

$$J = \frac{q}{p_s - p_{wf}} = \frac{250}{1850 - 1387} = 0.54 \text{ bbl/(day)(psi)}$$

The potential of the well is Jp_s, or 1000 bbl/day, so the IPR is as shown in Fig. 4-46.

Either method 1 or method 2 of Sec. 4-5 could be used, but in this problem method 2 only will be employed: that is, the THP will be determined at various rates, using the correct flowing BHPs.

Values of q and the THP, obtained from Table 4-6, are plotted in Fig. 4-46. Evidently, from this figure the flow rates against a THP of 170 psig are:

$3^1/_2$-in. tubing	430 bbl/day
$2^3/_8$-in. tubing	500 bbl/day
1.9-in. tubing	515 bbl/day

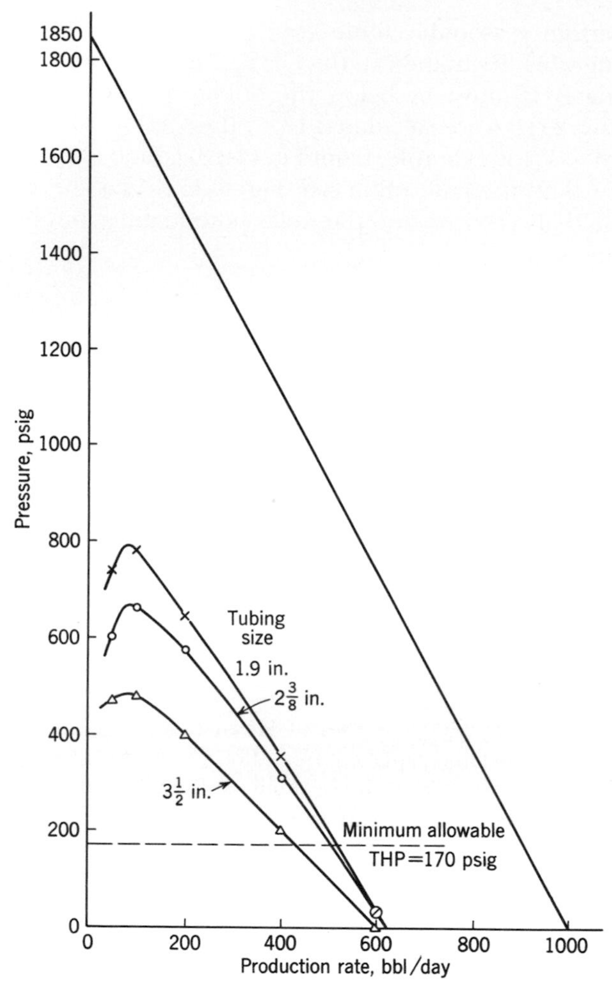

Fig. 4-46 Example 4-6: Determination of optimum tubing size.

However, the curve for 2³/₈-in. tubing is nearly as good as that for 1.9-in. tubing, and 2³/₈-in. tubing is more convenient in a well because it has greater strength and the larger diameter allows a greater selection of tools to be run into the hole. Thus, in practice in a case like this, 2³/₈-in. tubing would probably be rerun into the hole.

The following example illustrates a somewhat different way of using the pressure-distribution curves.

Example 4-7 The completion data of a well, producing from a reef limestone structure, are as follows:

TABLE 4-6 EXAMPLE 4-6: Determination of THP at Various Rates for Various Tubing Sizes*

q, bbl/day	p_{wf}, psi	Equiv. Depth of p_{wf}, ft	Equiv. Depth of THP, ft	THP, psi
1.9-in. tubing:				
50	1760	10,400	5400	740
100	1660	11,200	6200	780
200	1480	10,500	5500	650
400	1100	8,000	3000	360
600	740	5,300	300	30
2⅜-in. tubing:				
50	1760	8,900	3900	600
100	1660	9,500	4500	660
200	1480	9,600	4600	580
400	1100	7,500	2500	310
600	740	5,300	300	30
3½-in. tubing:				
50	1760	7,100	2100	470
100	1660	7,600	2600	480
200	1480	7,600	2600	400
400	1100	6,500	1500	200
600	740	5,000	0	0

* The p_{wf} values are determined by using the flow rate and the PI, and the equivalent depth of p_{wf} is taken from the pressure-distribution curves. Subtracting the tubing length (5000) from this figure gives the equivalent depth of the THP, and reference to the distribution curves results in the THP values shown in the last column.

Total depth	4052 ft
7-in. casing	Surface to 4020 ft
3½-in. tubing	Hung at 4000 ft

Casing-tubing packer installed just above tubing shoe

The well was flowing at 280 bbl/day of clean oil, a GOR of 600 cu ft/bbl, and a THP of 300 psi when it was decided to try the effects of an acidization treatment. During this treatment, 10,000 gal of acid was squeezed into the formation. A surface pressure of 3200 psi was needed to overcome the static reservoir pressure of 1800 psi and to achieve the desired injection rate of 2 bbl/min. After the treatment the well's production rate stabilized at 320 bbl/day of clean oil through 3½-in. tubing with a GOR of 1000 cu ft/bbl and a THP of 300 psi. Determine whether or not the treatment was successful, and give an explanation of the results obtained. What would have been the production rate of the well at a THP of 300 psi if, instead of the acidization treatment, the 3½-in. tubing string had been replaced with a 2⅜-in. string (on the assumption that the change could have been made without damaging the producing formation)?

The IPR

The first step is the calculation of the flowing BHP at various production rates and a GOR of 600 cu ft/bbl, using 3½-in. tubing and a THP of 300 psi. The results are shown in Table 4-7 and are plotted in Fig. 4-47 (crosses).

The actual production rate of the well before acidization was 280 bbl/day

TABLE 4-7 EXAMPLE 4-7: Calculation of Pressure at Foot of 3½-in. Tubing, Various Flow Rates, GLR of 600 cu ft/bbl

q, bbl/day	Equiv. Depth of THP of 300 psi, ft	Equiv. Depth of Well, ft	p_{wf}, psi
50	1600	5600	1150
100	2000	6000	1030
200	2300	6300	970
400	2800	6800	950
600	2700	6700	950

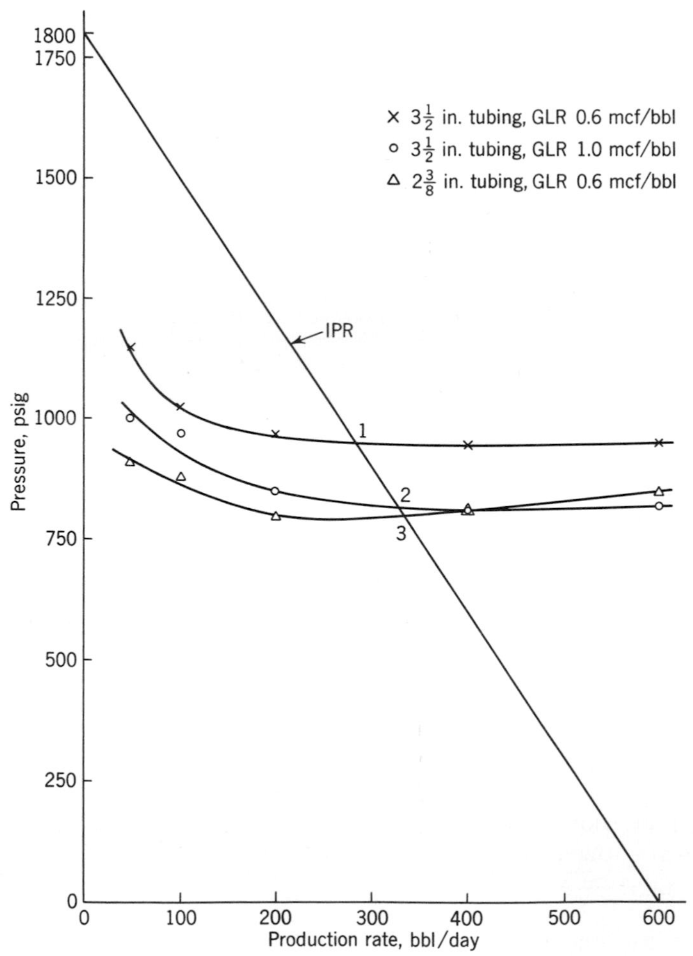

Fig. 4-47 Example 4-7: Evaluation of formation-stimulation treatment.

TABLE 4-8 EXAMPLE 4-7: Calculation of Pressure at Foot of 3½-in. Tubing, Various Flow Rates, GLR of 1000 cu ft/bbl

q, bbl/day	Equiv. Depth of THP of 300 psi, ft	Equiv. Depth of Well, ft	p_{wf} psi
50	2000	6000	1000
100	2800	6800	970
200	3100	7100	850
400	3500	7500	810
600	3300	7300	820

through $3^1/_2$-in. tubing. This point is marked 1 on Fig. 4-47. The static well pressure is given as 1800 psi, and so the IPR can be drawn as shown (on the assumption that it is a straight line).

The Well's Potentialities after Treatment

The flowing BHPs at various production rates through $3^1/_2$-in. tubing with a THP of 300 psi and a GOR of 1000 cu ft/bbl are calculated in Table 4-8, and the results are plotted in Fig. 4-47 (circles).

Effect of Treatment

The actual production rate after acidization was 320 bbl/day through $3^1/_2$-in. tubing. This point is marked 2 on Fig. 4-47, and it can be seen that, within the limits of accuracy of the method, it lies on the original IPR. Thus the acidization was completely unsuccessful in improving the oil productivity of the well. Indeed, the only real effect of the acid treatment was to increase the GOR and gas production of the well (possibly by acidizing into a gas stringer), and the additional gas has enabled the well to flow at a slightly higher rate.

Effect of Running $2^3/_8$-in. Tubing Instead of Acidizing

The flowing BHPs at various production rates through $2^3/_8$-in. tubing with a THP of 300 psi and a GOR of 600 cu ft/bbl are calculated in Table 4-9 and the results are plotted in Fig. 4-47 (triangles). The curve cuts the IPR at a production rate of 330 bbl/day, approximately, so a change in the tubing size would have been just as effective as—and a lot cheaper than—acidization.

TABLE 4-9 EXAMPLE 4-7: Calculation of Pressure at Foot of 2⅜-in. Tubing, Various Flow Rates, GLR of 600 cu ft/bbl

q, bbl/day	Equiv. Depth of THP of 300 psi, ft	Equiv. Depth of Well, ft	p_{wf}, psi
50	2400	6400	910
100	2800	6800	880
200	3000	7000	800
400	2900	6900	810
600	2800	6800	860

REFERENCES

1. Poettmann, F. H., and P. G. Carpenter: "The Multiphase Flow of Gas, Oil and Water through Vertical Flow Strings with Application to the Design of Gas-Lift Installations," *API Drill. Prod. Practice,* 1952, p. 257.
2. Baxendell, P. B.: "Producing Wells on Casing Flow, an Analysis of Flowing Pressure Gradients," *Trans. AIME,* **213:**202 (1958).
3. Baxendell, P. B., and R. Thomas: "The Calculation of Pressure Gradients in High-Rate Flowing Wells," *J. Petrol. Technol.,* **13**(10):1023 (1961).
4. Ros, N. C. J.: Simultaneous Flow of Gas and Liquid as Encountered in Well Tubing, *J. Petrol. Technol.,* **13**(10):1037 (1961).
5. Gilbert, W. E.: "Flowing and Gas-Lift Well Performance," *API Drill. Prod. Practice,* 1954, p. 126.
6. Brown, Kermit E.: *The Technology of Artificial Lift Methods,* vol. 1, Petroleum Publishing Company, Tulsa, Okla., 1977.
7. Aziz, K., et al.: *Gradient Curves for Well Analysis and Design,* Canadian Institute of Mining Special vol. 20, Montreal, Que., 1978.
8. Versluys, J.: "Mathematical Development of the Theory of Flowing Wells," *Trans. AIME,* **86:**192 (1930).
9. Tek, M. Rasin: "Multiphase Flow of Water, Oil, and Natural Gas through Vertical Flow Strings," *J. Petrol. Technol.,* **13**(10):1029 (1961).
10. Shaw, S. F.: *Gas-lift Principles and Practices,* Gulf Publishing Company, Houston, Tex., 1939.
11. Babson, E. C.: "The Range of Application of Gas-Lift Methods," *API Drill. Prod. Practice,* 1939, p. 266.
12. *Handbook of Gas Lift,* Garrett Oil Tools, Division of U.S. Industries, Inc., New York, 1959.
13. Rawlins, E. L., and M. A. Schellhardt: *Back-Pressure Data on Natural Gas Wells and Their Application to Production Practices,* U.S. Bureau of Mines Monograph 7, 1936.
14. Katz, Donald L., et al.: *Handbook of Natural Gas Engineering,* McGraw-Hill Book Company, Inc., New York, 1959.

5 Choke Performance: Well Performance History

5-1 INTRODUCTION

A necessary component in the study of flowing wells is a knowledge of the behavior of the choke used to control the production rate and ensure stability. The choke is most commonly installed at the wellhead, but other possibilities are a down-hole choke, often used as a safety measure in offshore wells against the possibility that the wellhead might suffer damage, or a choke installed downstream of the wellhead in the gathering lines. In certain circumstances it may be decided to produce the well without any flow-line restriction.

The case of a wellhead choke is the only one discussed here, although some general comments applying to other situations are made in Sec. 5-9. The reason for this decision is that there is already a considerable literature on chokes and on the detailed analysis of choke performance; the intention here is to outline general procedures, an understanding of which should enable the engineer to address successfully the particular problems and circumstances involved in a given situation.

Once the principal features of choke performance are understood, it is possible to incorporate these into the framework already discussed of inflow and vertical lift performance and so arrive at a picture that brings together the main features that determine flowing well behavior. Some aspects of this behavior are reviewed in Secs. 5-3 through 5-8.

5-2 CHOKE PERFORMANCE

In this section the performance of a choke (or bean) installed at the wellhead will be considered. It is standard oil field practice to choose the choke on a flowing well in such a way that small variations in the downstream pressure (that is, in the flow-line pressure, caused, for instance, by the use of a dump separator) do not affect the THP and thus the well's performance. This implies fluid flow through the choke at velocities greater than that of sound, and it has been found, under the range of conditions met in oil field work, that this requirement is satisfied if the THP is at least double the average flow-line pressure. It can be shown theoretically, assuming a knife-edge choke and making several simplifying assumptions with regard to the pressure versus volume characteristics of the oil and gas, that

$$p_{tf} = \frac{C\,R^{0.5}\,q}{S^2} \tag{5-1}$$

where p_{tf} = THP, psia
R = GLR, mcf/bbl
q = gross liquid rate, bbl/day
S = choke size, $^1/_{64}$ in.
C = constant (about 600 in the system of units defined above)

Using production data from Ten Section Field in California, Gilbert (Ref. 1) obtained the empirical formula

$$p_{tf} = \frac{435R^{0.546}q}{S^{1.89}} \tag{5-2}$$

where now p_{tf} is in psig.

In what follows, Gilbert's formula and the nomogram he presented (Fig. 5-1) will be used in the worked examples unless otherwise stated. The left-hand section of this nomogram represents the performance of a $^{10}/_{64}$-in. bean, and the right-hand section is a means of correcting results for other bean sizes.

Example 5-1 A well is producing 100 bbl/day gross with a GLR of 700 cu ft/bbl. If the bean size is $^1/_4$ in., calculate the THP from Gilbert's nomogram (Fig. 5-1) and from the theoretical formula shown in Eq. (5-1).

Using the Nomogram

Enter the left-hand chart at 100 bbl/day, and drop vertically to a GLR of 700 cu ft/bbl. Move horizontally on to the right-hand chart until the line intersects that for a $^{10}/_{64}$-in. bean. Move vertically once more to the diagonal line corresponding to a $^{16}/_{64}$-in. bean. Read the THP from the scale at the left-hand side of the right-hand chart. In this case, the result is 190 psig.

Using the Formula

Inserting the value 600 for the constant C in Eq. (5-1), the expression becomes

$$p_{tf} = \frac{600R^{0.5}q}{S^2} \text{ psia}$$

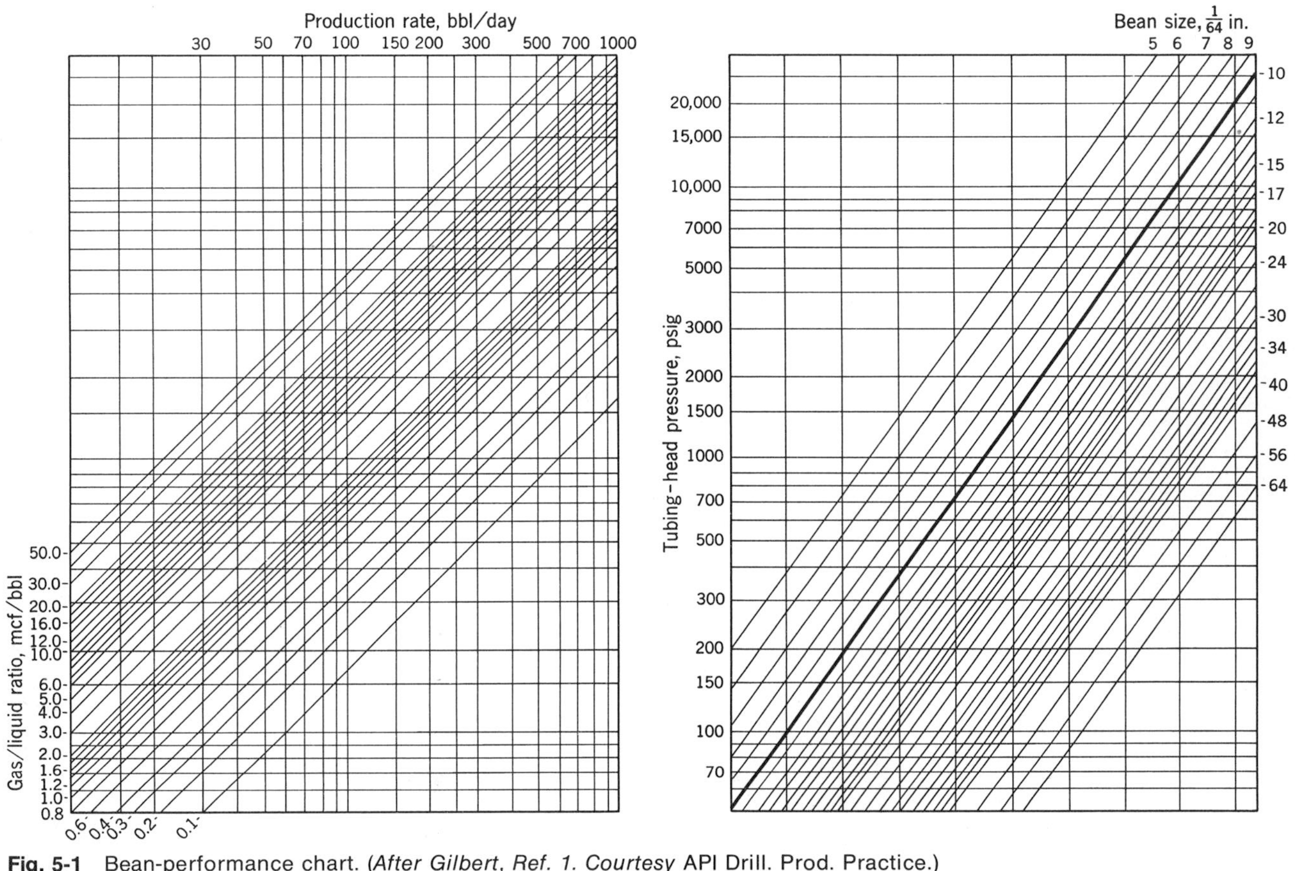

Fig. 5-1 Bean-performance chart. (*After Gilbert, Ref. 1. Courtesy* API Drill. Prod. Practice.)

In this example, $R^{0.5} = 0.7^{0.5} = 0.837$

$$q = 100$$
$$S^2 = (16)^2 = 256$$

Thus,

$$p_{tf} = \frac{600 \times 0.837 \times 100}{256} = 196 \text{ psia}$$
$$= 181 \text{ psig}$$

Example 5-2 A well is producing through a 1/4-in choke at 100 bbl/day with a THP of 150 psi. What is the GLR as calculated from Fig. 5-1 and from Eq. (5-2)? What would be the calculated GLR if, all other things being equal, the choke size were 17/64 in.?

Enter the right-hand chart of Fig. 5-1 at 150 psi and run across to the diagonal line corresponding to a 16/64-in. choke. Move vertically to the line for a 10/64-in. bean and then horizontally to the left-hand chart to the intersection with the vertical 100-bbl/day line. Read off the GLR using the diagonal grid. This procedure gives

$$\text{GLR} = 0.44 \text{ mcf/bbl}$$

A similar procedure for a 17/64-in. choke leads to the result

$$\text{GLR} = 0.55 \text{ mcf/bbl}$$

Use of Eq. (5-2) directly, rather than through Fig. 5-1, results in the values

1/4-in. choke: GLR = 0.456 mcf/bbl
17/64-in. choke: GLR = 0.561 mcf/bbl

Example 5-2 illustrates the large variation in the calculated GLR that results from only a small change in the supposed choke size and points up the danger of using the bean-performance formula to calculate GLRs. In symbols, the reason for this sensitivity can be seen easily from Eq. (5-1), which can be written in the form

$$R = \frac{(p_{tf})^2}{(Cq)^2} S^4$$

illustrating the fact that the GLR is dependent on the fourth power of the bean size.

It should be remembered too that unless chokes are frequently renewed, the effects of gas and/or sand cutting or of asphalt or wax deposition will result in a bean distorted in shape as well as out of gauge. The severity of these effects can be checked from time to time from measurements of the production rate, GLR, and THP, by using the facts that $p_{tf}S^{1.89}/R^{0.546}q$ should be about 435, from Eq. (5-2), or $p_{tf}S^2/R^{0.5}q$ should be about 600, from Eq. (5-1).

Example 5-3 Using the data of Example 4-4, what size of bean is required in the flow line to hold a THP of 100 psi? What would be the production rate on a 1/4-in. bean?

As was seen in Example 4-4, the flow rate with a THP of 100 psi is 270 bbl/day of oil and 30 bbl/day of water, or a gross rate of 300 bbl/day. The GLR is 0.27 mcf/bbl. If these data are used in Fig. 5-1, S is seen to be 32, so that the bean size is $^1/_2$ in.

To determine the production rate on a $^1/_4$-in. bean, note that p_{tf} and q are unknowns. Substituting 0.27 for R and 16 for S in Eq. (5-2) gives the result

$$p_{tf} = \frac{435 \times 0.27^{0.546} q}{16^{1.89}} = 1.13q$$

This is a straight line through the origin of the pressure versus rate graph. A second point on this line may be found by letting q be, say, 600 bbl/day. Then p_{tf} is 678 psig. Drawing this line in on Fig. 5-2, which is identical to Fig. 4-23, it is seen that it cuts the THP curve where q is 210 bbl/day and the THP is 235 psi. These values define the well's performance on a $^1/_4$-in. bean.

Other simplified empirical formulas modeled after that of Gilbert [Eq. (5-2)] are in use—for example Achong (Ref. 2) correlated data obtained from

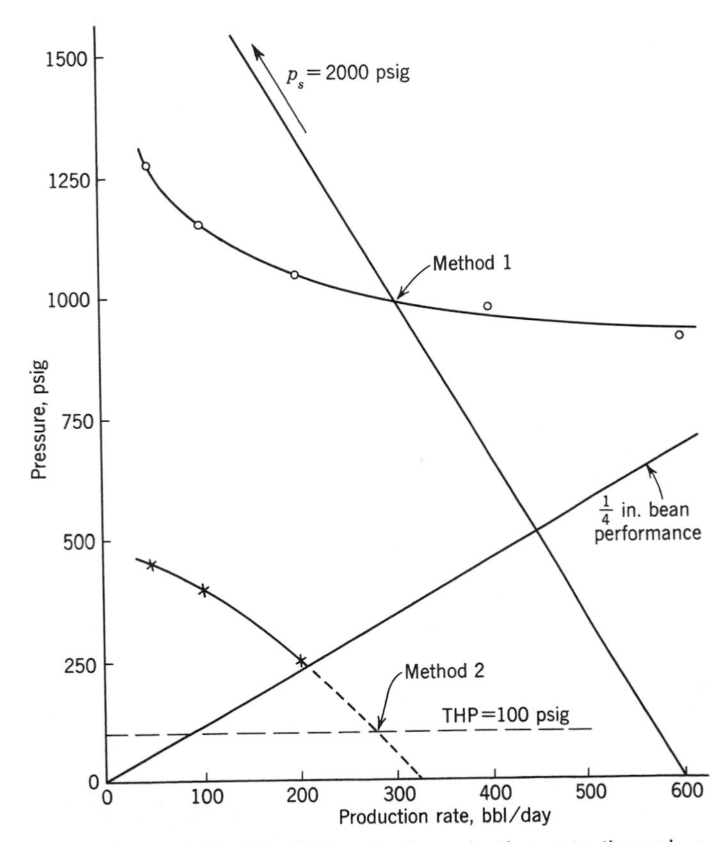

Fig. 5-2 Example 5-3: Forecast of production rate through a predetermined size of choke.

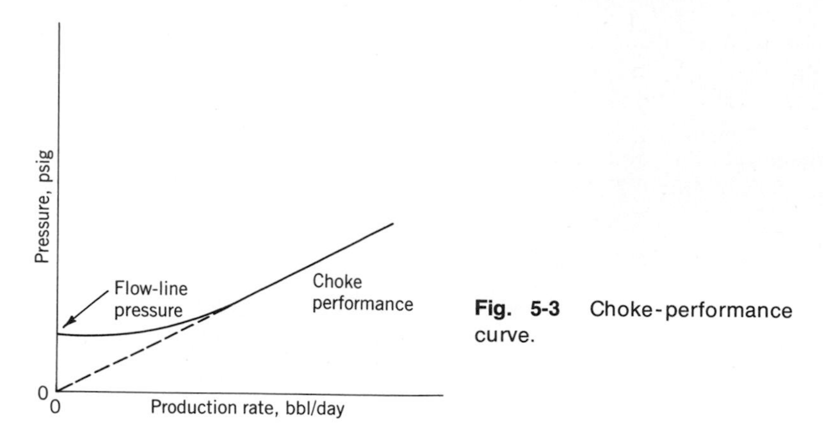

Fig. 5-3 Choke-performance curve.

wells producing from the Lake Maracaibo fields in Venezuela through Cameron positive-type chokes to obtain the formula

$$p_{tf} = \frac{340\, q\, R^{0.65}}{S^{1.88}} \tag{5-3}$$

where the symbols have their earlier meanings.

It should be noted that each of Eqs. (5-1), (5-2), and (5-3) may be written in the form

$$p_{tf} = Aq \tag{5-4}$$

that is, the THP is proportional to the production rate. This is true only under conditions of supersonic flow through the choke; at low flow rates the pressure upstream of the bean depends on downstream conditions. In the limit, when the flow rate is zero, the upstream and downstream pressures are equal (Fig. 5-3).

More complex expressions are available in the literature, and in use within individual companies, for two-phase flow performance through chokes. These expressions include refinements that make allowances for such variables as the specific gravity of the liquid and of the gas, the upstream temperatures, and the compressibility factor of the gas. One of the best known is that derived by Poettmann and Beck (Ref. 3) from the work of Ros (Ref. 4).

5-3 STABLE AND UNSTABLE FLOWING CONDITIONS

The flow from an oil well—from the formation, through the tubing, and up to the flow-line choke—is never completely stable. The liquid flow rate, the GLR, and the formation pressure all vary continuously. So long as the limits on these fluctuations remain narrow, the flow system may be regarded as

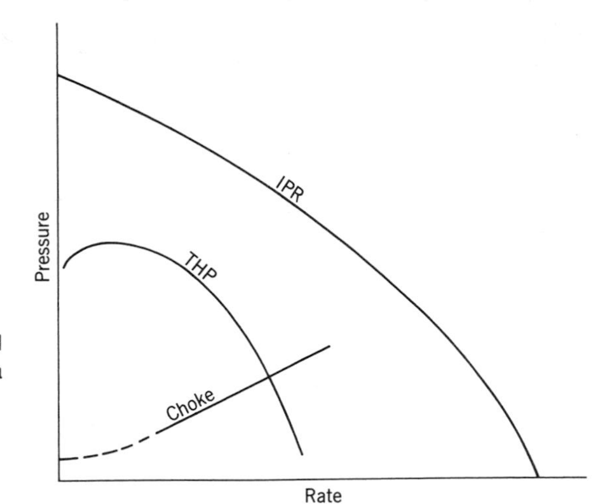

Fig. 5-4 Typical IPR, THP, and choke-performance curves for a flowing well.

stable. On the other hand, any tendency for the variations to grow in size is an indication of incipient instability and can have serious consequences, with the well either dying or—at the other extreme and far more rarely—building up excessively high surges in production rate.

In analyzing stability of flow it is necessary to keep in mind the IPR, THP, and choke-performance curves of a flowing well (Fig. 5-4), the typical curve of pressure loss in the tubing as a function of production rate (Fig. 5-5), and the fact that the THP curve of Fig. 5-4 is the difference between the IPR curve and the curve shown in Fig. 5-5. It must also be remembered that the function of a flow-line choke is to regularize flow, that is, to hold the flow rate—and consequently the THP—as steady as possible.

In Sec. 4-7 attention was drawn to the shape of the pressure-loss curve (Fig. 5-5), which falls rapidly with increasing rate at low flow rates and then slowly achieves a very flat minimum. The flow rate at which this minimum is attained falls with increasing GLR, and the rate at which the curve rises, at

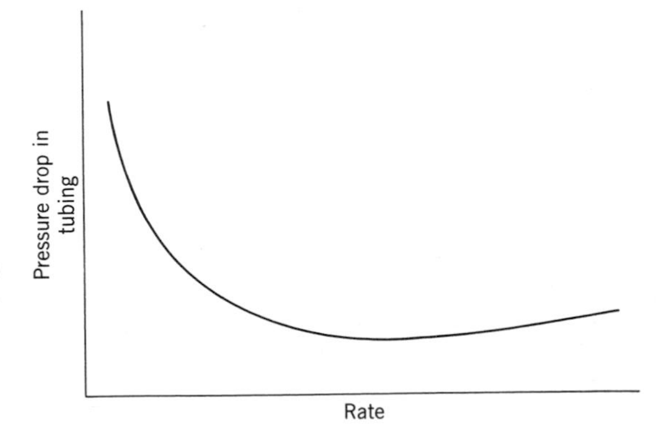

Fig. 5-5 Pressure loss in the tubing as a function of production rate.

flow rates above that at which the minimum occurs, increases with increasing GLR (Fig. 5-6).

A typical situation is that in which the well is being flowed at a rate at which the "pressure-loss-in-the-tubing" curve of Fig. 5-5 is relatively flat, dropping or rising slowly. For the sake of argument, consider the case in which this curve is dropping slowly at the flow point (Fig. 5-7).

Suppose that, for some reason, there is a variation in the free GLR from the formation, and that this variation persists for a few seconds. As a result, the pressure loss in the tubing will change; assume that it drops by an amount δp—point B_2 on Fig. 5-7. Since the choke is maintaining a (reasonably) constant THP, it follows that the drop in the pressure loss in the tubing must be accompanied by a rise in the inflow pressure from the formation—point A_2 on the IPR curve, where the vertical (pressure) difference between A_1 and A_2 is also δp.

The flow rate from the formation is reduced, and so the pressure loss in the tubing rises (B_3). The inflow pressure falls so that the new flowing position from the formation is A_3, where the vertical (pressure) difference between A_1 and A_3 equals that between B_1 and B_3. This leads to a new point (B_4) and so on.

Stability will occur when the points A_2, A_3, A_4, and so forth tend to converge toward A_1 (the points B_3, B_4, B_5, and so forth will of course simultaneously converge toward B_1). Instability will be a danger if the A points tend to become more widely separated.

In order to study this in greater detail, let m and M be the slopes of the IPR curve and the pressure-loss curve at the points A_1 and B_1, respectively (Fig. 5-7). Since B_2B_1 equals δp, the horizontal (flow rate) distance A_2A_1 is $\delta p/m$. It follows that the vertical (pressure) distance B_1B_3 is $\delta p\ (M/m)$; this is also the vertical (pressure) distance A_1A_3.

Thus the horizontal (flow rate) distance A_1A_3 is

$$\delta p \frac{1}{m}\frac{M}{m}$$

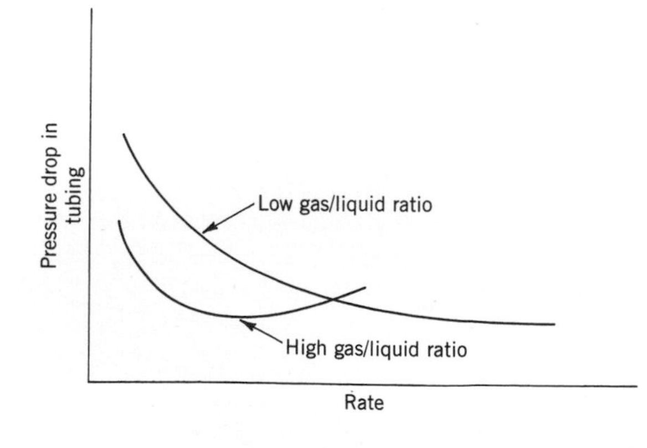

Fig. 5-6 Influence of GLR on the pressure loss in the tubing plotted as a function of production rate.

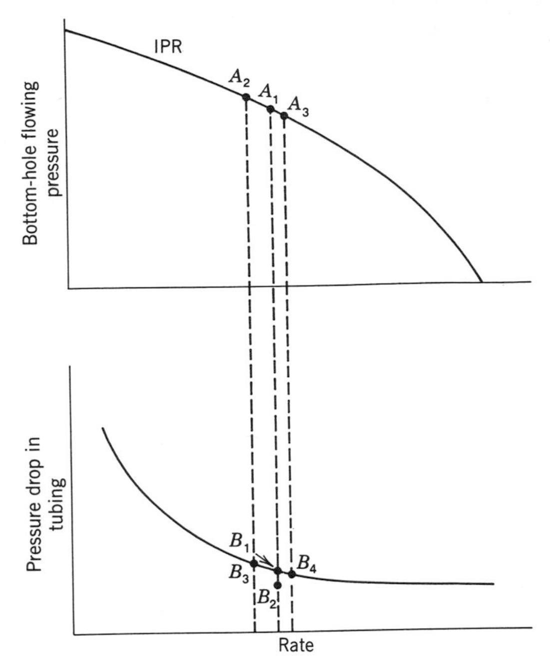

Fig. 5-7 Analysis of stability of flow.

so that the vertical (pressure) distance B_1B_4 is

$$\delta p \left(\frac{M}{m}\right)^2$$

Continuing this line of argument, the horizontal (flow rate) distance A_1A_4 is

$$\delta p \frac{1}{m} \left(\frac{M}{m}\right)^2$$

the horizontal distance A_1A_5 is

$$\delta p \frac{1}{m} \left(\frac{M}{m}\right)^3$$

and so on.

It follows that if $|M/m| > 1$, the points A become wider apart (instability), while if $|M/m| < 1$, the points A converge to A_1 (stability). In words, the system is stable if the slope of the pressure-loss curve (Fig. 5-5) is numerically less than the slope of the IPR. Otherwise the system is potentially unstable.

At the low-rate end, the slopes of these two curves are equal when the curve of their difference—that is, the THP curve of Fig. 5-4—is horizontal. Hence any attempt to flow a well at a point on the THP curve to the left of its maximum is dangerous, and the well is liable to die.

At the high-rate end, it is possible to imagine situations in extremely prolific, high GLR wells in which the curve of pressure loss in the tubing will be numerically steeper than the IPR (which will of necessity be flat in a high-volume producer). Under such circumstances, attempts to obtain steady, high-rate production by conventional means may fail, with excessive surges in production rate being encountered.

5-4 EFFECTS OF CHANGES IN CHOKE SIZE

In those cases in which the casing-tubing annulus is not packed off and the flowing BHPs are below the bubble point, the sequence of events that follows a reduction in the choke size in a flowing well is as described below.

The reduced orifice at the surface results in an increased back pressure on the tubing, which causes a decreased flow rate through the tubing. The back pressure on the formation does not change instantaneously, however, because the CHP plus the pressure due to the annulus gas column is momentarily unaltered. The formation continues for a short while to produce at the old rate, and the difference between this production and what can now pass up the tubing must go into the annulus. This liquid production into the annulus causes the flowing BHP to rise, first, because the liquid column itself exerts a back pressure and, second, because the gas that initially occupied the entire annulus volume is compressed. The rise in the flowing BHP results in a lower formation production rate, and so on until equilibrium is attained,[1] that is, until a balance is reached between the formation inflow (governed by the flowing BHP, which is itself determined by the pressure exerted by the fluids in the annulus), the vertical two-phase pressure loss for this particular inflow rate, and the bean performance.

At this stage, gas begins to displace the oil that has accumulated in the annulus; while this is taking place, the free GLR in the tubing is reduced, not only because of the loss of gas into the annulus but also because of the production of oil (associated with no free gas) from the annulus into the tubing. During this period of reduced GLR in the tubing, the pressure loss in the tubing will increase (unless the well is producing at above-optimum GLRs) and the THP will fall. Thus, immediately after the well is beaned back, there is a critical period during which the THP falls and there is real danger that the well may die. It follows, therefore, that for wells which have no casing-tubing packer, which are producing with flowing BHPs below the optimum, and which have GLRs approaching the lower limit for natural flow, considerable care should be exercised in beaning back. The desired bean change should be made over a period of several days, a number of intermediate bean sizes being

[1] There is always the attendant danger that if the annulus liquid builds up too rapidly, it will "overshoot," raising the BHP so much and altering the tubing flow rate so drastically that the well dies during this phase of the process.

used so that no sharp increase in CHP is ever recorded. Unfortunately, this is precisely the class of well that develops heading problems, one of the (temporary) cures of which is to bean back; the subject of heading is discussed in Sec. 5-8.

It is easily seen that beaning up presents no immediate problem; the decreased flowing BHP allows some of the annulus gas to escape into the tubing, thus raising the tubing GLR and reducing the vertical two-phase flow pressure loss, provided GLRs are below the optimum (see, however, Sec. 5-8).

If the well is completed with a casing-tubing packer at or near the foot of the tubing, there is little or no "reservoir" of free gas available in the well itself, and the danger that has just been discussed is either not present or is at worst a marginal consideration. An exception to this general conclusion may arise in the case of a fractured pay zone, for example, a relatively impermeable fractured limestone. In such circumstances it is possible that a gas-filled fracture close to the well bore may play the role of the casing-tubing annulus. Even if the producing formation is, say, a permeable sandstone, some degree of instability may be encountered after a large increase or decrease in choke size if the produced fluids are volatile in the sense that the gas solubility is high and gas moves readily into and out of solution. Such instability would result from the time lag between a constriction change at the surface and the reaction of the fluid flow to that change. During that time, pressures rise or fall in the vicinity of the well bore, with consequent absorption or release of free gas and possible "overshoot" in both directions.

5-5 EFFECT OF STATIC PRESSURE ON FLOWING WELL EFFICIENCY

In this section it will be shown that, other factors remaining constant, two-phase vertical-flow efficiency declines as the static reservoir pressure drops; in other words, as the static pressure falls, an increasing percentage of the total pressure drop from formation to tubing head is due to the pressure drop in the tubing. This decreasing efficiency is reflected in the production rate attainable from the well through a given choke size, so production-rate-decline curves based on past flowing performance will frequently give meaningless results. To illustrate this point, consider the following example:

Example 5-4 A well producing from a pay zone between 5000 and 5020 ft is completed with $2^7/_8$-in. tubing hung at 5000 ft. At what rate will the well flow against a $^1/_2$-in. bean when the static BHP is (1) 2500, (2) 2000, (3) 1500, and (4) 1300 psig, assuming that the GLR of the well varies as shown in Table 5-1 and that the PI remains constant at 0.3 bbl/(day)(psi)?

To solve the problem, construct the THP and IPR curves for the well at each of the four static pressures given; draw in the choke-performance lines at the corresponding GLRs, and read off the flowing rates. Results of these calculations are shown in Fig. 5-8, which indicates that

TABLE 5-1 GLR as Function of Static Pressure

Static BHP, psig	GLR, scf/bbl
2500	600
2000	700
1500	400
1300	300

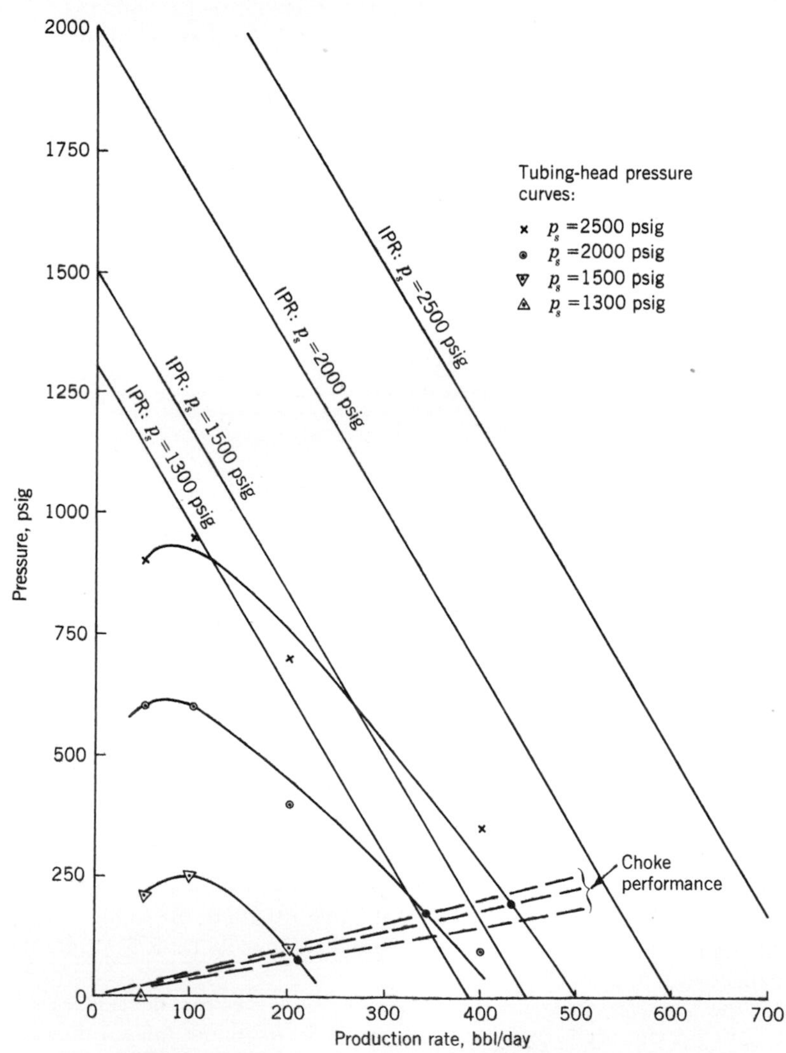

Fig. 5-8 Example 5-4: Production forecast, flowing well, constant choke size.

1. At 2500 psig static: q = 430 bbl/day = 57.4% potential
2. At 2000 psig static: q = 345 bbl/day = 57.5% potential
3. At 1500 psig static: q = 210 bbl/day = 46.6% potential
4. At 1300 psig static: q = 0 bbl/day = 0% potential

These figures point up declining vertical lift efficiency: a similar result would have been obtained if the GLR had been assumed to be constant throughout.

In Fig. 5-9 the potential of the formation and the actual production rate are plotted as functions of reservoir pressure. The figure shows that, in this particular example, the production-rate decline curve has a lower slope than the formation potential over the pressure range from 2500 to about 1700 psig. The situation between 1700 and 1300 psig is entirely different. Overall, it is clear that the production decline is no measure of formation potential decline. Moreover, changes in the PI as the pressure declines will further complicate the picture, so that it is to be expected that extrapolation of the production-rate decline curve of a flowing well will give misleading results. In

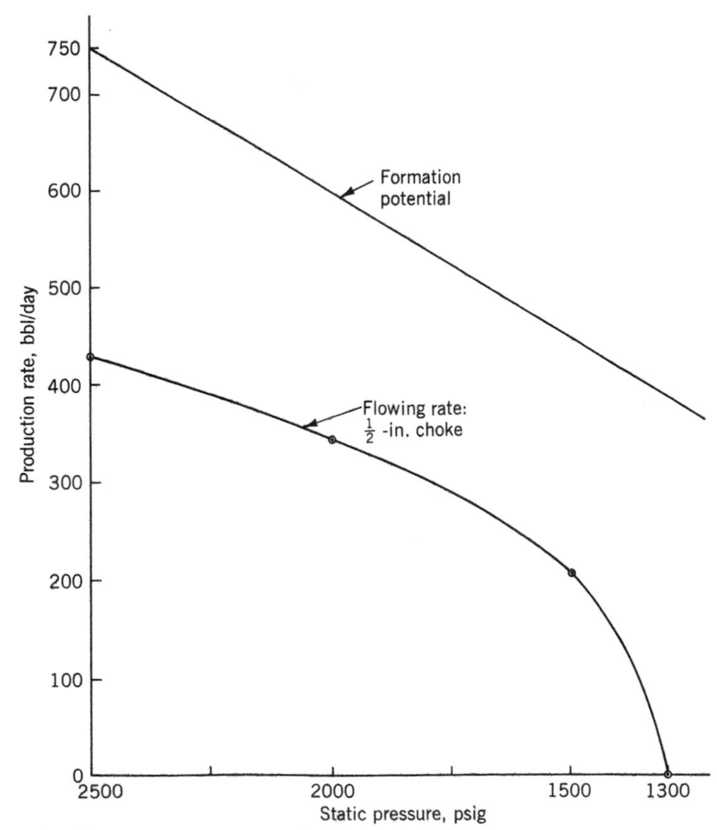

Fig. 5-9 Comparison between the potential of a well and its flowing production rate.

place of applying production-rate-decline curve analyses, it is recommended that a method similar to that illustrated in the next section be used wherever long-term forward predictions of fair to good accuracy are required. For short-term predictions, or for long-term estimates in which only orders of magnitude are of importance, production-rate decline curves form a quick and simple way of making estimates (Chap. 2).

5-6 PRODUCTION FORECAST FOR A POOL

It was illustrated in the preceding section that extrapolation of decline curves does not give too reliable a picture of the future possibilities of a well or pool. The following example outlines a method based on individual well performance that should lead to more accurate results. One major drawback to the use of this approach, or similar approaches, is the present lack of precise knowledge concerning IPR decay (Sec. 3-4); an increase in our understanding of this process is perhaps one of the most important challenges facing modern production engineering. A second drawback to the method outlined—the computational work involved—carries little weight in light of the availability of computers and the importance of accurate forecasting. In the example shown, arbitrary extrapolations of the curves of pressure and GOR against cumulative recovery have been made. This was done in order to avoid loss of continuity of the procedure outlined and burdening the reader with more detail than is absolutely necessary. In an actual field case it would be necessary to turn to one of the reservoir engineer's techniques (such as Tarner's or Muskat's method for solution gas-drive reservoirs) to derive more trustworthy curves on which to base the predictions (Sec. 1-8).

TABLE 5-2 Well Static Pressures and PIs

Well	Static BHP, psig	PI, bbl/(day)(psi)
A	2350	0.22
	1820	0.19
	1710	0.09
	1420	0.14
B	2100	0.06
	1730	0.07
	1550	0.05
C	2100	0.19
	1660	0.15
	1400	0.12
D	2100	0.11
	1770	0.09
	1420	0.07

TABLE 5-3 Pool Static Pressure at Various Cumulative Productions

Cumulative Oil Production from Pool, 10^6 bbl	Average Static Pressure at Datum Level, psig
0.031	2100
0.353	1720
0.669	1410

In the example used to illustrate the proposed method, certain production data are available from the four wells draining a small pool. The data are shown in Table 5-2. Wells A and C are flowing against 100-psig THP through 4500 ft of $2^3/_8$-in. tubing. Well B is pumping and well D, which recently died, has just been put to the pump. Initial reservoir pressure, as measured in the discovery well, well A, was 2350 psig at the datum level of 4120 ft subsea. Since production started from the pool, three pressure surveys have been run and the results are listed in Table 5-3; the pool's GORs at various points in the producing history are shown in Table 5-4.

Current water production is zero, and it is thought that the recovery mechanism will be depletion-drive. A volumetric estimate indicates that the volume of oil initially in place in the pool was 10 million stock-tank bbl. The pool currently has an average static pressure of 1410 psig and a cumulative oil production of 669,000 bbl. Assuming that wells A and C are allowed to flow until they die and are then put to pump, the problem is to determine the future production history of each well and the cumulative production to be expected from the pool.

It will be assumed that the IPRs are straight lines, that there is a constant flowing BHP (intake pressure) of 150 psig during the pumping phase, and that a well reaches its economic limit when the production is down to 5 bbl/day.

TABLE 5-4 GOR History of Pool

Cumulative Oil Production, 10^6 bbl	Pool's GOR, scf/bbl
0.120	209
0.229	208
0.307	214
0.402	220
0.471	242
0.533	240
0.565	255
0.602	298
0.641	353
0.669	365

Static Pressure Behavior

Plot the static pressure as a function of the cumulative oil production[2] (Fig. 5-10) from Table 5-3. The manner in which the static pressure drops indicates that the bubble-point pressure of the crude is a little over 2000 psi and was reached at a cumulative oil production of about 40,000 bbl.

PI Decline

Prepare a PI decline curve that will fit the available information in the best possible manner. When the symbol J_i is used for the PIs at pressures above the bubble point, the indicated bubble-point pressure of 2000 psig, taken in conjunction with the information of Table 5-2, leads to the conclusions

$$(J_i)_A = 0.22$$
$$(J_i)_B = 0.06$$
$$(J_i)_C = 0.19$$
$$(J_i)_D = 0.11$$

Now determine (J/J_i) and plot this value as a function of pressure.

When this is done on semilog paper, the result is as shown in Fig. 5-11. With the exception of one value from well A and the well B results, the points lie reasonably well on a straight line.[3] It will be noted that "adjusted values"

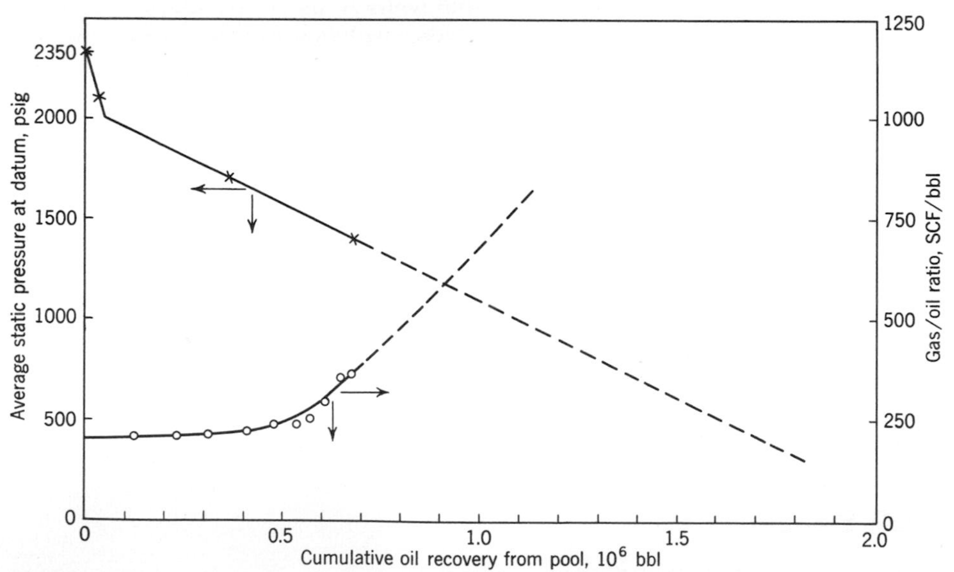

Fig. 5-10 Production forecast: assumed pressure and GOR history.

[2] See comments at the beginning of this section.
[3] See comments at the beginning of this section.

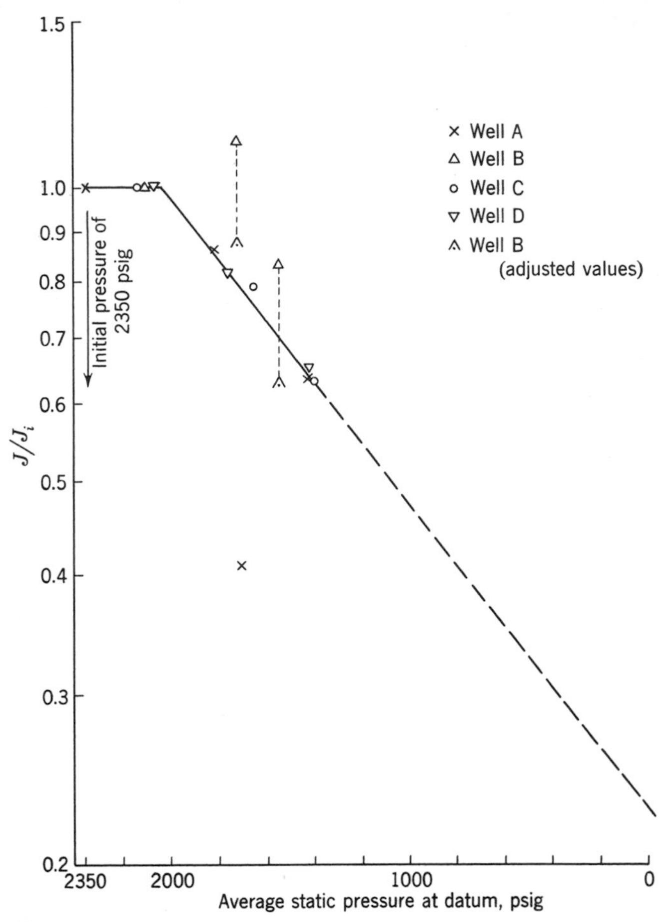

Fig. 5-11 Production forecast: PI data.

for well B have also been plotted in Fig. 5-11. These are in quite good agreement with the trend defined by the readings from the other three wells. They have been obtained by assuming that the initial PI of well B was 0.08 bbl/(day)(psi), rather than 0.06 as measured, the difference being due, possibly, to faulty readings, to failure to allow the well to clean up before taking the first pressure survey, or to some other similar cause.

GOR Behavior

Plot the GOR versus cumulative oil recovery trend from Table 5-4 (Fig. 5-10). This can be extrapolated over a relatively short distance without too great inaccuracy.[4]

[4] See comments at the beginning of this section.

Estimated Well Potentials and Pumping Rates

Determine the future PIs, potentials, and pumping production rates of the four wells. These values can be assigned with the help of Figs. 5-10 and 5-11, and the necessary calculation steps are shown in Tables 5-5, 5-6, and 5-7.

TABLE 5-5 Establishing a PI Decline

Static Pressure, psig	Cumulative Oil from Pool, 10^3 bbl (Fig. 5-10)	$\frac{J}{J_i}$ (Fig. 5-11)	PIs bbl/(day)(psi)			
			A	B*	C	D
1500†	575	0.680	0.150	0.054	0.129	0.075
1410	669	0.635	0.140	0.051	0.121	0.070
1300	780	0.590	0.130	0.047	0.112	0.065
1200	880	0.550	0.121	0.044	0.104	0.061
1000	1090	0.475	0.105	0.038	0.090	0.052
800	1300	0.410	0.090	0.033	0.078	0.045
600	1500	0.350	0.077	0.028	0.067	0.039
400	1710	0.305	0.067	0.024	0.058	0.034
300	1820	0.285	0.063	0.023	0.054	0.031
200	1920	0.265	0.058	0.021	0.050	0.029

* Initial PI value J_i taken to be 0.08 bbl/(day)(psi).

† Included to make interpretation in Figs. 5-12 and 5-13 easier.

TABLE 5-6 Variations in Wells' Potentials with Pool Static Pressure

Static Pressure, psig	Potential, bbl/day			
	A	B	C	D
1500	150.0*	81.0	193.5	112.5
1410	197.4	71.9	170.6	98.7
1300	169.0	61.1	145.6	84.5
1200	145.2	52.8	124.8	73.2
1000	105.0	38.0	90.0	52.0
800	72.0	26.4	62.4	36.0
600	46.2	16.8	40.2	23.4
400	26.8	9.6	23.2	13.6
300	18.9	6.9	16.2	9.3
200	11.6	4.2	10.0	5.8

* Production rate of well A at a drawdown of 1000 psi when the pool static pressure is 1500 psig.

TABLE 5-7 Future Pumping Production Rates

Static Pressure, psig	Static Pressure Less 150, psig	Pumping Production Rate, bbl/day			
		A	B	C	D
1500	1350	202.5	72.9	174.2	101.3
1410	1260	176.4	64.3	152.5	88.2
1300	1150	149.5	54.1	128.8	74.8
1200	1050	127.1	46.2	109.2	64.1
1000	850	89.3	32.3	76.5	44.2
800	650	58.5	21.5	50.7	29.3
600	450	34.7	12.6	30.2	17.6
400	250	16.8	6.0	14.5	8.5
300	150	9.5	3.5	8.1	4.7
200	50	2.9	1.1	2.5	1.5

Future Flowing Performance of Well A

In Fig. 5-12, the IPR of well A is plotted at reservoir static pressures of 1500, 1410, 1300, 1200, and 1000 psig (if it is seen later that more IPR lines are needed, they can be drawn in as required). The information required for drawing these lines is summarized in Table 5-6. The GORs corresponding to

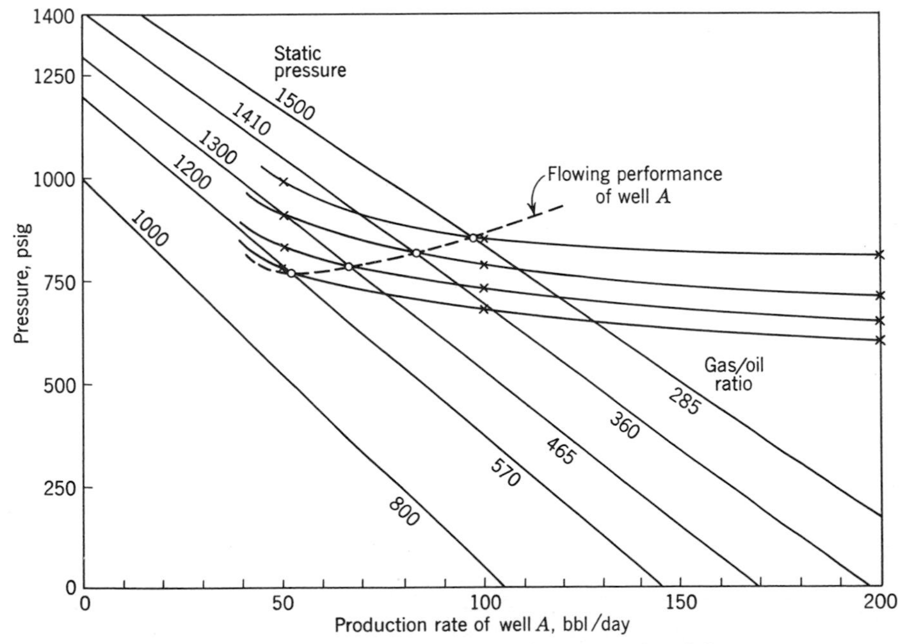

Fig. 5-12 Production forecast: future flowing performance of well A.

TABLE 5-8 Predicted GORs for the Pool

Static Pressure, psig	Pool GOR, scf/bbl
1500	285
1410	360
1300	465
1200	570
1000	800

the various pressures may be obtained from Fig. 5-10 and are shown in Table 5-8.

Use of the pressure-distribution curves for flow through $2^3/_8$-in. tubing (Fig. 4-18) and a fixed THP of 100 psig permits determination of the rate of flow of well A at each of the assumed static pressures by plotting the pressure at the foot of the tubing against the production rate and picking out the point of intersection with the corresponding IPR. The calculations are shown in Table 5-9, and the resultant curves are plotted in Fig. 5-12. The well will die when the flowing performance curve (dashed line on Fig. 5-12) becomes tangential to the corresponding IPR line; evidently this occurs for well A when the static pressure is slightly lower than 1200 psi. To simplify the calculations, it will be assumed that well A will die when the static pressure has

TABLE 5-9 Future Flowing Performance, Well A ($2^3/_8$-in. Tubing; THP = 100 psig)

Static Pressure, psig	GOR, scf/bbl	Production Rate, bbl/day	Equiv. Depth of 100-psi THP, ft	Equiv. Depth of Tubing Shoe, ft*	p_{wf} psig
1500	285	50	500	5000	990
		100	500	5000	850
		200	700	5200	810
1410	360	50	700	5200	910
		100	600	5100	790
		200	700	5200	710
1300	465	50	900	5400	830
		100	900	5400	730
		200	800	5300	650
1200	570	50	1000	5500	780
		100	1000	5500	690
		200	1000	5500	610

* Equivalent depth of tubing shoe equals the equivalent depth of 100-psi THP plus the tubing length (that is, 4500 ft).

TABLE 5-10 Future Flowing Rates, Well A

Reservoir pressure, psig	Flowing Rate, bbl/day
1410	82
1300	66
1200	51 (well dies)

dropped to 1200 psi. A summary of the flowing rates achieved by well A is given in Table 5-10.

Future Flowing Performance of Well C

An analysis similar to that for well A results in Fig. 5-13, and the flowing production rates of well C at various pressures are listed in Table 5-11.

Future Performance of Pool

By utilizing the information shown in Tables 5-5, 5-7, 5-10, and 5-11, it is possible to set up Tables 5-12 and 5-13, which between them summarize the future performance of the pool. The calculation has been terminated at a reservoir pressure of 400 psi, a cumulative production of 1,710,000 bbl (or 17.1 percent recovery), and a future life of just under 20 years, since production rates after this point appear to be so low that any further prediction is very

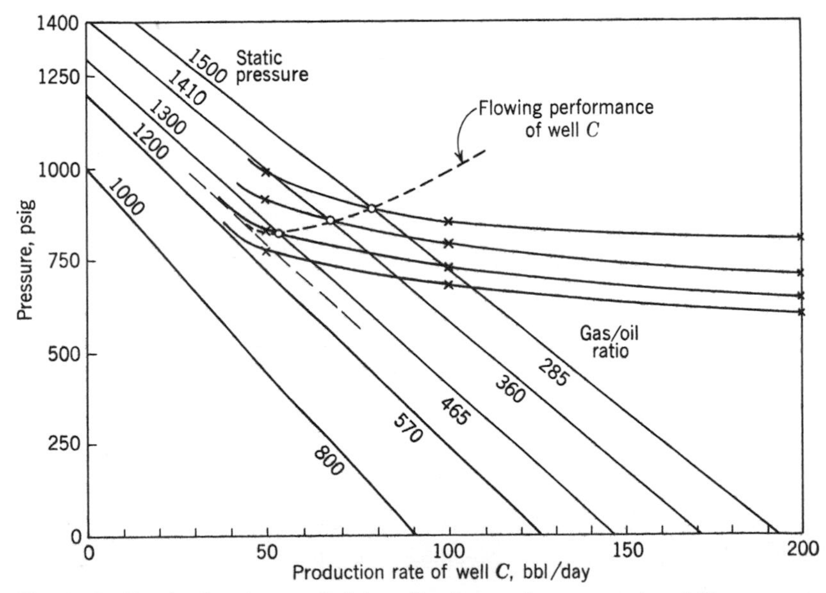

Fig. 5-13 Production forecast: future flowing performance of well C.

TABLE 5-11 Future Flowing Rates, Well C

Reservoir Pressure, psig	Flowing Rate, bbl/day
1410	66
1300	53
1250	45 (well dies)

TABLE 5-12 Predicted Production-Rate Performance of Pool against Pressure

Static Pressure, psig	Cum. Prod. from Pool, 10^3 bbl	Increase in Cumulative, 10^3 bbl	Production Rates, bbl/day				
			A	B	C	D	Total
1410	669		82.0	64.3	66.0	88.2	300.5
1300	780	111	66.0	54.1	53.0	74.8	247.9
1250	830	50	(58.5)	(50.2)	45.0 }	(69.5)	223.2 }
					118.8* }		297.0 }
1200	880	50	51.0 }	46.2	109.2	64.1	270.5 }
			127.1 }				346.6 }
1000	1090	210	89.3	32.3	76.5	44.2	242.3
800	1300	210	58.5	21.5	50.7	29.3	160.0
600	1500	200	34.7	12.6	30.2	17.6	95.1
400	1710	210	16.8	6.0	14.5	8.5	45.8
300	1820	110	9.5	3.5	8.1	4.7	25.8
200	1920	100	2.9	—	2.5	—	5.4

* This figure of 118.8 bbl/day is calculated by obtaining the value of J/J_i from Fig. 5-11, and thus the J value. Use of a pump-intake pressure of 150 psig results in the calculated rate of 118.8 bbl/day.

TABLE 5-13 Predicted Production-Rate Performance of Pool against Time

Pressure Step, psig	Average Production Rate, bbl/day					Incremental Time, months	Cumulative Time, months
	A	B	C	D	Total		
1410–1300	74.0	59.2	59.5	81.5	274.2	13.3	13.3
1300–1250	62.3	52.2	49.0	72.2	235.7	7.0	20.3
1250–1200	54.8	48.2	114.0	66.8	283.8	5.8	26.1
1200–1000	108.2	39.3	92.9	54.2	294.6	23.5	49.6
1000–800	73.9	26.9	63.6	36.8	201.2	34.3	83.9
800–600	46.6	17.1	40.5	23.5	127.7	51.5	135.4
600–400	25.8	9.3	22.4	13.1	70.6	97.8	233.2
400–300	13.2	—	11.3	6.6	13.1	116.4	*
300–200	6.2	—	5.3	—	11.5	286.1	*

* Omitted, as time becomes unrealistically long.

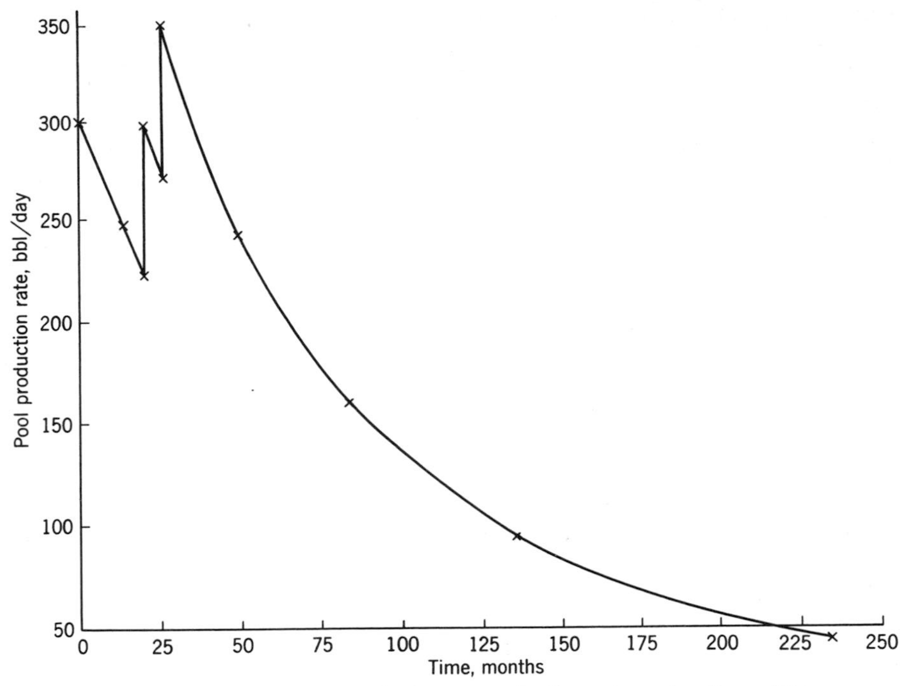

Fig. 5-14 Production forecast: pool's future production rate as function of time.

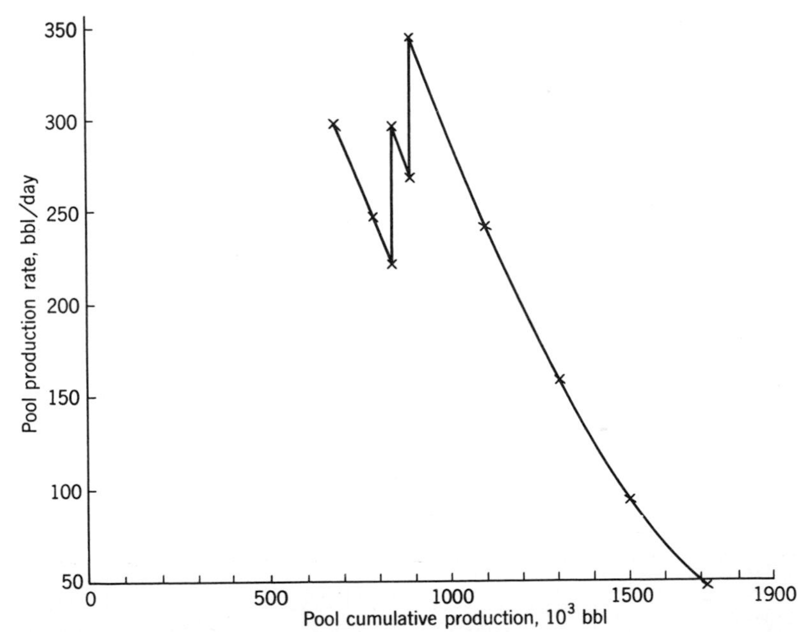

Fig. 5-15 Production forecast: pool's future production rate as function of cumulative production.

unreliable. Plots of future pool production rates against time and cumulative production are shown in Figs. 5-14 and 5-15, respectively.

Future Performance of Individual Wells

By using the information of Tables 5-12 and 5-13, curves showing the future production-rate histories of the individual wells may be drawn (Fig. 5-16). Table 5-13 may also be used to calculate the future cumulative production of each well as a function of reservoir pressure. The results are shown in Table

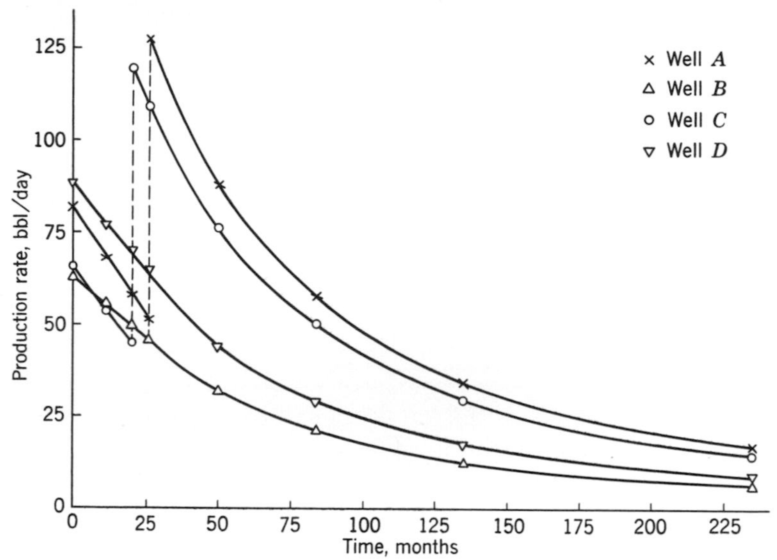

Fig. 5-16 Production forecast: individual well future production rates as functions of time.

TABLE 5-14 Predicted Cumulative Production as Function of Pressure

Reservoir Pressure, psig	Future Cumulative Production, 10^3 bbl				
	A	B	C	D	Total
1410	0	0	0	0	0
1300	30.0	23.9	24.1	33.0	111.0
1250	43.2	35.0	34.5	48.3	161.0
1200	52.9	43.5	54.5	60.1	211.0
1000	130.0	71.6	120.7	98.7	421.0
800	207.1	99.7	186.9	137.3	631.0
600	280.1	126.5	250.3	174.1	831.0
400	356.9	154.1	317.0	213.0	1041.0

5-14, and curves of production rate against well cumulative for the four wells are drawn in Fig. 5-17. It is of interest to note from Fig. 5-17 that, although in the early stages of pumping the production rate declines of individual wells appear to be exponential (the plot of production rate against cumulative is a straight line), this trend disappears later in the pumping life and the production rates of the wells are in general higher than would be indicated by an extrapolation of the early, straight-line portions of the curves. The curves for wells B and D in Fig. 5-17 are unexpected and illustrate the type of departure from formalized decline curves that may be obtained as the result of a detailed analysis.

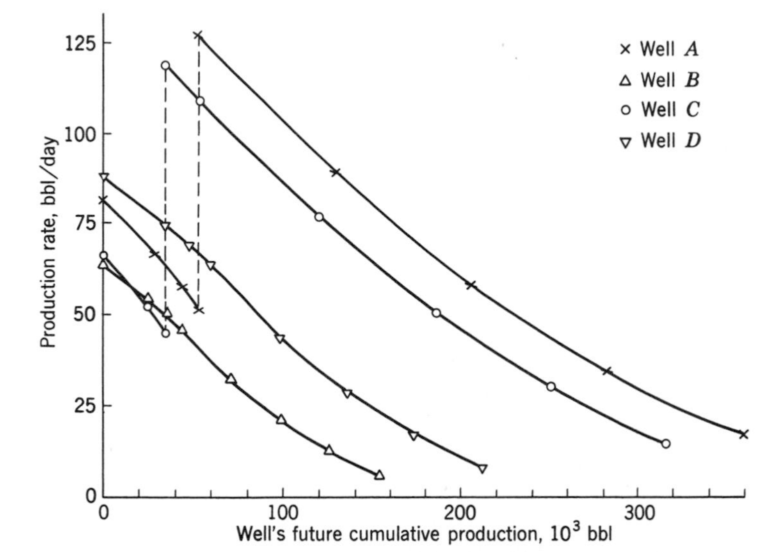

Fig. 5-17 Production forecast: individual well future production rates as functions of the future cumulative productions of the respective wells.

5-7 EFFECT OF A SMALL CHANGE IN THE PRODUCING GLR

For a flowing well, the flowing BHP less the pressure loss in the tubing equals the THP; that is,

$$p_{wf} - \Delta p = p_{tf}$$

Since

$$p_{tf} = \frac{C\,R^{0.5}}{S^2}\,q$$

from Eq. (5-1), it follows that

$$p_{wf} - \Delta p = \frac{C\,R^{0.5}}{S^2}\,q$$

or

$$p_{wf} - \Delta p = \frac{C\,R^{0.5}}{S^2}\,J\,(p_s - p_{wf})$$

from Eq. (3-2). This may be rewritten in the form

$$p_{wf}\left(1 + \frac{C\,R^{0.5}}{S^2}\,J\right) = \frac{C\,R^{0.5}}{S^2}\,Jp_s + \Delta p \tag{5-5}$$

The manner in which the flowing BHP varies with small changes in the GLR may be found by determining the rate of change dp_{wf}/dR from Eq. (5-5). If this equation is differentiated then, with respect to R,

$$\frac{dp_{wf}}{dR}\left(1 + \frac{C\,R^{0.5}}{S^2}\,J\right) + \frac{1}{2R}\,\frac{C\,R^{0.5}}{S^2}\,Jp_{wf} = \frac{1}{2R}\,\frac{C\,R^{0.5}}{S^2}\,Jp_s + \frac{d}{dR}\,\Delta p$$

or

$$\frac{dp_{wf}}{dR}\left(1 + \frac{C\,R^{0.5}}{S^2}\,J\right) = \frac{1}{2R}\,\frac{C\,R^{0.5}}{S^2}\,q + \frac{d}{dR}\,\Delta p$$

from Eq. (3-2), so that finally,

$$\frac{dp_{wf}}{dR}\left(1 + \frac{C\,R^{0.5}}{S^2}\,J\right) = \frac{1}{2R}\,p_{tf} + \frac{d}{dR}\,\Delta p \tag{5-6}$$

from Eq. (5-1).

Suppose that the well is producing at a GLR below the optimum. A small increase in GLR will cause the pressure loss in the tubing to drop (Sec. 4-7), so that $d(\Delta p)/dR$ is negative. Equation (5-6) shows that if $p_{tf}/2R$ is less than $-d(\Delta p)/dR$, then dp_{wf}/dR will be negative; that is, the pressure at the foot of the tubing will drop and the production rate will increase when the GLR increases. Conversely, if $p_{tf}/2R$ is greater than $-d(\Delta p)/dR$, the pressure at the foot of the tubing will rise and the production rate will decrease when the GLR increases. The precise point of balance between these two cases is dependent on the shape of the pressure-distribution curves for two-phase vertical flow, but, in general, the smaller the bean installed at the wellhead, the greater the THP p_{tf} and the greater the probability that the intake pressure will increase as the GLR increases. Moreover, when the GLR is less than, but close to, the critical, the value of $d(\Delta p)/dR$ will be small, so it is quite possible that the right-hand side of Eq. (5-6) will be positive, that is, that the intake pressure will increase with the GLR.

Inspection of the pressure-distribution curves, Figs. 4-11 to 4-20 inclusive, confirms that the value of $-d(\Delta p)/dR$ increases both as the tubing size decreases and as the liquid flow rate decreases, so that the smaller the tubing size and the smaller the production rate, the more likely it is that the flowing BHP will decrease as the GLR increases. Furthermore, the magnitude of $-d(\Delta p)/dR$ increases as the depth of the well increases, so the deeper the well, the more probable it is that the intake pressure will decrease as the GLR increases. The following two examples illustrate orders of magnitude.

Example 5-5 A 2000-ft well completed with a 3½-in. tubing is flowing 600 bbl/day, GLR 1.0 mcf/bbl, against a THP of 600 psi.

Referring to Fig. 4-20, Δp = 300 psi at the given rate and GLR. At a GLR of 1.2 mcf/bbl, Δp = 260 psi at 600 bbl/day.

Thus,
$$\frac{d}{dR}\Delta p = \frac{-40}{0.2} = -200 \text{ psi/mcf}$$

Also,
$$\frac{p_{tf}}{2R} = \frac{600}{2} = 300 \text{ psi/mcf}$$

Therefore,
$$\frac{d}{dR}\Delta p + \frac{1}{2R}p_{tf} = -200 + 300$$
$$= 100 \text{ psi/mcf}$$

and it follows from Eq. (5-6) that dp_{wf}/dR is positive.

Example 5-6 A 5000-ft well completed with 2⅜-in. tubing is flowing 200 bbl/day, GLR 1.0 mcf/bbl, against a THP of 200 psi.

Referring to Fig. 4-18, Δp = 470 psi at the given rate and GLR. At a GLR of 1.2 mcf/bbl, Δp = 430 psi at 200 bbl/day.

Thus,
$$\frac{d}{dR}\Delta p = \frac{-40}{0.2} = -200 \text{ psi/mcf}$$

Also,
$$\frac{p_{tf}}{2R} = \frac{200}{2} = 100 \text{ psi/mcf}$$

Therefore,
$$\frac{d}{dR}\Delta p + \frac{1}{2R}p_{tf} = -200 + 100$$
$$= -100 \text{ psi/mcf}$$

and it follows from Eq. (5-6) that dp_{wf}/dR is negative.

Thus, in summing up, it can be said that for wells sufficiently deep, completed with small-diameter tubing, producing at low rates, with low THPs, and with GLRs well below the optimum, small decreases in intake pressure will accompany small increases in the GLR. However, as the well depth decreases or as the other variables mentioned increase, this reaction to an increased GLR may be reversed. These results are of importance in an understanding of the phenomenon of heading, which is the subject of the next section.

5-8 THE HEADING CYCLE

Suppose that a well flowing at a low rate and with a relatively low GLR (that is, far below the optimum) is suddenly beaned up. It will be assumed that the new flow position is one in which a small decrease in the flowing BHP accom-

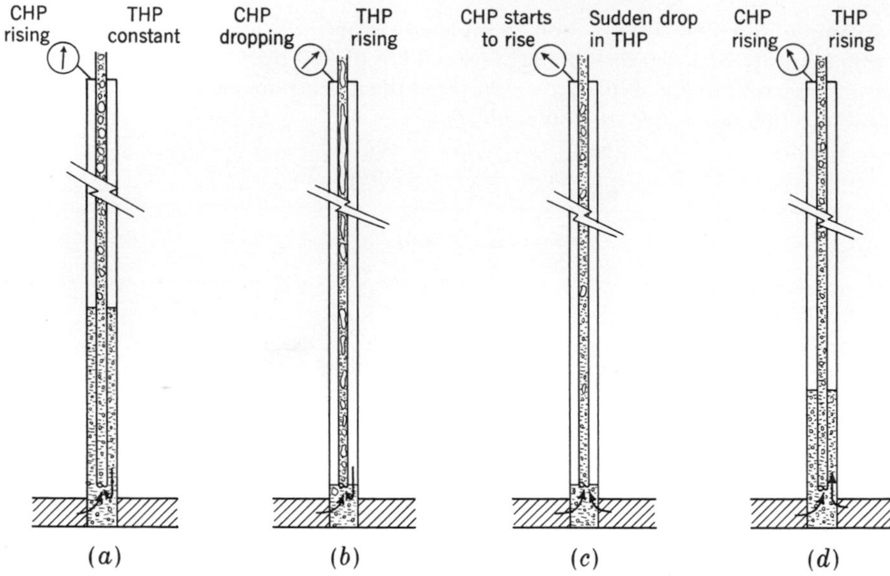

Fig. 5-18 The heading cycle. (*Adapted from Gilbert, Ref. 1. Courtesy* API Drill. Prod. Practice.) (*a*) Annulus liquid gradually displaced; (*b*) gas production from annulus; (*c*) annulus gas production terminates; (*d*) liquid moves into annulus.

panies a small increase in the GLR (Sec. 5-7); that the casing-tubing annulus is open at the lower end of the tubing; and that the flowing BHP in the new flow position is far enough below the bubble-point pressure to allow some of the free gas produced by the formation to escape entrainment and to move into the casing-tubing annulus.

Since the new flowing BHP is lower than the old, some of the gas stored in the annulus is produced up the tubing, temporarily raising the producing GLR. This fact, taken in conjunction with the first of the assumptions, implies that the well will flow temporarily at a higher rate than can be sustained with the natural GLR of the formation. As soon as this movement of surplus gas out of the annulus comes to a stop, however, the flow rate that can be accommodated through the tubing drops and the surplus liquid production from the formation moves into the annulus. As a result, the flowing BHP rises, and the liquid rate from the formation drops. This action continues until the flowing BHP has risen to such an extent that the production from the formation can be handled by the tubing and surface bean. Conditions in the well are then as shown in Fig. 5-18*a*.

At this stage some of the free gas produced by the formation passes into the annulus (which follows from the second and third of the initial assumptions), displacing the liquid there. Thus the annulus liquid level slowly drops, and this is shown at the surface by a steady increase in CHP. When the annulus liquid level reaches the tubing shoe, the net movement of free gas into the annulus becomes zero; the GLR of the fluid passing up the tubing is increased slightly. The first assumption then implies that the intake pressure drops or, to put it another way, that the liquid flow rate increases slightly. This permits some gas production from the annulus, which further raises the tubing GLR and hence, in all probability, the production rate. As a result of this action, the CHP drops steadily while the THP rises (Fig. 5-18*b*).

But there is a limited volume of gas stored in the annulus. When the annulus pressure has dropped sufficiently, free-gas production from the annulus stops and once again the surplus liquid production moves into the annulus. The CHP rises during this period because the annulus gas is being compressed. The THP, on the other hand, first falls sharply because of the decrease in both gas and liquid throughput rates at the surface and then steadily increases as the flowing BHP increases.

This sequence of events is known as the *heading cycle.* The types of THP and CHP behavior to be expected during the cycle are shown in Fig. 5-19.

A clear understanding of the conditions that must be fulfilled before

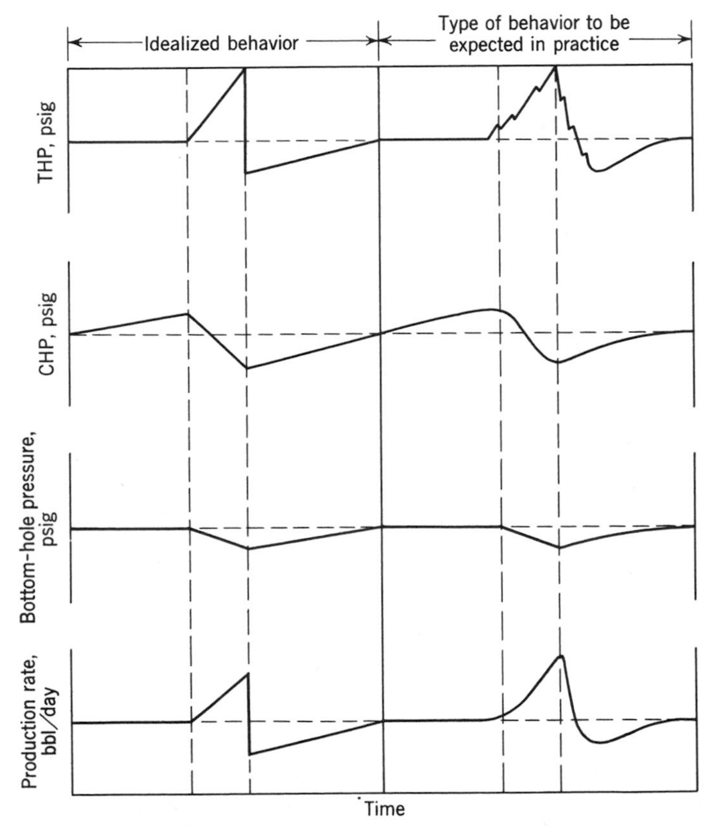

Fig. 5-19 Behavior of THP, CHP, flowing BHP, and production rate during heading cycle.

heading can occur points the way to control and also indicates the type of well in which heading may occur. First, the condition that the casing-tubing annulus must be open at the lower end to permit accumulation of free gas in this volume implies that the use of a casing-tubing packer eliminates annulus heading. If it is not felt desirable to install such a packer, a similar effect may be obtained by the use of an inverted swab rubber or some similar device placed at the tubing shoe to divert gas into the tubing.

Normally, little or no control can be exercised over the condition that there must be free gas present in the fluid stream at the tubing shoe, although in a few isolated cases it may be possible to cut the free-gas volume sufficiently to ensure entrainment of the gas bubbles with the oil by simply beaning back the well to raise the flowing BHP.

The third condition is that the new flow position must be one in which small decreases in the intake pressure accompany small increases in the GLR. If this were not so, then when the annulus fluid level reached the tubing shoe and all the formation free gas passed up the tubing, the resultant small increase in GLR would be accompanied by a slight increase in flowing BHP, which would preclude the blow-round of annulus gas. It is evident that under such circumstances, equilibrium would rapidly be established and the well would flow steadily. It follows (Sec. 5-7) that heading can be controlled to some extent by beaning back the well to raise the THP or by running a larger-sized tubing string. Both these remedies have their dangers, however, for heading action is most likely to occur in wells producing at low rates with low GORs against low THPs. Thus, beaning back the well may very possibly kill it (see Sec. 5-4); on the other hand, it may not be possible to bring the well back into production after running a larger-diameter tubing string, either because the formation has been damaged in the work-over or because the larger tubing is an inefficient flow string (Sec. 4-8).

Although it has here been assumed that the heading cycle started as a result of beaning up the well, it is clear that if the conditions leading to annulus heading are present, any slight alteration can start the cycle, which, once established, will gradually build up and, unless controlled in some way, will eventually kill the well during one of the periods when liquid is moving up into the annulus.

In certain types of formation, heading action may occur even though the casing-tubing annulus is packed off. Such a phenomenon is known as *formation heading;* since the occurrence of serious fluctuations in the producing GLR, production rate, and THP implies a comparatively large volume available for the accumulation of free gas, which periodically spills over, it follows that formation heading can take place only in those formations in which such "caverns" are present. A fissured or cavernous limestone or dolomite reservoir, for example, satisfies this requirement, and formation heading may take place when a pay section of this type is being produced. Conversely, it is not to be expected that a porous sandstone section will exhibit such behavior (see Sec. 5-4).

The presence of annulus heading is most readily noticed from the fluctuations in the CHP. However, since the period of the heading cycle may vary from 1 or 2 hr to as long as a day, it may be difficult to diagnose the occurrence of heading in a well from spot CHP readings taken at irregular intervals of a week or two. Because of the inefficiency of the heading cycle as a producing mechanism (see below) and because a heading well is liable to die, it is important to diagnose heading as soon as possible after its onset. To such an end the installation of a maximum-minimum pressure gauge on the casinghead may be justified; such a gauge indicates the range of CHP variation that has occurred since it was last read. If this range is large, say, more than 100 psi, the annulus pressure should be checked carefully for a few days to establish whether or not heading is actually taking place. In such a way heading can be discovered soon after its onset and the necessary countermeasures taken.

The inefficiency of heading as a producing mechanism stems from the inefficient use of the formation gas that is involved. This inefficiency may be illustrated by an example.

Example 5-7 A 4000-ft well completed with a $2^3/_8$-in. tubing has a static pressure of 1500 psi and a PI of 0.4 bbl/(day)(psi). It is producing through a $^{24}/_{64}$-in. bean. What is the daily production rate if the well produces for 22 hr at a GLR of 0.1 mcf/bbl and for 2 hr at a GLR of 2.0 mcf/bbl? What would be the production rate if the well produced steadily at the same daily GLR, the latter now remaining constant throughout the day?

To solve the first part of the problem, make a plot of THP against rate for the cases of 0.1 and 2.0 mcf/bbl GLR. The calculations are shown in Tables 5-15 and 5-16 and plotted in Fig. 5-20. The bean-performance formula, Eq. (5-2), is used to prepare the bean-performance lines. The equation gives a value of $0.303q$ for p_{wf} when the GLR is 0.1 mcf/bbl and $1.55q$ when the GLR is 2.0 mcf/bbl. The intersections of the bean-performance lines with the respective THP curves give the flow rates at the two GLRs under consideration.

It follows that the well will produce for 22 hr at a rate of 230 bbl/day, GLR 0.1 mcf/bbl, THP 70 psi, and for 2 hr at a rate of 295 bbl/day, GLR 2.0 mcf/bbl, THP 450 psi.

Thus, the daily production is

$$230 \times \frac{22}{24} + 295 \times \frac{2}{24} = 236 \text{ bbl/day}$$

TABLE 5-15 EXAMPLE 5-7: THP as a Function of Production Rate (GLR of 0.1 mcf/bbl)

Production Rate, bbl/day	Flowing BHP, psi	Equiv. Depth p_{wf}, ft	Equiv. Depth THP, ft	THP, psi
50	1370	5200	1200	300
100	1250	5100	1100	260
200	1000	4700	700	120
400	500	2700		
600	0			

TABLE 5-16 EXAMPLE 5-7: THP as a Function of Production Rate (GLR of 2.0 mcf/bbl)

Production Rate, bbl/day	Flowing BHP, psi	Equiv. Depth p_{wf}, ft	Equiv. Depth THP, ft	THP, psi
50	1370	16,700	12,700	940
100	1250	17,600	13,600	890
200	1000	14,300	10,300	680
400	500	6,100	2,100	170
600	0			

while the daily gas production is

$$(230 \times 100)\frac{22}{24} + (295 \times 2000)\frac{2}{24} = 70{,}300 \text{ scf/day}$$

so that the average (formation) GLR is 298 scf/bbl.

To solve the second part of the problem, the THP versus rate curve is plotted

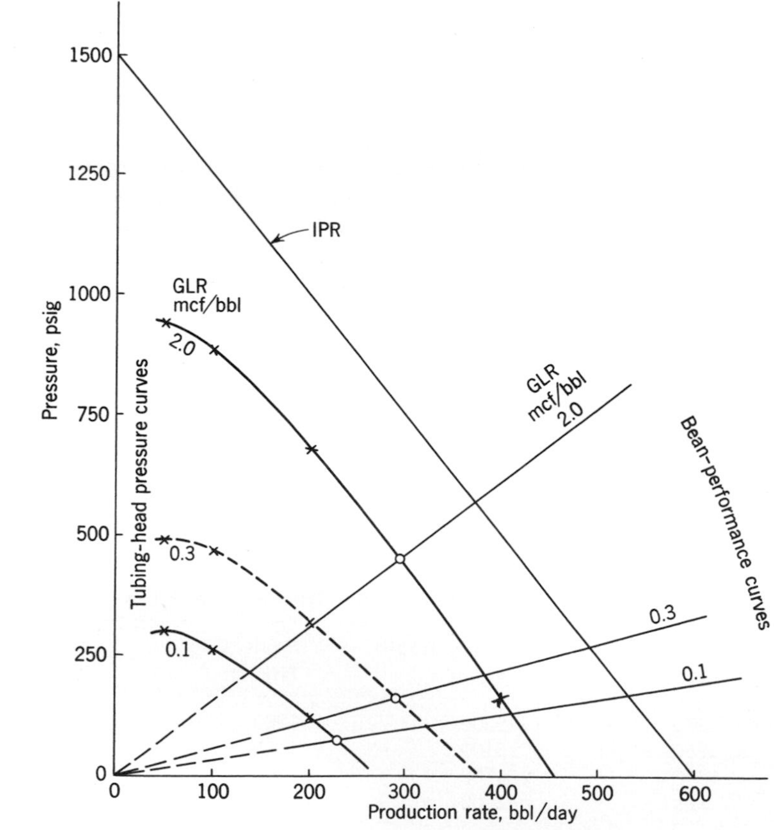

Fig. 5-20 Example 5-7: Performance of a heading well.

TABLE 5-17 EXAMPLE 5-7: THP as a Function of Production Rate (GLR of 0.30 mcf/bbl)

Production Rate, bbl/day	Flowing BHP, psi	Equiv. Depth of p_{wf}, ft	Equiv. Depth of THP, ft	THP, psi
50	1370	6800	2800	490
100	1250	7000	3000	470
200	1000	6400	2400	320
400	500	3600		
600	0			

with use of a GLR of 0.30 mcf/bbl. This is shown as a broken line in Fig. 5-20, and the calculations are summarized in Table 5-17. The bean-performance equation gives

$$p_{tf} = 0.556q$$

and when this line is drawn in on Fig. 5-20, it is seen that a steady flow rate of 290 bbl/day at a GLR of 0.30 mcf/bbl and a THP of 160 psi may be maintained on a $^{24}/_{64}$-in. choke.

Thus, even on the same bean, the well will maintain a higher flow rate under steady conditions. Moreover, if a bean change is made and the THP is maintained under steady conditions at 70 psi (equal to the minimum THP under unsteady conditions on the $^{24}/_{64}$-in. choke), a flow rate of about 340 bbl/day may be achieved.

In general, the principles illustrated by this example hold good and a greater production rate will be achieved, for the same overall GLR, from a well on steady production than from a heading well. Together with this increased production rate, which stems from a more effective use of the formation gas, goes a reduced likelihood that the well will head up and die in the immediate future. It follows that the diagnosis and regulation of the heading condition may be an important factor in optimizing the performance from an individual well and from a field.

5-9 DOWN-HOLE AND FLOW-LINE CHOKES: GENERAL COMMENTS

Control systems, other than wellhead chokes, may be visualized in terms of the pressure versus rate diagrams that have formed the basis of the work so far.

Figure 5-21 illustrates the case in which there is a wellhead choke, and in which the variation of pressure with rate at the downstream end of the gathering line is the critical variable—perhaps the pressure versus throughput rating of a separator.

Curve 1 is the IPR from which, using the pressure distribution curves, the THP (curve 2) is derived. If supersonic flow through the wellhead choke is a

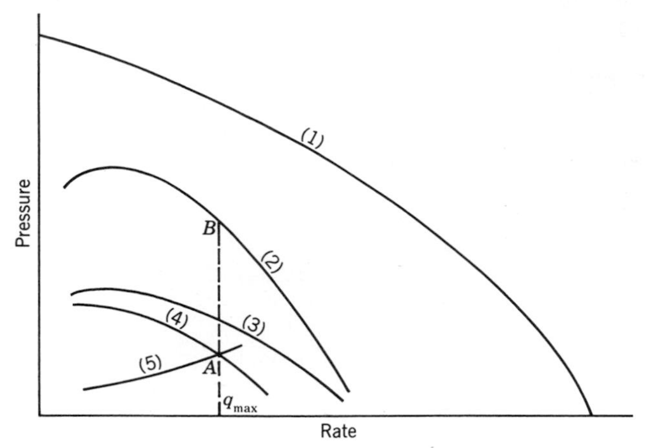

Fig. 5-21 Pressure versus rate curves for flowing well: (1) IPR; (2) THP; (3) maximum pressure downstream of choke (supersonic flow); (4) downstream end of gathering line. Curve 5 defines separator pressure versus rate performance.

requirement, then curve 3 is obtained. This is the maximum practical pressure versus rate curve downstream of the choke to ensure supersonic flow, and at each rate the corresponding pressure on curve 3 is one-half that on curve 2. A pressure loss correlation for the gathering system is used to obtain curve 4, which gives the maximum allowable pressure at the downstream end of the gathering line from the point of view of the well system.

Curve 5, the pressure versus throughput rating of the separator, is superimposed, and the point of intersection *A* gives the maximum-rate operating condition. The THP, point *B*, corresponds to this maximum rate, and the choke-performance equation can be used to determine the appropriate choke size.

A smaller choke, and a lower rate, are consistent with the well's performance characteristics and the equipment, but a higher rate would not permit adherence to the requirement for supersonic flow through the choke. Under certain circumstances the decision might, of course, be made to waive that requirement and to attempt a higher rate of production.

Figure 5-22 illustrates a more complex case, in which a choke is installed in the tubing, not on the bottom, but some way up the string. The pressure versus rate distribution grid (Sec. 4-6) is used to define the pressure versus rate curve 1 immediately upstream of the down-hole choke. The choke-performance correlation gives the curve 2, immediately downstream of the choke. Pressure-distribution curves then identify the pressure loss up the balance of the tubing so that the THP curve 3 is known (in effect, curve 2 is being used as the IPR at the depth at which the choke is installed). From that point forward, the procedure is similar to that outlined above (Fig. 5-21).

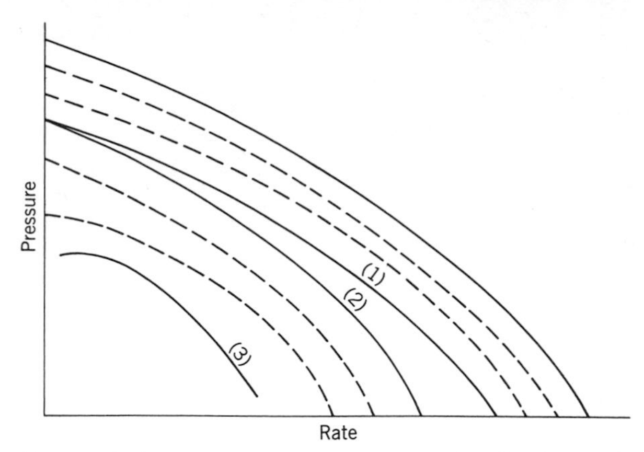

Fig. 5-22 Pressure versus rate curves for flowing well; down-hole choke.

REFERENCES

1. Gilbert, W. E.: "Flowing and Gas-Lift Well Performance," *API Drill. Prod. Practice,* 1954, p. 126.
2. Achong, Ian B.: *Revised Bean Performance Formula for Lake Maracaibo Wells,* University of Zulia, Maracaibo, Venezuela, 1974.
3. Poettmann, Fred H., and Richard L. Beck: "New Charts Developed to Predict Gas-Liquid Flow through Chokes," *World Oil,* March 1963.
4. Ros, N. C. J.: "Simultaneous Flow of Gas and Liquid as Encountered in Well Tubing," *J. Petrol. Technol.,* **13**(10):1037 (1961).

The Principles of Gas Lift

6

6-1 INTRODUCTION

When a well comes to the end of its natural flow life, the question arises as to what method should be used to keep it in production. One solution is to go to sucker rod pumping (Chap. 9) or some other pumping technique and another is to supply energy in the form of gas to help lift the formation liquids up through the tubing. In this chapter optimum-flow gas lift is discussed first; modifications to this system are then considered which enable the daily volume of injected gas, and so the compressor horsepower needed, to be substantially reduced with only a slight decrease in production rate. These modifications are discussed by means of examples. Example 6-2 illustrates continuous optimum-flow gas lift. Example 6-4 examines the relationship between input gas pressure and horsepower requirements, the effect of a limitation on the total daily volume of supply gas, and the use of macaroni tubing. In Example 6-5 the limitation imposed is that of an available compressor. In Example 6-6 it is assumed that input gas is available at a certain pressure (perhaps from a nearby gas well) and the question of the attainable gas-lift production rate is posed.

Finally, the problem of *kicking off* a dead well is considered, and it is shown that much higher compressor output pressures are required for kickoff than for steady production. Attempts to overcome this difficulty have led to the development of gas-lift valves that are run as part of the tubing string. These valves permit the introduction of gas (which is injected down the annulus[1]) into the static fluid column at intermediate depths in order to initiate flow. Several valves are usually included in the flow string and are located in such a way that the injected gas is able to enter the tubing at progressively lower depths during the kickoff process. In this way the flowing BHP can be reduced to the point at which the surface gas-injection pressure is sufficient to inject gas around, or close to, the foot of the tubing (see Chap. 7).

As in earlier chapters the IPR is assumed to be a straight line in the examples, but this is not a limitation on the methods discussed.

6-2 ILLUSTRATIVE EXAMPLES

Example 6-1 A well producing from 5000 to 5040 ft is completed with 2⅞-in. tubing set at 5000 ft. Production is clean, the well's PI is 0.5 bbl/(day)(psi), and the GLR is 300 cu ft/bbl. What will be the well's flow rate against a THP of 100 psi if the static BHP is (*a*) 1400 and (*b*) 1300 psi?

Calculate the flowing BHP at various rates, using a THP of 100 psi. The results, which are listed in Table 6-1, are plotted in Fig. 6-1 together with the IPRs appropriate to static pressures of 1400 and 1300 psi, respectively. It is evident that although the well is capable of flowing 150 bbl/day when the reservoir pressure is 1400 psi, nevertheless the well will die before the reservoir pressure drops to 1300 psi; in fact, it appears that the well will die when the pressure reaches approximately 1350 psi.

But a drop in reservoir pressure as cumulative production increases is typical of all but a very few oil fields. Artificial lift is therefore needed to keep the well producing, and the choice is between pumping and gas lift. In this chapter the problem of continuous gas lift is considered. Intermittent gas lift is discussed in Chap. 8, and pumping in Chap. 9.

Example 6-2 Consider the well of Example 6-1 when the static BHP is 1350 psi. The GLR of 300 cu ft/bbl is well below the optimum, so if it could be increased by

TABLE 6-1 EXAMPLE 6-1: Determination of Flowing BHP at Various Rates (GLR 300 cu ft/bbl)

q, bbl/day	Equivalent Depth of THP of 100 psi, ft	Equivalent Depth of Well, ft	p_{wf}, psi
50	500	5500	1250
100	700	5700	1150
200	800	5800	1050
400	800	5800	975
600	800	5800	910

[1] An alternative is to inject gas into the tubing and to produce the well via the casing-tubing annulus. This method, which has its main application in prolific producers, does not differ in principle from the system here described, namely, gas injection into the annulus and production via the tubing.

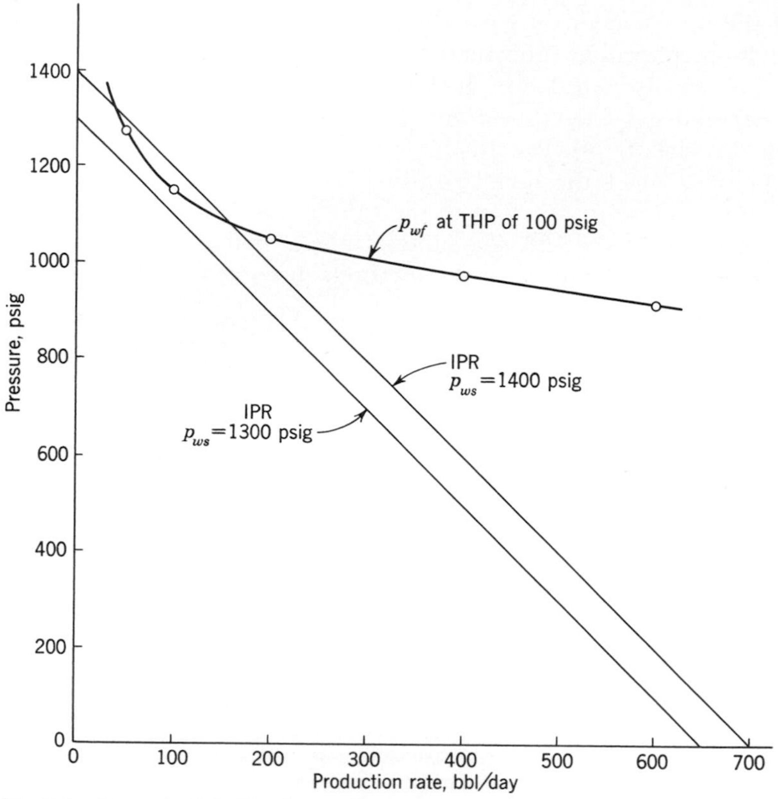

Fig. 6-1 Example 6-1: Flowing well about to die.

some means, the two-phase-flow pressure loss in the tubing would be reduced and the well would continue to produce. Since the pressure loss in the tubing is a minimum at the optimum GLR, it follows that to find the well's maximum possible rate of production it is necessary to use the optimum GLR at each stage.

One method is to calculate the flowing BHP at various production rates, using a THP of 100 psi and the optimum GLR at each rate. The calculations are shown in Table 6-2 and the results plotted in Fig. 6-2.

TABLE 6-2 EXAMPLE 6-2: Determination of Flowing BHPs at Various Rates, Using Optimum GLRs

q, bbl/day	Opt. GLR, mcf/bbl	Equiv. Depth of THP of 100 psi, ft	Equiv. Depth of Well, ft	p_{wf} at Opt. GLR, psi
50	8.8	5000	10,000	230
100	6.3	4000	9,000	270
200	4.3	3400	8,400	320
400	3.25	2200	7,200	380
600	2.4	1800	6,800	430

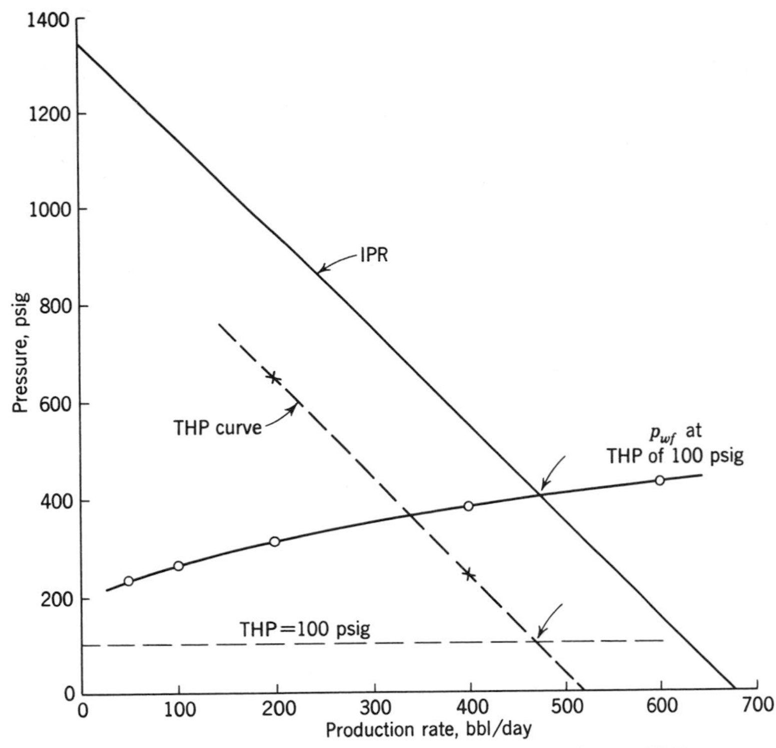

Fig. 6-2 Example 6-2: Gas-lift rate attainable using optimum GLR.

A second method is to calculate the THPs using the correct flowing BHPs (Table 6-3 and Fig. 6-2).

Evidently, since the second method gives only two points, the first is the more reliable in this case, and it appears that if the GLR could be raised to the optimum, the well would produce at a rate of 475 bbl/day.

To determine the optimum GLR at a production rate of 475 bbl/day, a plot is made of optimum GLR against rate (from the data shown in Table 6-3) and the GLR value at 475 bbl/day is obtained by interpolation. From Fig. 6-3 it appears that this GLR is 2.9 mcf/bbl.

TABLE 6-3 EXAMPLE 6-2: Determination of THP from IPR at Various Rates, Using Optimum GLRs

q, bbl/day	Opt. GLR, mcf/bbl	p_{wf}, psi	Equiv. Depth of p_{wf}, ft	Equiv. Depth of THP, ft	THP, psi
50	8.8	1250	Off graph		
100	6.3	1150	Off graph		
200	4.3	950	19,200	14,200	650
400	3.25	550	9,800	4,800	240
600	2.4	150	2,800		

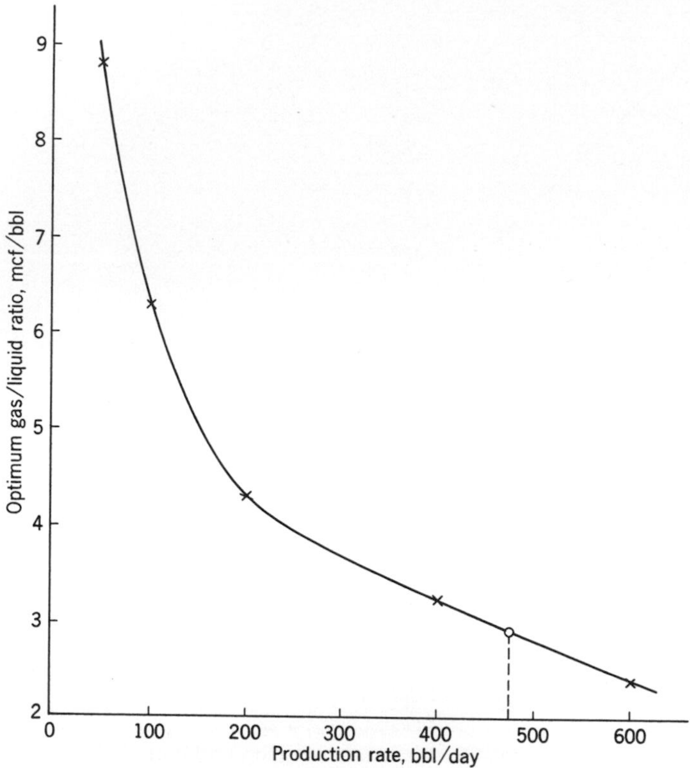

Fig. 6-3 Example 6-2: Determination of optimum GLR (2⅞-in. tubing).

The daily volume of injection gas required may then be found as follows:

$$\text{Total daily volume of gas} = 475 \times 2900 \text{ scf}$$
$$\text{Daily gas volume supplied by formation} = 475 \times 300 \text{ scf}$$
$$\text{Injection gas required daily} = 475 \times 2600 \text{ scf}$$
$$= 1.235 \times 10^6 \text{ scf}$$

This additional gas volume may be supplied to the well by injecting gas into the annulus and around the tubing shoe, thus raising the GLR in the tubing from the formation GLR of 300 cu ft/bbl to the optimum of 2900 cu ft/bbl. A rough estimate of the surface injection pressure required may be made as follows.

When the formation is producing 475 bbl/day, the flowing BHP is found from the IPR. Referring to Fig. 6-2, it is seen that this pressure is 400 psi. Thus the surface gas-injection pressure is 400 psi (the flowing BHP) *less* the pressure due to the gas in the annulus *plus* the friction loss due to gas movement down the annulus. If it is assumed that the friction-loss term and the pressure due to the gas column in the annulus roughly cancel, the value of the flowing BHP, that is, 400 psi in this example, may be taken as a first approximation to be the required surface gas-injection pressure.

6-3 COMPRESSOR HORSEPOWER REQUIREMENTS

It can be shown (see, for example, Katz et al., Ref. 1, page 316) that the power required to compress adiabatically a volume V_0 cu ft/sec of perfect gas (measured at pressure p_0 lb/ft^2 and temperature $T_0 K$) from pressure p_1 to pressure p_2 is

$$p_0 V_0 \frac{k}{k-1} \frac{T_1}{T_0} \left[\left(\frac{p_2}{p_1} \right)^{(k-1)/k} - 1 \right] \text{ ft-lb/sec}$$

where k is the ratio of the specific heat of the gas at constant pressure to the specific heat at constant volume and T_1 is the suction-gas temperature in kelvins.

For reasonably dry oil field gases (that is, gases containing a relatively small percentage of the heavier hydrocarbons), k is 1.25 approximately (Ref. 2), so the above expression gives

$$\text{Power} = 5 p_0 V_0 \frac{T_1}{T_0} \left[\left(\frac{p_2}{p_1} \right)^{0.2} - 1 \right] \text{ ft-lb/sec} \qquad (6\text{-}1)$$

If p_0 is atmospheric pressure, that is, 14.7×144 lb/ft^2, and T_0 and T_1 are equal and are both 520 K (60°F), while M is the gas rate measured in mcf/day at standard conditions, then

$$M = \frac{V_0 \times 60 \times 60 \times 24}{1000}$$

and Eq. (6-1) reduces to

$$\text{Power} = 5 \times 14.7 \times 144 \times \frac{1000}{60 \times 60 \times 24} M \left[\left(\frac{p_2}{p_1} \right)^{0.2} - 1 \right] \quad \text{ft-lb/sec}$$

$$= 0.223 M \left[\left(\frac{p_2}{p_1} \right)^{0.2} - 1 \right] \quad \text{hp} \qquad (6\text{-}2)$$

where M = gas rate, mcf/day at standard conditions
p_1 = compressor input pressure, psia
p_2 = compressor output pressure, psia

To make allowances for the fact that the suction temperature T_1 may be above 60°F, for supercompressibility effects, and for overall plant efficiency and to allow a safety margin in the calculated value of the required output pressure, it is suggested that the brake horsepower of the compressor be obtained by increasing the figure derived from Eq. (6-2) by about one-third.

Example 6-3 What would be the horsepower of the compressor required to gas-lift the well of Examples 6-1 and 6-2 at 475 bbl/day, assuming gas were available at 50 psi?

M = 1235, from Example 6-2; p_2 = 400 psig = 415 psia, from Example 6-2; and p_1 = 50 psig = 65 psia. Equation (6-2) gives

$$hp = 0.223 \times 1235 \left[\left(\frac{415}{65} \right)^{0.2} - 1 \right]$$

so the compressor should have a brake horsepower of approximately $^4/_3 \times 124$, that is, 165.

6-4 MINIMIZING COMPRESSOR HORSEPOWER REQUIREMENTS

The cost of a compressor installation naturally goes up with the horsepower, and it is reasonable to question whether it is possible to reduce substantially the compressor horsepower and yet maintain an oil production rate close to the optimum. The answer to this question, and the various methods utilized, can best be illustrated by means of examples (6-4, 6-5, and 6-6).

Example 6-4 Well data:

Productive interval	6507–6551 ft
Tubing depth	6500 ft
Static pressure at 6500 ft	2000 psig
Flow-line pressure	100 psig

The well is currently flowing 200 bbl/day of clean oil with a GOR of 600 cu ft/bbl and a CHP of 1225 psig.

1. If it were decided to gas-lift this well, what would be the maximum gas-lift rate through $2^3/_8$-in. tubing, assuming a THP of 250 psig? What would be the required horsepower of the compressor if input gas were available at (*a*) 15 psig and (*b*) 100 psig?
2. If the supply gas were limited to 180 mcf/day, what would be the maximum production rate on gas lift through $2^3/_8$-in. tubing at a 250-psig THP? What horsepower would be required in this case if gas were available at 15 psig?
3. What would the well make on gas lift through $3^1/_2$-in. tubing at 6500 ft and 1.9-in. tubing (inside the $3^1/_2$-in.) to depths of 1000, 3000, and 5000 ft (THP of 250 psig in each case)? Determine the horsepower of the compressor in each of these cases if input gas is available at 15 psig.

1. The solution to the first part of the problem is similar to that of Example 6-2. The flowing BHPs at various production rates have been calculated by assuming a THP of 250 psi, and the results are shown in Table 6-4 and Fig. 6-4. Also shown in Fig. 6-4 is the IPR, derived as follows:

   If the casing is assumed to be full of gas, the value of p_{wf} when the CHP is 1225 psig is given by

$$p_{wf} = (1225 + 15) \left[1 + \frac{(6.5)^{1.5}}{100} \right]$$

   from Eq. (4-18). Thus p_{wf} is 1445 psia, or 1430 psig. So two points are known on the IPR, namely,

TABLE 6-4 EXAMPLE 6-4: Determination of Flowing BHPs at Various Rates, Using Optimum GLRs

q, bbl/day	Opt. GLR, mcf/bbl	Equiv. Depth of THP of 250 psi, ft	Equiv. Depth of Well, ft	p_{wf} at Opt. GLR, psi
50	5.0	7500	14,000	530
100	3.55	6000	12,500	620
200	2.5	4700	11,200	710
400	1.8	3300	9,800	830
600	1.47	2900	9,400	920

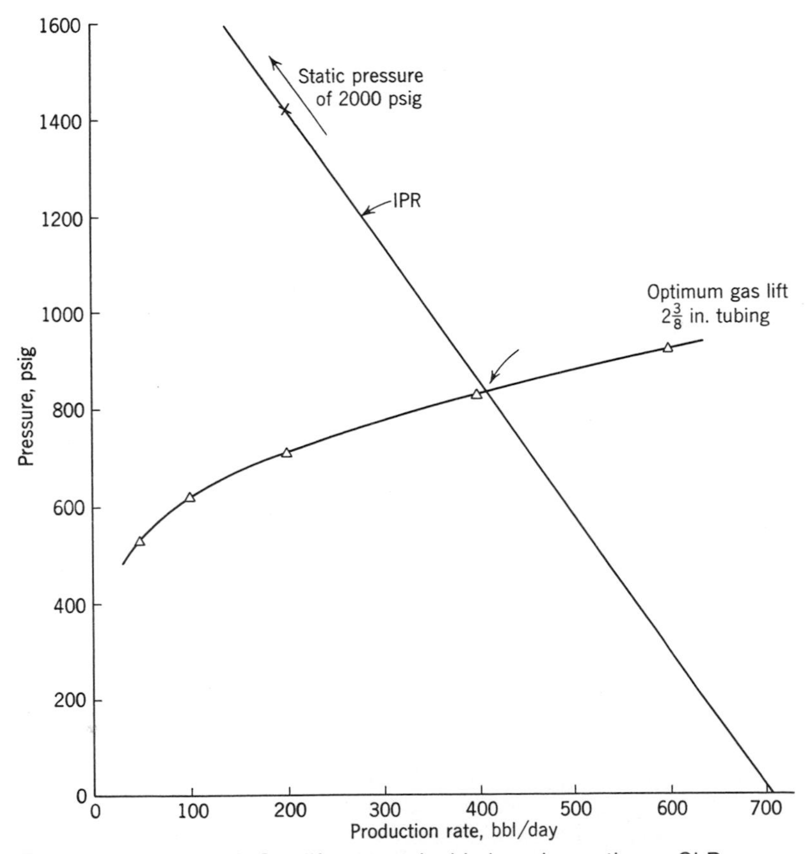

Fig. 6-4 Example 6-4: Gas-lift rate attainable by using optimum GLR.

$$p = 2000 \qquad q = 0$$

and

$$p = 1430 \qquad q = 200$$

It will be assumed that the IPR is a straight line through these two points (but see Sec. 3-4).

From Fig. 6-4 the optimum rate is 410 bbl/day, and the corresponding optimum GLR is 1.8 mcf/bbl (Fig. 6-5). Thus

$$\text{Supply gas} = 410\,(1.8 - 0.6)\ \text{mcf/day} = 492\ \text{mcf/day}$$

So in Eq. (6-2), M is 492 and p_2 is 835 psig (see Sec. 6-3 and Fig. 6-4), or 850 psia.

a. When $p_1 = 15$ psig $= 30$ psia,

$$\text{hp} = 0.223 \times 492\,[(^{850}/_{30})^{0.2} - 1]$$
$$= 104 \quad \text{(theoretical)}$$
$$= 139 \quad \text{(practical)}$$

b. When $p_1 = 100$ psig $= 115$ psia,

$$\text{hp} = 0.223 \times 492\,[(^{850}/_{115})^{0.2} - 1]$$
$$= 54 \quad \text{(theoretical)}$$
$$= 72 \quad \text{(practical)}$$

Thus one answer to the question of how to reduce the horsepower required is to use the highest-pressure supply gas available.

2. To answer this part of Example 6-4, construct the curve of flowing BHP

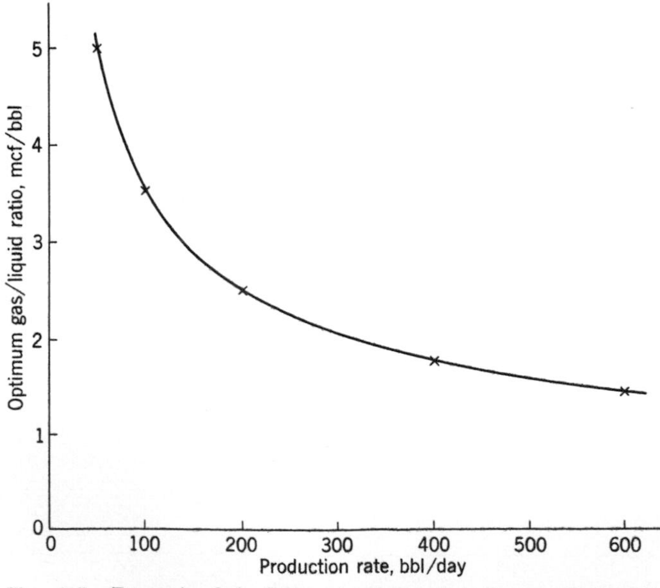

Fig. 6-5 Example 6-4: Determination of optimum GLR (2⅜-in. tubing).

against rate at an input gas rate of 180 mcf/day by following the steps outlined in Table 6-5. The results are plotted in Fig. 6-6, from which it is apparent that under the stated conditions the maximum production rate is 380 bbl/day.

In this case,

$$M = 180 \text{ mcf/day}$$
$$p_1 = 15 \text{ psig} = 30 \text{ psia}$$
$$p_2 = 920 \text{ psig (Fig. 6-6)} = 935 \text{ psia}$$
$$\text{hp} = 40 \quad \text{(theoretical)}$$
$$= 53 \quad \text{(practical)}$$

Comparing this result with that of the first part of this example, it is seen that, by accepting a small cut in the production rate (from 410 to 380 bbl/day), the horsepower requirements are reduced from 139 to 53 and the volume of supply gas needed is reduced from 492 to 180 mcf/day.

3. This part of Example 6-4 may be solved by calculating the pressure p^* (see Fig. 6-7) in two ways: first, by $3\frac{1}{2}$-in.-tubing performance at a GLR of 0.6 mcf/bbl from 6500 ft to 1000, 3000, and 5000 ft, respectively, and second, by 1.9-in. tubing performance at optimum GLR, the length of this small-diameter tubing being, in turn, 1000, 3000, and 5000 ft and the THP being 250 psig.

   The calculations involved in the first of these ways of obtaining p^* are listed in Table 6-6; those involved in the second are in Table 6-7. The results are plotted in Fig. 6-8, the full lines summarizing the first method of obtaining p^* and the broken lines applying to the second method.

   Obtaining the optimum GLRs in the 1.9-in. tubing at the various rates involved from Fig. 6-9, it is possible to summarize the results of the last part of Example 6-4 as shown in Table 6-8.

Comparing these results with those of the first and second parts of the example, it can be seen that both the horsepower and the volume of supply gas required are reduced in each case but that the oil production rate has been substantially cut. The method of gas lifting with the use of a macaroni string is not to be recommended unless the market situation or government regulations restrict the production rate of the well or unless a low-powered com-

TABLE 6-5 EXAMPLE 6-4: Determination of Flowing BHPs at Various Rates, Using a Restricted Input Gas Rate

Prod. Rate, bbl/day	Supply GLR, mcf/bbl	Total GLR, mcf/bbl	Equiv. Depth of THP of 250 psi, ft	Equiv. Depth of Well, ft	p_{wf} at Total GLR, psi
50	3.6	4.2	7200	13,700	580
100	1.8	2.4	5100	11,600	660
200	0.9	1.5	4100	10,600	780
400	0.45	1.05	3000	9,500	930
600	0.3	0.9	2800	9,300	1020

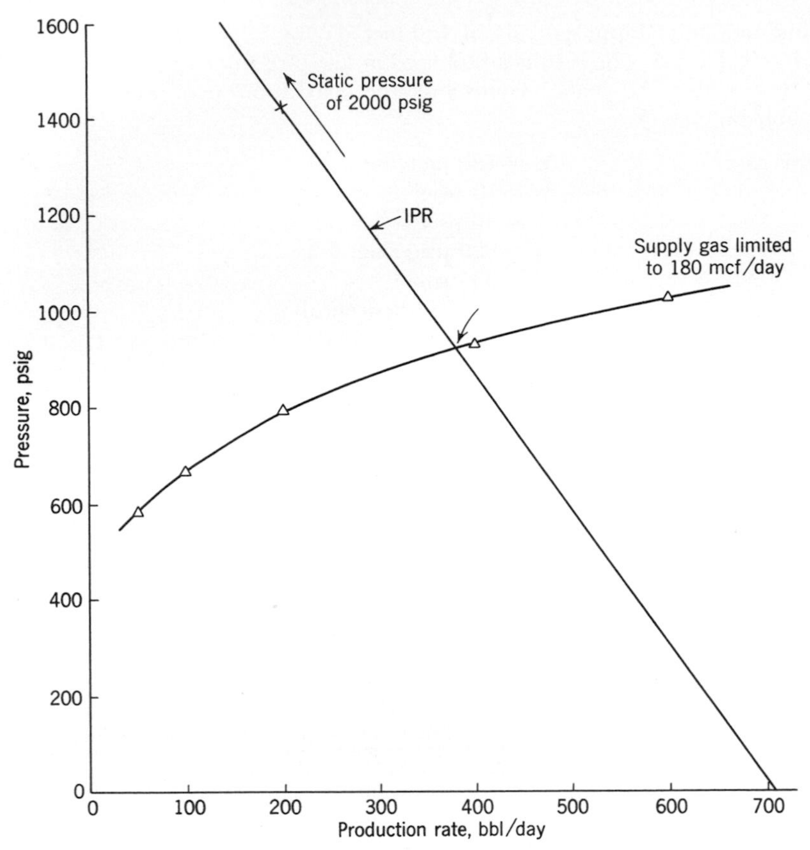

Fig. 6-6 Example 6-4: Gas-lift rate attainable if supply gas is limited.

pressor is available and it is felt that the purchase of a larger one is not justifiable.

It might be asked how much the well of Example 6-4 would produce through $2^3/_8$-in. tubing using a 32-hp compressor. Would the production rate be greater or less than the 275 bbl/day obtainable with $3^1/_2$-in. tubing and 3000 ft of 1.9-in. tubing?

Example 6-5 The well of Example 6-4 is to be gas-lifted through 6500 ft of $2^3/_8$-in. tubing with a compressor of 32 brake hp. Gas is available at a pressure of 15 psig. What will be the well's production rate on gas lift, assuming a THP of 250 psig?

From Eq. (6-2) and the discussion of Sec. 6-3, the following relationship holds:

$$32 = \frac{4}{3} \times 0.223M \left[\left(\frac{p_{wf} + 15}{30} \right)^{0.2} - 1 \right]$$

where M is the gas-injection rate in mcf/day and p_{wf} is the pressure at the foot of

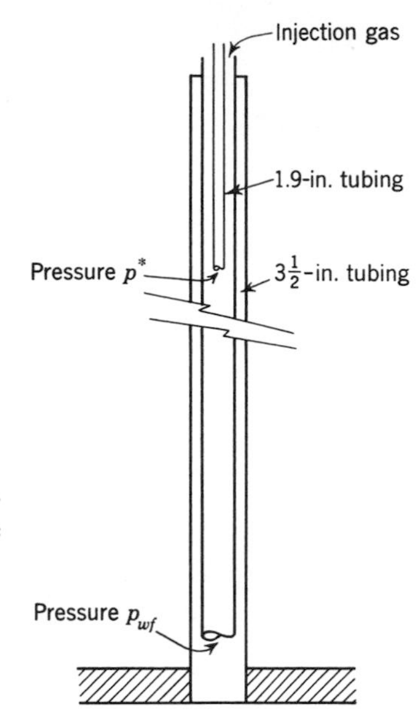

Fig. 6-7 Example 6-4: Gas lifting with the use of concentric tubing strings.

the tubing in psig. Hence,

$$M = \frac{107.6}{\left(\frac{p_{wf} + 15}{30}\right)^{0.2} - 1} \tag{6-3}$$

Plotting the IPR once again, it is possible to set up Table 6-9, the results of which are shown in Fig. 6-10. This figure illustrates that with a 32-brake-hp compressor the well would produce 355 bbl/day against a THP of 250 psig.

Thus, for this particular well, gas lifting with a 32-brake-hp compressor would give a greater production rate through $2^3/_8$-in. tubing than through

TABLE 6-6 EXAMPLE 6-4: Determination of Pressure at Foot of 1.9-in. Tubing Run inside 3½-in. Tubing, Using Vertical-Flow Pressure Losses for Flow in 3½-in. Tubing at GLR of 600 cu ft/bbl

Prod. Rate, bbl/day	p_{wf} from IPR, psi	Pressure p^*, psi, When 1.9-in. Tubing Is at		
		1000 ft	3000 ft	5000 ft
50	1860	520	960	1400
100	1710	520	900	1350
200	1430	410	720	1110
400	860	70	290	590
600	300			120

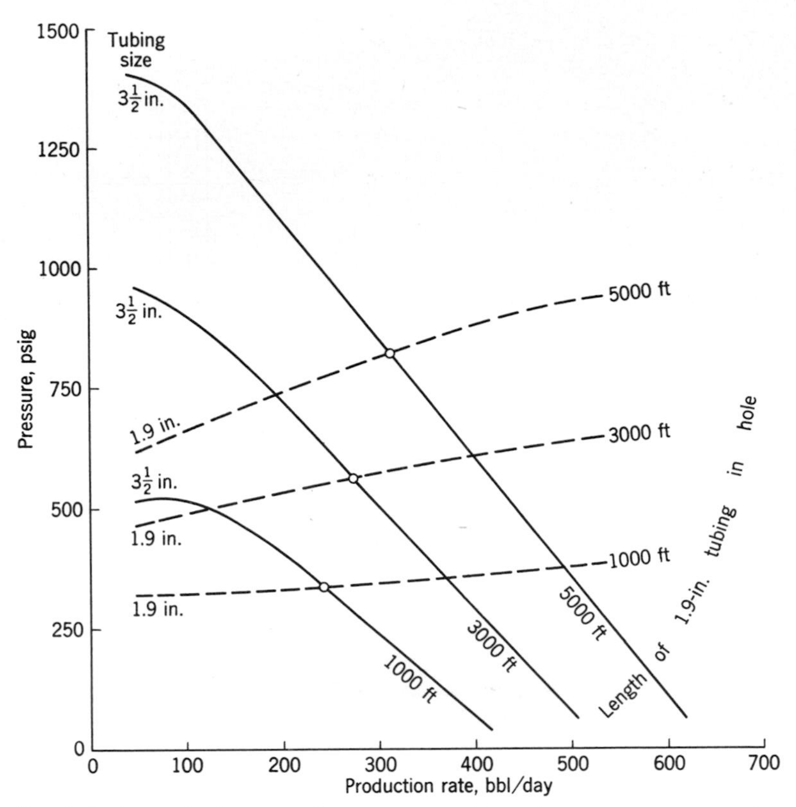

Fig. 6-8 Example 6-4: Gas-lift rates attainable with the use of concentric tubing strings.

TABLE 6-7 EXAMPLE 6-4: Vertical Two-Phase Flow Pressure Losses in 1.9-in. Tubing at Various Rates, Using Optimum GLRs

			1000 ft		3000 ft		5000 ft	
Prod. Rate, bbl/day	Opt. GLR, mcf/bbl	Equiv. Depth 250-psi THP, ft	Equiv. Depth of p^*, ft	p^*, psi	Equiv. Depth of p^*, ft	p^*, psi	Equiv. Depth of p^*, ft	p^*, psi
50	2.6	4800	5800	320	7800	470	9800	620
100	1.8	3900	4900	320	6900	490	8900	670
200	1.27	3000	4000	330	6000	530	8000	740
400	0.88	2500	3500	360	5500	610	7500	880
600	0.72	2200	3200	400	5200	660	7200	950

TABLE 6-8 EXAMPLE 6-4: Summary of Results of Final Part

	Length of 1.9-in. Tubing Used		
	1000 ft	**3000 ft**	**5000 ft**
Production rate, bbl/day, from Fig. 6-8	240	275	315
Pressure at foot of 1.9-in., psig, from Fig. 6-8	340	560	820
Optimum GLR, mcf/bbl, from Fig. 6-9	1.16	1.08	1.01
Supply GLR, mcf/bbl	0.56	0.48	0.41
Supply gas, mcf/day	134.3	132.0	129.1
Compressor input pressure, psia	30	30	30
Compressor output pressure, psia	355	575	835
Horsepower, from Eq. (6-2)	19	24	27
Practical brake horsepower	25	32	36

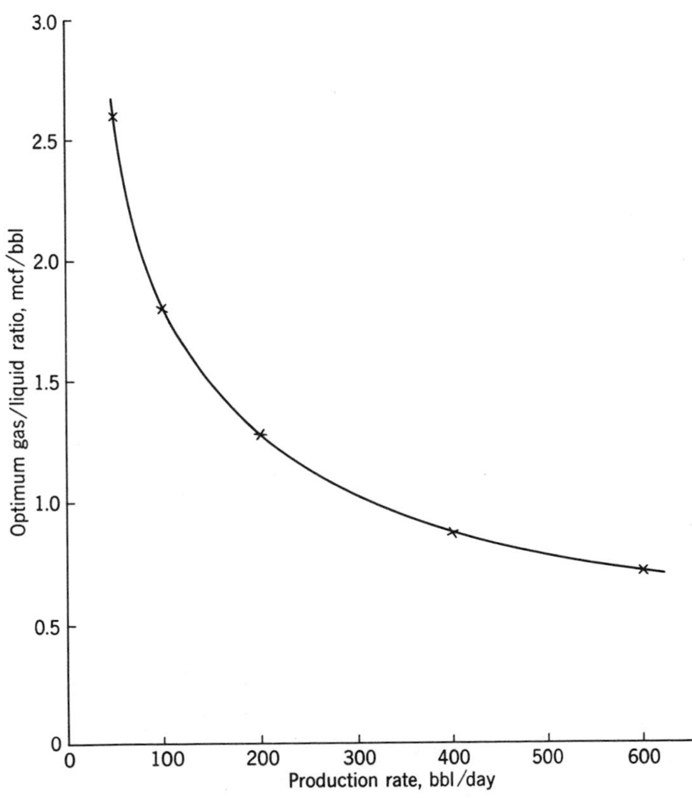

Fig. 6-9 Example 6-4: Determination of optimum GLRs (1.9-in. tubing).

TABLE 6-9 EXAMPLE 6-5: Vertical Two-Phase Flow Pressure Losses at Various Rates When Compressor Horsepower Is Restricted

Prod. Rate, bbl/day	p_{wf}, psig	*M* from Eq. (6-3), mcf/day	Injected GLR, mcf/day	Total GLR, mcf/bbl	Equiv. Depth of p_{wf}, ft	Equiv. Depth of Tubing Head, ft	THP, psig
50	1860	81.5	1.630	2.230	22,500	16,000	1170
100	1710	86.2	0.862	1.462	18,000	11,500	900
200	1430	91.9	0.460	1.060	13,800	7,300	630
400	860	111.9	0.280	0.880	8,500	2,000	160
600	300	179.1	0.299	0.899	3,200		

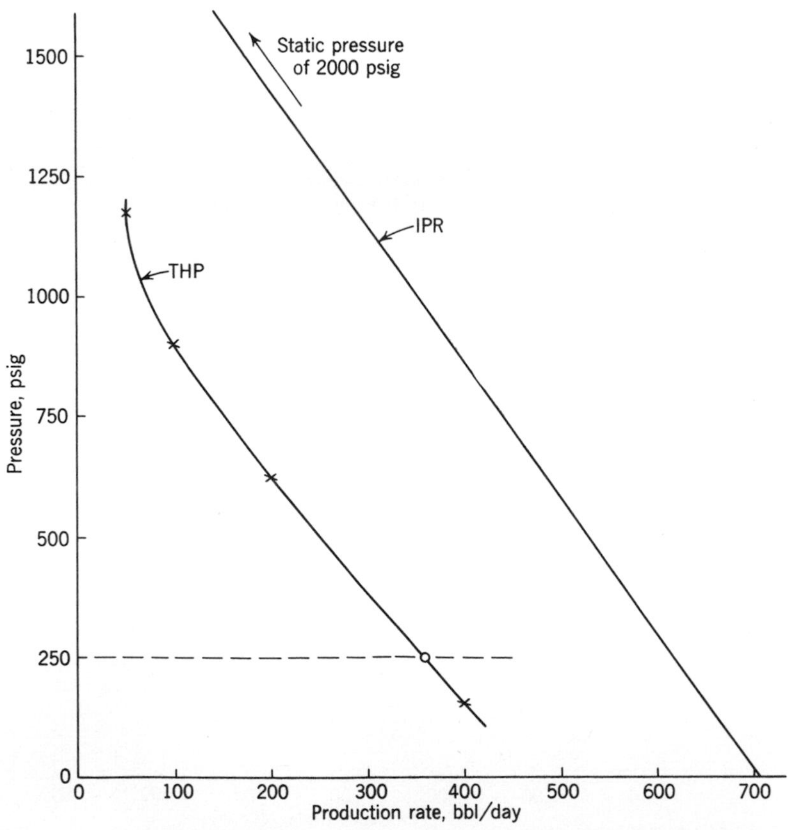

Fig. 6-10 Example 6-5: Gas-lift rates attainable with restricted compressor size.

$3^1/_2$-in. with 3000 ft of concentric 1.9-in. tubing (see Table 6-8). However, one great advantage of using the concentric tubing is that, in the absence of gas-lift valves for kickoff purposes, the requirement on the maximum compressor output pressure is less severe in the case of the $3^1/_2$-in. to 1.9-in. arrangement than it is with $2^3/_8$-in. tubing alone. Suppose, for instance, that in Examples 6-4 and 6-5 the specific gravity of the oil were 0.9. Then if the well died, the formation static pressure of 2000 psig would support a column of oil of length

$$\frac{2000}{0.9 \times 0.433} = 5130 \text{ ft}$$

so that the fluid level would be 1370 ft from the surface (Fig. 6-11).

If gas has to be blown around the foot of the tubing at 6500 ft to start the well, the compressor output pressure must be sufficient to overcome the pressure exerted by 5130 ft of dead oil; that is, it must be greater than 2000 psi. On the other hand, to blow around the foot of 1.9-in. tubing located at 3000 ft only requires a compressor output pressure of 1170 psi.

As a final example, the case will be considered in which injection gas is available (perhaps from a nearby gas well) at a certain pressure.

Example 6-6 Well data are as follows:

Production interval	9050–9200 ft
$2^7/_8$-in. tubing	9100 ft
Static pressure at 9100 ft	3000 psig
PI	0.333 bbl/(day)(psig)
GOR	450 cu ft/bbl
Water cut	zero

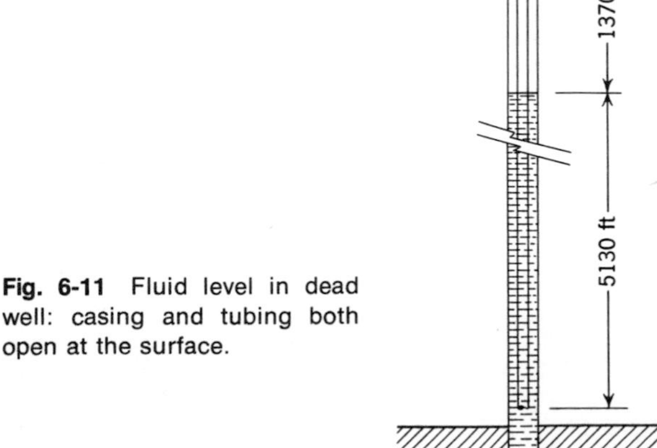

Fig. 6-11 Fluid level in dead well: casing and tubing both open at the surface.

TABLE 6-10 EXAMPLE 6-6: Pressure versus Rate Curves at 9100 ft and at 7100 ft, 5100 ft, and 3100 ft, Obtained by Assuming Natural GLR from the Tubing Shoe to the Intermediate Depth

Prod. Rate, bbl/day	p_{wf}, psig	Equiv. Depth of p_{wf}, ft	Equivalent Depth of 7100 ft, ft	Equivalent Depth of 5100 ft, ft	Equivalent Depth of 3100 ft, ft	Pressure at 7100 ft, psig	Pressure at 5100 ft, psig	Pressure at 3100 ft, psig
50	2850	11,200	9,200	7200	5200	2150	1550	1000
100	2700	12,100	10,100	8100	6100	2100	1600	1050
200	2400	11,800	9,800	7800	5800	1850	1350	875
400	1800	10,000	8,000	6000	4000	1300	850	500
600	1200	7,900	5,900	3900	1900	825	450	200

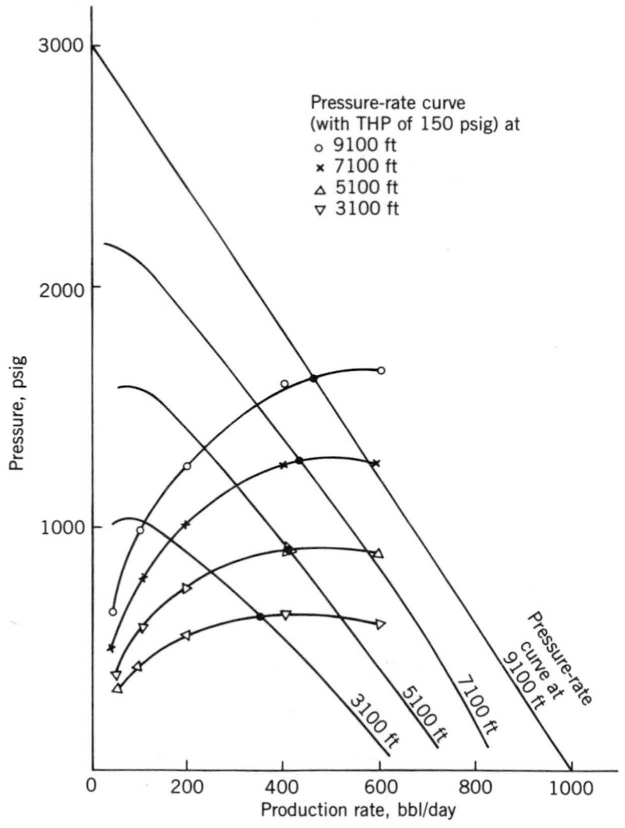

Fig. 6-12 Example 6-6: Gas-lift rates attainable with varying depths of gas blow-round.

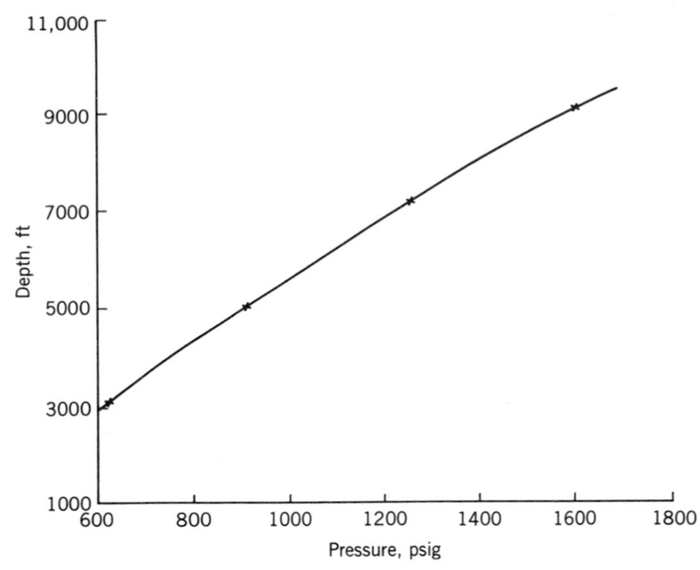

Fig. 6-13 Example 6-6: Pressure at gas injection point as function of depth.

1. Determine the natural rate of flow with a THP of 150 psig.
2. If 240,000 cu ft/day of high-pressure gas is available for gas lifting, at what rate could the well be gas-lifted while maintaining the THP at 150 psig? What gas-injection pressure would be required to maintain this rate?
3. If the high-pressure gas were only available at 700 psig, what gas-lift production rate could be attained at the same THP, and how deep would be the lowest gas-injection point?

1. This is a standard calculation, the answer to which is about 330 bbl/day.
2. The calculation is similar to that illustrated in Table 6-5 and Fig. 6-6 (Example 6-4), and it is found that the well could be gas-lifted at a rate of 460 bbl/day. At that rate the value of p_{wf} is 1620 psig, so 1600-lb gas would be required.
3. It is clear that the lowest point of injection of 700-lb gas will be several thousand feet off bottom. The first step in the solution is to plot a pressure-rate-depth grid (Sec. 4-6). Figure 6-12 shows the pressure versus rate curves at 9100 ft (the tubing shoe) and at 7100 ft, 5100 ft, and 3100 ft following the calculations of Table 6-10. The curves at 7100, 5100, and 3100 ft have been determined from the IPR on the assumption that flow at the natural GLR takes place from the tubing shoe to that particular depth. This is because the next step in the calculation will be to assume that each of these intermediate depths, in turn, is the lowest point at which gas injection occurs.

Now determine the value of the pressure as a function of rate at each of the four indicated depths, assuming a THP of 150 psig and a GLR made up of both natural and injected gas. This is done in Table 6-11, and the resulting curves are plotted on Fig. 6-12. The points of intersection—as shown—of corresponding curves give the rates at which the well would produce if all the injected gas were

TABLE 6-11 EXAMPLE 6-6: Pressure versus Rate Curves at 9100 ft, 7100 ft, 5100 ft, and 3100 ft Assuming a THP of 150 psig and All the Injected Gas Introduced at the Depth in Question

Prod. Rate, bbl/day	Supply GLR, mcf/bbl	Total GLR, mcf/bbl	Equiv. Depth THP, psig	Equiv. Depth of				Pressure at Depth of			
				9100 ft, ft	7100 ft, ft	5100 ft, ft	3100 ft, ft	9100 ft, psig	7100 ft, psig	5100 ft, psig	3100 ft, psig
50	4.8	5.25	5700	15,800	13,800	11,800	9800	650	500	400	325
100	2.4	2.85	4300	14,400	12,400	10,400	8400	1000	800	600	400
200	1.2	1.65	3200	13,300	11,300	9,300	7300	1250	1000	750	550
400	0.6	1.05	2300	12,400	10,400	8,400	6400	1600	1250	900	625
600	0.4	0.85	1700	11,800	9,800	7,800	5800	1650	1250	875	600

TABLE 6-12 EXAMPLE 6-6: Rate and Pressure in Tubing Opposite Injection Point if Injection Gas All Introduced at an Intermediate Depth (from Fig. 6-12)

Depth of Gas Injection Point, ft	Production Rate, bbl/day	Pressure in Tubing Opposite Injection Point, psig
9100	460	1600
7100	430	1260
5100	400	900
3100	350	620

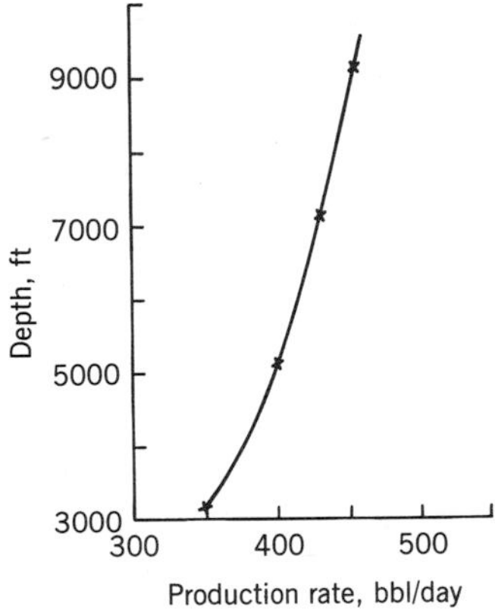

Fig. 6-14 Example 6-6: Attainable gas-lift production rate as function of depth of gas injection point.

introduced into the tubing at the appropriate depth, as well as the pressures in the tubing at these depths.

The results are listed in Table 6-12 and cross-plotted in Figs. 6-13 and 6-14, from which it is seen that 700-psig gas could be introduced at a depth of 3600 ft with a resultant production rate of 365 bbl/day.

6-5 THE PLACE OF GAS LIFT IN A WELL'S PRODUCING HISTORY

Although modifications to the gas-lift technique exist which permit it to be used on small producers and down to a well's economic limit (for example, intermittent lift and chamber lift; see Chap. 8), gas lifting remains essentially a high-production-rate lifting method. As such, the gas-lift phase in a well's history is generally fairly short (say up to 5 years) and comes between the initial natural flow period and the final artificial-lift installation, which will probably be some form of pump. Gas lift, then, has its greatest applications where market and government restrictions are not the limiting factors in deciding a well's production rate. And, since a pump will generally have to be installed eventually in any case, the economic justification for putting in a gas-lift system frequently rests on the economic advantages to be gained by bringing production, and so income, forward in time as much as possible. It follows that the installation of gas lift is a good example of an acceleration project (Sec. 12-9).

The fact that gas lift, as it has been discussed here, cannot be used down to final (economic) depletion of a well is readily understood when it is realized that the injection gas itself exerts some back pressure against the formation, so that the flowing BHP can never be reduced to the level attained, say, in a pumped-off well (Secs. 9-5 and 10-3).

Gas lift may also have its place in water-flood operations, where BHPs are kept high artificially and where large gross volumes need to be lifted from the wells. However, the use of gas lift under these conditions will depend to a considerable extent on the availability and cheapness of gas, because other high-rate methods (for example, down-hole centrifugal pumping) are at their peak efficiency in such conditions of low formation GLR.

6-6 THE NEED FOR GAS-LIFT VALVES

In Sec. 6-4 it was seen that for the well discussed in Examples 6-4 and 6-5, a pressure of some 2000 psig would be needed to blow gas around the foot of the $2^3/_8$-in. tubing after the well had died, whereas it is seen from Fig. 6-10 that a pressure of slightly less than 1000 psig is required to keep the well on steady production at 355 bbl/day. This flowing pressure would be even lower with a compressor of greater power, which would be able to inject more gas daily and so bring the GLR in the tubing even closer to the optimum. This type of difference always exists if gas has to be blown around the tubing shoe to kick the well off, and so the output pressure of the compressor required to bring a dead well back into production in such circumstances is always considerably in excess of the pressure required to keep the well producing.

One solution to this problem is to have a portable *kickoff compressor,* but this has the major disadvantage of not being automatic, with the result that there may be considerable delays in bringing some wells back into production after a general or partial shutdown. If the scale of operations does not warrant the purchase and maintenance of a portable compressor, a *pinhole collar* or *permanently open port* is a cheap and simple way of overcoming the kickoff problem. This method utilizes a special collar containing a choke that can be inserted in the tubing string at some predetermined depth (even simpler than a special collar is a hole punched in the tubing, but frequently this is only a temporary solution to the problem, as the hole may become gas-cut and too large to permit any sort of control to be exercised on the way in which gas is injected into the tubing). Referring to Fig. 6-15, the pinhole collar operates as follows.

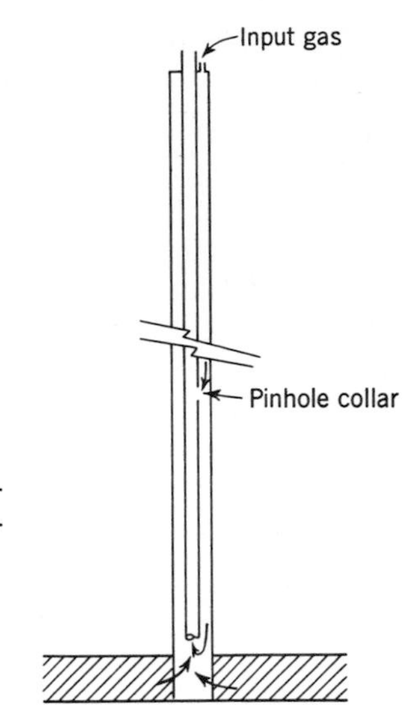

Fig. 6-15 Diagrammatic representation of operation of pinhole collar.

Gas is injected into the annulus of the off-production well, and the pinhole collar is set at such a depth that the gas-injection pressure can just overcome the static head due to the dead oil column above the collar under off-production conditions. Gas injection into the tubing first occurs at this intermediate depth (Fig. 6-15). This injected gas lightens the oil column between the pinhole collar and the surface and so reduces the pressure at the foot of the tubing. Thus the formation will produce some oil and gas, the fluid column up to the pinhole collar will be lightened, and so on. The intake pressure at the foot of the tubing is gradually reduced, which enables the injection gas to replace oil in the annulus, and the installation is so designed that eventually injection gas blows around the foot of the tubing.

Since the pressure at the tubing shoe is always greater than the pressure in the tubing at the pinhole collar, gas injection around the foot of the tubing necessarily implies simultaneous gas injection into the tubing at the collar. Alternatively, it is possible for gas injection to occur at the intermediate depth without taking place around the foot of the tubing. In neither case is it possible to have optimum GLR conditions over the entire tubing string. Thus a well cannot be produced at its optimum rate by means of gas lift with the use of a pinhole collar. In fact, it may be said that the pinhole collar installation is, generally, inefficient in the use of gas.

A natural development of the pinhole collar was a port that could be opened and closed as required. Early examples included the Nixon valve (Ref. 3), which could be operated by a wireline tool, and the Acme valve (Ref. 3), which was opened and closed by varying the tension in the tubing string with use of a surface jack.

Requirements of greater efficiency and more automatic control have led to the development of valves that either operate according to well conditions (the *differential valve*) or that can be operated by adjusting the surface gas-injection pressure (the *pressure-charged valve*). These valves and the design of gas-lift tubing strings are discussed in the next chapter.

REFERENCES

1. Katz, Donald L., et al.: *Handbook of Natural Gas Engineering,* McGraw-Hill Book Company, Inc., New York, 1959.
2. Gilbert, W. E.: "Flowing and Gas-Lift Well Performance," *API Drill. Prod. Practice,* 1954, p. 126.
3. Shaw, S. F.: *Gas-Lift Principles and Practices,* Gulf Publishing Company, Houston, Tex., 1939.

Gas-Lift Valves and String Design

7

7-1 INTRODUCTION

Unless a pinhole collar or a series of gas-lift valves or some similar device is placed in the flow string, much higher gas-injection pressures are needed to start (kick off) a nonflowing well than are needed for the steady-flow state (Sec. 6-6). One objective, therefore, in the installation of a string of gas-lift valves is to be able to bring a well back into production easily and cheaply after a shutdown, without having to resort either to a high-pressure gas-injection system or to portable compressors. A second objective that must be satisfied by a gas-lift string is that of stability, and a third is the ability to compensate automatically for variations in the pressure of the gas-injection system.

The subject of gas-lift string design is a highly complex one, the placement of the valves in the optimum positions in the string being dependent on a large number of variables, such as the gravity and temperature of the injected gas, tubing and casing sizes, well temperature at various depths, gas-injection pressure, and the well inflow performance among others, in addition to many different design features built into the various types of gas-lift valve available on the market. The leading manufacturers of these valves publish manuals on oil production by gas lift (see, for example, Refs. 1 and 2), and within the scope of the present book it is not possible to do more than summarize the main features of the two basic types and to indicate by examples some of the major considerations in the design of gas-lift strings. The principle behind the *differential valve* is that it is controlled in its operation by well conditions, while the *pressure-charged bellows valve* is operated by changes in the gas-injection pressure.

7-2 DIFFERENTIAL VALVES

The principal features of the differential valve are illustrated diagrammatically in Fig. 7-1. The valve stem is attached to a coil spring, which, in the absence of other forces, holds the stem clear of the valve seat so that the valve is normally open. The spring setting may be regulated by an adjustable nut. When the valve is open, external pressure is applied to the valve stem in two ways: via the open port in the top of the valve so that the full pressure of the injection gas at the depth at which the valve is placed is applied over the valve-stem area, opposing the spring and acting to close the valve; and via the choke situated in the valve wall.

When the valve is open, and provided that the gas-injection pressure is greater than the pressure in the tubing at the depth of the valve, gas flows through the choke and into the tubing. The presence of the choke results in a pressure drop, so the pressure applied to the lower end of the valve stem is less than that applied at the upper end. When the gas flow through the choke becomes sufficiently great, the difference in the gas pressures applied to the upper and lower ends of the stem, multiplied by the stem cross-sectional area, becomes greater than the force exerted by the spring and the valve closes. Typically, the valve may be set to close when the pressure difference across the stem is in the range of 100 to 150 psi (the precise value of the closing pressure is determined by the setting of the adjustable nut).

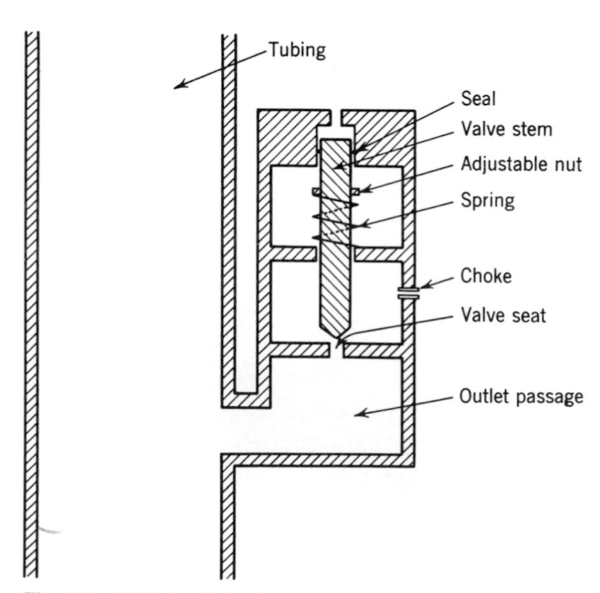

Fig. 7-1 Differential valve.

Since the pressure in the tubing is even less than the pressure of the injected gas downstream of the choke, the valve will not reopen until the tubing pressure has undergone a substantial rise (or the gas-injection pressure a substantial drop), so that the closing action is positive and there is no tendency for the valve to chatter.

For gas flow through a choke, the downstream pressure will continue to influence the throughput rate (that is, the gas velocity through the choke will be less than that of sound) provided[1]

$$\frac{p_2}{p_1} > \left(\frac{2}{k+1}\right)^{k/(k-1)} \tag{7-1}$$

where p_1 = upstream pressure, psia
p_2 = downstream pressure, psia
k = ratio of the specific heat at constant pressure to the specific heat at constant volume of the gas

Taking 1.25 to be a reasonable value for k for lean gas (Ref. 4), it appears that as long as

$$p_2 > 0.555p_1 \tag{7-2}$$

the downstream pressure will influence the flow rate through the choke. Taking a spring setting of 150 psi, the maximum attainable value of $p_1 - p_2$ before the valve closes is 150 psi; hence, the value of p_2 when the valve is on the point of closing is given by p^*, say, where

$$p^* = p_1 - 150 \tag{7-3}$$

Substituting this into Eq. (7-2), it appears that even the minimum downstream pressure (that is, when the valve is about to close) influences the flow rate through the choke as long as

$$p_1 - 150 > 0.555p_1$$

or

$$p_1 > 337 \text{ psia}$$

For a 100-psi spring setting, this figure would be 225 psia.

As pointed out below, the automatic feature of the differential valve is dependent on the gas flow rate through the choke being less than the sonic velocity (in other words, the flow is *subcritical*); the spring setting must be tailored to the gas-injection pressure to ensure that the pressure in the flow stream in the tubing at the valve depth is a controlling factor in the gas throughput of the choke.

Assuming that the GLR is below the optimum, so that the addition of extra gas reduces the pressure loss in the tubing under two-phase vertical flow, the pressure in the tubing decreases once gas begins to pass through the valve. This allows more gas to pass through the choke, and the pressure in the tubing decreases further. This process continues until the pressure drop

[1] See, for instance, Binder, Ref. 3, p. 299.

across the choke is equal to the spring setting, at which time the valve closes. However, at this time, provided that the valves are spaced correctly in the string, the pressure conditions will permit gas injection into the next valve down, and so on.

Although this type of valve is of comparatively simple construction and is relatively cheap and although it has the advantage that its operation is governed by conditions in the flowing column, it has one major drawback. To ensure that the conditions in the flowing column govern the valve operation, it is necessary to limit the pressure drop across the choke. As a result, there is a limit to the rate at which injection gas can be passed into the tubing via a differential valve, and hence optimum-flow conditions are usually not achieved in the tubing above the valve. Thus the full value of each valve is not attained; a different type of valve that will permit GLRs to reach the optimum (or, at least, to approach this value) may be spaced more widely down the tubing string than the differential type. For this reason differential valves are not usually used in deep wells, say, below 4000 ft. They find their greatest application in wells in which large increases in production rate accompany relatively small increases in GLR or in wells in which low production is lifted from shallow depths.

Before an example is given to illustrate the design of a string of differential valves, it is necessary to discuss the flow of gas through chokes in rather more detail so that the correct choke sizes may be installed in the various valves in the string.

7-3 GAS FLOW THROUGH CHOKES

For the adiabatic frictionless flow of an ideal gas through an orifice it can be shown (see, for example, Binder, Ref. 3, pages 298 ff.) that

$$W = A \left\{ \frac{2kg}{k-1} \frac{p_1}{v_1} \left[\left(\frac{p_2}{p_1} \right)^{2/k} - \left(\frac{p_2}{p_1} \right)^{1+1/k} \right] \right\}^{1/2} \tag{7-4}$$

where W = weight of gas passing per unit time
A = cross-sectional area of the choke
k = ratio of the specific heat of the gas at constant pressure to its specific heat at constant volume
g = acceleration due to gravity
p_1 = upstream pressure
p_2 = downstream pressure
v_1 = specific volume at pressure p_1

Converting this expression to oil field units and introducing a *discharge coefficient* C to allow for the fact that the minimum area of the flow stream will be somewhat less than A, the resulting equation is

$$Q = 155.5C\, Ap_1 \left\{ \frac{2g}{GT} \frac{k}{k-1} \left[\left(\frac{p_2}{p_1} \right)^{2/k} - \left(\frac{p_2}{p_1} \right)^{1+1/k} \right] \right\}^{1/2} \tag{7-5}$$

where Q = gas flow rate, mcf/day at standard conditions
A = choke cross-sectional area, sq in.
C = discharge coefficient
G = specific gravity of the gas (relative to air)
T = inlet temperature, °R

The pressures p_1 and p_2 are given in psia. Substituting the following average numerical values:

$$C = 0.86$$
$$k = 1.25$$
$$G = 0.6$$
$$T = 520°\text{R } (60°\text{F})$$

Eq. (7-5) reduces to

$$Q = 136Ap_1 \left[\left(\frac{p_2}{p_1}\right)^{1.6} - \left(\frac{p_2}{p_1}\right)^{1.8} \right]^{1/2} \tag{7-6}$$

or

$$Q = 136Ap_1 F\left(\frac{p_2}{p_1}\right) \tag{7-7}$$

where

$$F\left(\frac{p_2}{p_1}\right) = \left[\left(\frac{p_2}{p_1}\right)^{1.6} - \left(\frac{p_2}{p_1}\right)^{1.8} \right]^{1/2}$$

The function $F(p_2/p_1)$ for various values of p_2/p_1 is plotted in Fig. 7-2.

The maximum value of $F(p_2/p_1)$ illustrated in Fig. 7-2 corresponds to a condition in which the gas velocity through the orifice is equal to the velocity

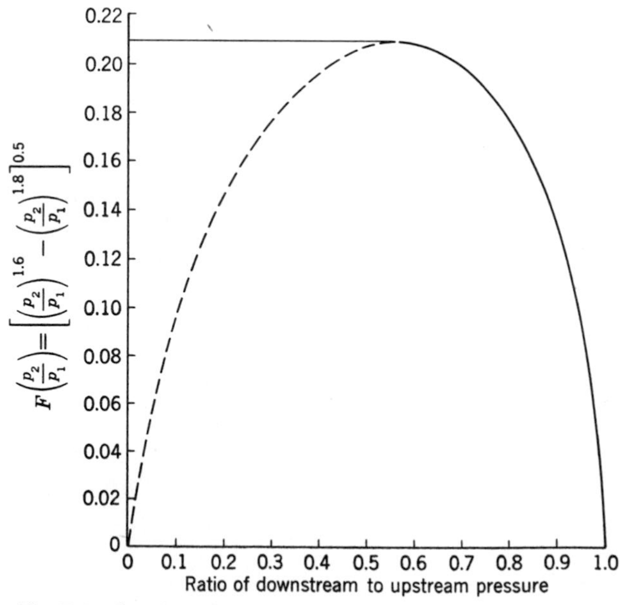

Fig. 7-2 Gas flow through a constriction.

of sound. For values of the ratio of the downstream to the upstream pressure less than the value corresponding to this maximum, the downstream pressure cannot be transmitted back through the orifice and the flow rate through the choke becomes independent of the downstream pressure. This is the so-called *critical-flow range,* over which the value of the flow rate through the choke is proportional to the upstream pressure. For this reason the left-hand arch of the curve in Fig. 7-2 has been shown as a broken line, and a full horizontal line has been drawn through the maximum value. The value of the gas flow rate through the choke in mcf/day at standard conditions can be found from Fig. 7-2 by reading the value attained by the full line corresponding to the value of the downstream/upstream pressure ratio and multiplying this figure by $136Ap_1$.

7-4 EXAMPLE OF STRING DESIGN: DIFFERENTIAL VALVES

Suppose that a string of differential valves is to be designed for use in the following well:

Productive interval	7000–7030 ft
Static pressure	2000 psi at 7000 ft
Productivity index	0.4 bbl/(day)(psi)
Formation GLR	200 cu ft/bbl
Oil gravity	25° API

Gas of gravity 0.6 is available in unlimited quantities at a pressure of 550 psi. Differential valves of various choke sizes and of spring setting 100 psi are available.

The first step is to decide whether there is sufficient gas pressure to permit gas to be injected close to the tubing shoe (assumed to be at 7000 ft) when the well is producing steadily, and if not, at what (approximate) deepest level a valve can usefully be placed. In this example, if gas were being injected at 7000 ft, the pressure at that point could not be higher than about 500 psi. With a 500-psi back pressure at 7000 ft, the formation would produce

$$0.4(2000 - 500) = 600 \text{ bbl/day}$$

At this production rate, assuming optimum GLRs, the pressure-distribution curves show that a pressure of 500 psi supports a 10,000-ft flowing column in $3^1/_2$-in. tubing and a 7800-ft column in $2^7/_8$-in. tubing. Thus it might be decided to use a $2^7/_8$-in. tubing string, as it appears that gas will blow around either at or near the shoe at 7000 ft on steady production.[2]

Having decided, then, to run a 7000-ft string of $2^7/_8$-in. tubing, the next

[2] It will be seen at the end of the calculations that a $3^1/_2$-in. tubing would probably have been a better choice.

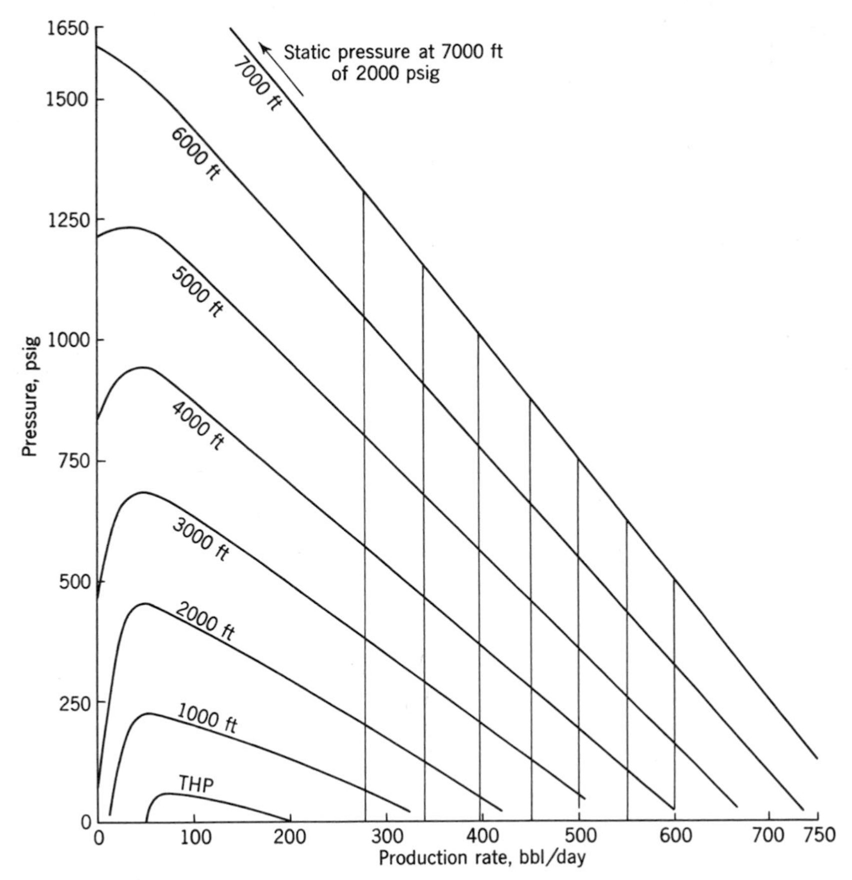

Fig. 7-3 Differential valve string design: pressure versus rate curves at different points in the tubing.

step is to position the first (top) valve. This may be done on the basis of static considerations alone.

Oil having a gravity of 25° API, that is, sp gr 0.9042, exerts a static pressure of 0.39 psi/ft. Thus when the well is off production, there is an oil column of length 2000/0.39, or 5130, ft in the tubing and the fluid level is 7000 − 5130, or 1870, ft below the surface.

Neglecting the weight of the gas column in the annulus and allowing a 50-psi safety margin, the top valve can be set at a depth at which the static column exerts a pressure of 500 psi. If this depth is D_1, then D_1 is calculated from the equation

$$0.39(D_1 - 1870) = 500$$

or

$$D_1 = 3150 \text{ ft}$$

Now plots are made of the pressure versus rate curves in the tubing at various depths, using the natural formation GLR of 200 cu ft/bbl (Sec. 4-6).

TABLE 7-1 Differential Valve String Design: Pressure versus Rate Data at Various Depths

		Production Rate, bbl/day					
		0	**50**	**100**	**200**	**400**	**600**
p_{wf}, psi		2000	1875	1750	1500	1000	500
Equivalent depth p_{wf}, ft			7100	7300	6800	5400	3100
Equivalent depth, ft, of	6000 ft		6100	6300	5800	4400	2100
	5000 ft		5100	5300	4800	3400	1100
	4000 ft		4100	4300	3800	2400	100
	3000 ft		3100	3300	2800	1400	
	2000 ft		2100	2300	1800	400	
	1000 ft		1100	1300	800		
	Tubing head		100	300			
Flowing pressures, psi, at	6000 ft	1610	1550	1440	1200	760	330
	5000 ft	1220	1230	1150	940	550	160
	4000 ft	830	940	870	700	360	20
	3000 ft	440	680	620	490	200	
	2000 ft	50	450	400	300	50	
	1000 ft		225	200	130		
	Tubing head		20	50			

TABLE 7-2 Differential Valve String Design: Pressure versus Rate Data at the Depths of the Various Valves

		Production Rate, bbl/day					
		0	**50**	**100**	**200**	**400**	**600**
p_{wf}, psi		2000	1875	1750	1500	1000	500
Equivalent depth p_{wf}, ft			7100	7300	6800	5400	3100
Equivalent depth	3150 ft, ft		3250	3450	2950	1550	
Pressure at	3150 ft, psi	500	720	660	500	220	
Equivalent depth	3650 ft, ft		3750	3950	3450	2050	
Pressure at	3650 ft, psi	690	840	780	620	310	
Equivalent depth	4150 ft, ft		4250	4450	3950	2550	250
Pressure at	4150 ft, psi	890	1000	910	730	390	40
Equivalent depth	4700 ft, ft		4800	5000	4500	3100	800
Pressure at	4700 ft, psi	1100	1150	1050	850	500	120
Equivalent depth	5200 ft, ft		5300	5500	5000	3600	1300
Pressure at	5200 ft, psi	1300	1300	1200	1000	600	200
Equivalent depth	5800 ft, ft		5900	6100	5600	4200	1900
Pressure at	5800 ft, psi	1530	1480	1380	1150	730	290
Equivalent depth	6400 ft, ft		6500	6700	6200	4800	2500
Pressure at	6400 ft, psi	1770	1680	1570	1320	860	400

These are shown in Fig. 7-3, the depths chosen being 0 (that is, the surface), 1000, 2000, 3000, 4000, 5000, 6000, and 7000 ft. The information plotted in Fig. 7-3 is derived from Table 7-1, which is prepared from the pressure-distribution curves for $2^7/_8$-in. tubing and a GLR of 200 cu ft/bbl.

Next a plot is made of the pressure versus rate curve in the tubing at the depth of the first valve, namely, 3150 ft, using the natural formation GLR of 200 cu ft/bbl, since during kickoff, when injection gas is passing through the top valve only, the flow in the tubing string below this valve is at the formation GLR. The steps needed to make this plot are outlined in Table 7-2, and the results are shown in Fig. 7-4.

It is evident from Fig. 7-4 that the equilibrium flow rate that will be attained by gas injection into valve 1 is 275 bbl/day (when the flowing pressure

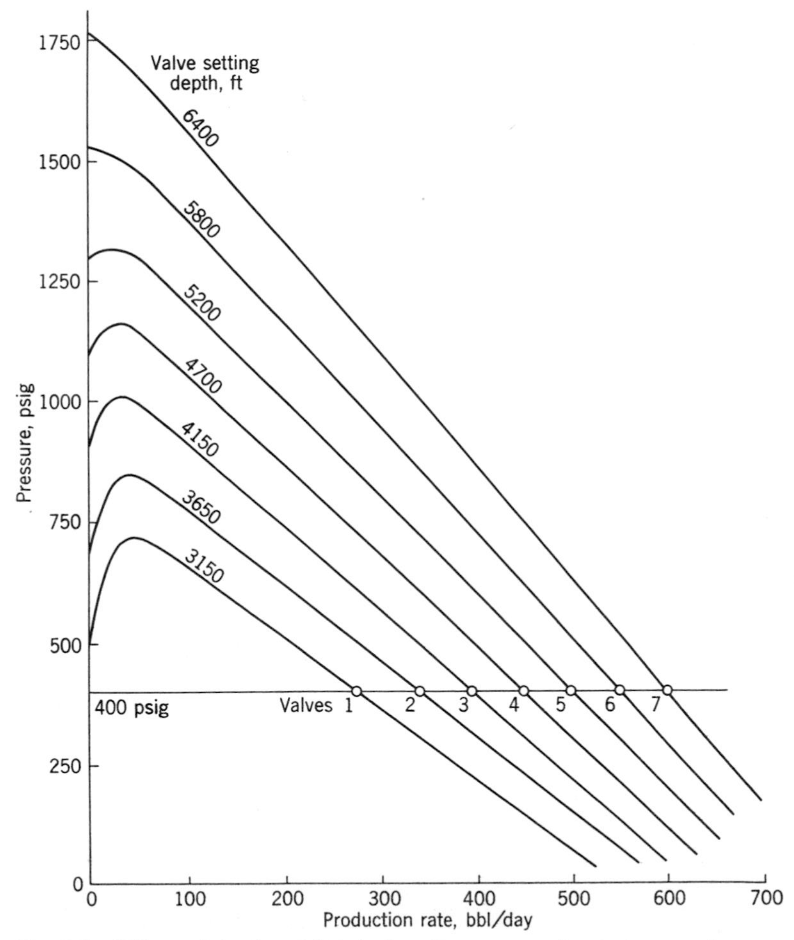

Fig. 7-4 Differential valve string design: flow rate attained when gas is passing through an intermediate valve in the string.

in the tubing at 3150 ft is the gas-injection pressure less the spring setting, that is, 500 − 100, or 400, psi).

To determine the depth at which valve 2 should be placed, it is necessary to find the point in the tubing that will have a pressure of 500 psi when gas is being injected into valve 1 and the pressure at 3150 ft is 400 psi. To do this, a cross plot is made of pressure against depth from Fig. 7-3 at the production rate of 275 bbl/day. The result is shown in Fig. 7-5, whence it is evident that valve 2 must be placed at 3650 ft. The pressure versus rate curve at this depth is plotted in Fig. 7-4 (see Table 7-2), from which it is seen that the flow rate from the well, when the pressure in the tubing at 3650 ft is 400 psi, will be 340 bbl/day.

Continuing this process, Table 7-2 and Figs. 7-4 and 7-5 are built up and the valve string comes out as shown in Table 7-3.

It remains to check whether in fact flowing pressures of 400 psi can be maintained—together with a positive THP—at the calculated setting depths of the various valves. As part of this check, the minimum rates at which the different valves must pass gas will be determined, and these will be used in the final stage to determine the required choke sizes.

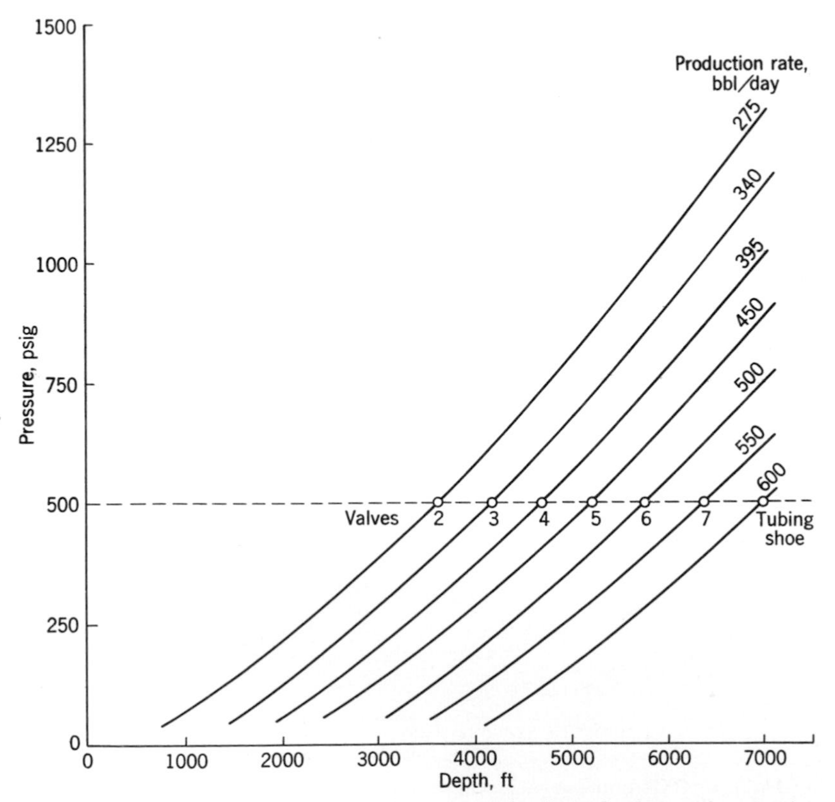

Fig. 7-5 Differential valve string design: determination of valve-setting depths.

TABLE 7-3 Differential Valve String Design: Details of Valve String

Valve	Depth, ft	Flow Rate When Pressure at This Depth is 400 psi, bbl/day
1	3150	275
2	3650	340
3	4150	395
4	4700	450
5	5200	500
6	5800	550
7	6400	600
Tubing shoe	7000	600*

* At 500 psi pressure.

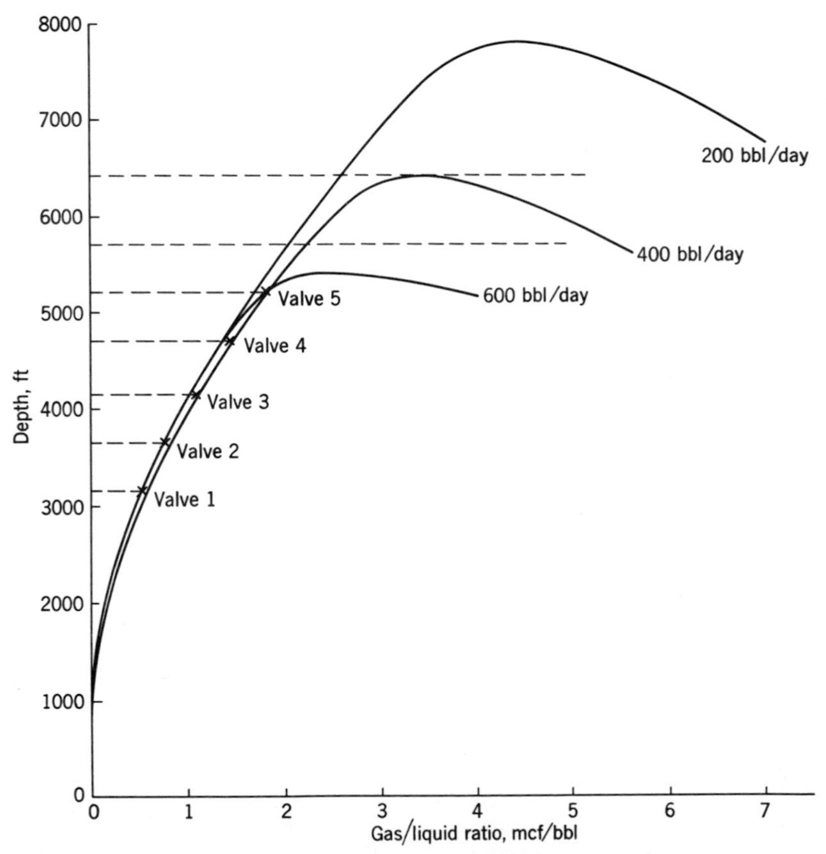

Fig. 7-6 Differential valve string design: a check to test whether the required flow conditions are possible.

Suppose that it is desired to operate with a THP of 50 psi. Since the flow rates in the tubing at the depths of the various valves all lie between 200 and 600 bbl/day (Table 7-3) and since the pressure at any one valve depth must drop to 400 psi before that valve closes, all that is required to determine whether the valve string as proposed in Table 7-3 will actually work is a plot of the type shown in Fig. 7-6. In this figure the depth at which the pressure in the tubing is 400 psi (assuming a THP of 50 psi) is plotted against GLR for production rates of 200, 400, and 600 bbl/day. The information is derived from the relevant pressure-distribution curves and is shown in Table 7-4. In Fig. 7-6, the positions (depth and corresponding liquid production rate) of the first five valves, as given in Table 7-3, have been plotted, but it is not possible to locate valves 6 and 7. It follows that valve 6 will open but the pressure in the tubing at this valve depth will never be reduced to 400 psi, even at the optimum GLR; hence, this valve will not close, and valve 7 will not come into operation.

To determine the minimum pressure attainable at 5800 ft (that is, the depth of valve 6) and the accompanying production rate, it is necessary to go back to the production rate versus pressure plot in the tubing at this depth

TABLE 7-4 Differential Valve String Design: Determination of the Depths at Which the Pressure in the Tubing Is Equal to 400 psi

Rate, bbl/day	GLR, mcf/bbl	Equiv. Depth of 50-psi THP, ft	Equiv. Depth of 400 psi, ft	Actual Depth of 400-psi Pressure with 50-psi THP, ft
200	0	128	1025	897
	0.2	350	2400	2050
	0.6	500	3800	3300
	1.2	800	5100	4300
	2.0	1200	6900	5700
	4.3	2000	9800	7800
	5.0	1600	9300	7700
	7.0	1100	7900	6800
400	0	128	1025	897
	0.2	350	2600	2250
	0.6	450	3600	3150
	1.2	700	4900	4200
	1.8	800	6000	5200
	3.25	1100	7500	6400
	4.0	1000	7300	6300
	5.5	900	6600	5700
600	0	128	1025	897
	0.2	250	2500	2250
	0.6	400	3700	3300
	1.2	600	5000	4400
	2.4	1000	6400	5400
	3.0	1000	6300	5300
	4.0	800	6000	5200

(curve *A*, say) and to superimpose the production rate versus pressure plot at the optimum GLR, assuming a THP of 50 psig (curve *B*, say). This is shown in Fig. 7-7. Curve *A* is derived from the information of Table 7-2, and curve *B* comes from the figures shown in Table 7-5, which has been prepared for $2^7/_8$-in. tubing. The intersection of curves *A* and *B* shows that the well will produce at 545 bbl/day, the pressure opposite valve 6 being 410 psi. The inset plot in Fig. 7-7 is also taken from the data of Table 7-5 and shows that, at a production rate of 545 bbl/day through $2^7/_8$-in. tubing, the optimum GLR is 2.6 mcf/bbl.

The gas-injection rate through each valve during kickoff, and hence the choke sizes required for the various valves, may be found from the information of Fig. 7-6 and Table 7-3; the calculation leading to these figures is given in Table 7-6.

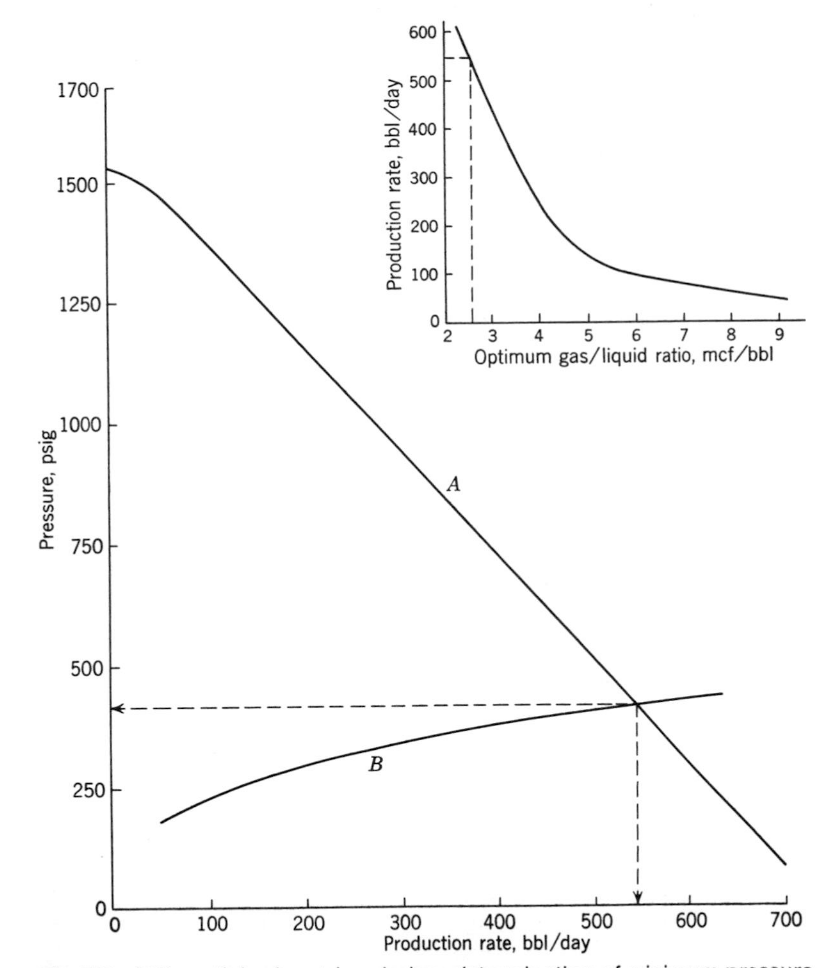

Fig. 7-7 Differential valve string design: determination of minimum pressure attainable at bottom valve. *Inset:* Production rate versus optimum GLR ($2^7/_8$-in. tubing).

TABLE 7-5 Differential Valve String Design: Determination of Maximum Gas-Lift Rate

Prod. Rate, bbl/day	Optimum GLR, mcf/bbl	Equiv. Depth of 50-psi THP at Opt. GLR, ft	Equiv. Depth of 5800 ft, ft	Pressure at 5800 ft at Opt. GLR, psi
50	8.8	2800	8600	190
100	6.3	2000	7800	230
200	4.3	2000	7800	300
400	3.25	1100	6900	370
600	2.4	1000	6800	430

For the differential valves under consideration the upstream pressure is 500 psi, and the closing downstream pressure for valves 1 to 5 inclusive is 400 psi. For valve 6, the minimum downstream pressure is 410 psi. Thus, utilizing Eq. (7-7) and Fig. 7-2, the bean sizes required in the various valves may be found as shown in Table 7-7.

So, in summary, if a $2^7/_8$-in. string of differential valves is to be used, the maximum gas-lift rate that can be expected is 545 bbl/day and six valves should be placed in the string as follows:

Valve	Depth, ft	Bean Size, in.
1	3150	$^7/_{64}$
2	3650	$^{10}/_{64}$
3	4150	$^{13}/_{64}$
4	4700	$^{16}/_{64}$
5	5200	$^{19}/_{64}$
6	5800	$^{25}/_{64}$

With this string, the lowest injection point would be valve 6. Injection gas required would be 1300 mcf/day of 500-psi gas.

It should be noted that the depths listed here make no allowance for safety factors; in practice, it might be advisable to place each valve two or three joints higher than the depths shown above. Finally, the choke sizes

TABLE 7-6 Differential Valve String Design: Daily Rate at Which Valves Must Be Capable of Passing Gas

Valve	Required GLR in Tubing, mcf/bbl	Injected GLR, mcf/bbl	Max. Prod. Rate at This Stage, bbl/day	Daily Gas Volume Injected through Valve, mcf
1	0.55	0.35	275	96.25
2	0.80	0.60	340	204.00
3	1.15	0.95	395	375.25
4	1.45	1.25	450	562.50
5	1.80	1.60	500	800.00
6	2.60	2.40	545	1308.00

TABLE 7-7 Differential Valve String Design: Determination of Choke Size Required in Each Valve

Valve	p_1, psia	p_2, psia	$\frac{p_2}{p_1}$	$F\left(\frac{p_2}{p_1}\right)$	A, sq in.	Choke Diameter, in.	Nearest Larger 1/64 in.
1	515	415	0.806	0.173	0.0079	0.100	7
2	515	415	0.806	0.173	0.0168	0.146	10
3	515	415	0.806	0.173	0.0309	0.198	13
4	515	415	0.806	0.173	0.0464	0.243	16
5	515	415	0.806	0.173	0.0659	0.290	19
6	515	425	0.825	0.167	0.1117	0.377	25

should be larger, if anything, than those calculated. This will lead to a greater daily rate of injection gas during the kickoff period and to THPs during this period which are higher than the 50 psi assumed. But if the choke sizes were too small, the valves would not pass sufficient gas and the THP would consequently fall below 50 psi; there would be a real danger that it might be impossible to kick off the well.

7-5 PRESSURE-CHARGED BELLOWS VALVES

The basic type of pressure-charged valve is the so-called *intermittent-flow valve,* which despite its name can be, and is, used for constant-flow as well as intermittent-flow conditions. The principal features of this valve are illustrated in Fig. 7-8.

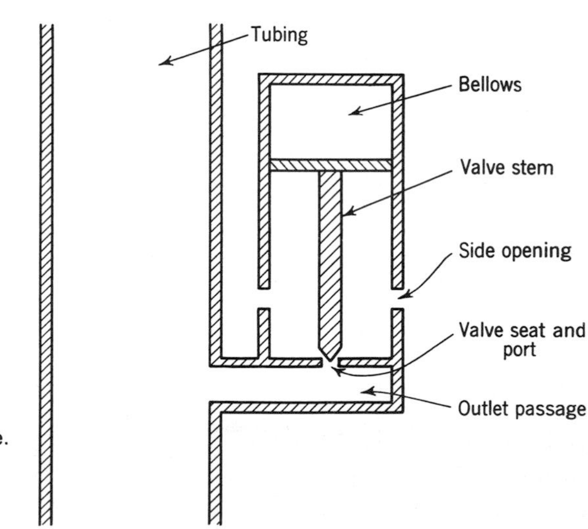

Fig. 7-8 Intermittent-flow valve.

The valve is kept normally closed by means of pressure-charged bellows. The force of the pressure charge on the bellows area is opposed by the force exerted by the injection-gas pressure on the bellows area less the valve-stem area and by the force exerted on the valve-stem area by the pressure in the tubing at the depth of the valve. If

p_i = gas-injection pressure, psi
p_b = bellows pressure charge, psi
p = pressure in the tubing, psi
A = bellows area, sq in.
B = valve-stem area, sq in.

then the valve will open when p_i rises to such a value that

$$p_i(A - B) + pB > p_bA \tag{7-8}$$

With the valve open, assuming that p_i is sufficiently large for gas injection to take place (p_i greater than p), gas moves into the tubing through the valve. Since the side openings are large and do not exert any appreciable choking effect, the pressure inside the valve body is equal to the gas-injection pressure p_i, and this now acts over the total bellows area, so that the force opposing the closing influence of the bellows charge is p_iA. But p_i is greater than p, so that

$$p_iA > p_i(A - B) + pB \tag{7-9}$$

It follows from Eqs. (7-8) and (7-9) that there is a positive opening action, and there will be no tendency for the valve to chatter. Once the valve is open, it may be reclosed only by the reduction of the gas-injection pressure to a value slightly less than the pressure charge p_b of the bellows. Similarly, once the valve is closed, the opening force is reduced even further, so the closing action is also positive.

It will be noted that the gas-injection pressure required to open the valve depends on the valve geometry, on the pressure charge, and on the pressure in the tubing, whereas the gas pressure at which the valve closes depends only on the pressure charge. Note, too, that the effective pressure charge in the bellows increases with increasing temperature, so that in the design of a gas-lift string involving pressure-charged valves, allowance should be made for the geothermal gradient, the magnitude of the allowance depending on the setting of the valve.

From the description of the action of the valve it is evident that when a series of these valves is run into a hole on a tubing string and gas pressure is applied until the top valve is open, all the lower valves are also open (unless the bellows pressure charge is increased considerably with increased valve setting depth, which is just the reverse of what is required in practice, as will be seen below). Gas flows through the valve seat and port of the top valve, the seat and port supplying the choking action, and into the tubing. The resultant increase in GLR of the fluid in the tubing above this top valve reduces the pressure loss in the tubing above the top valve, thus decreasing the pressure in the tubing at the depth of the valve and increasing the liquid production rate from the formation. With a correct choice of choke size associated with the valve seat and port, it is possible to attain the optimum GLR in the tubing above the valve. If the second valve is so placed that it is in the annulus gas column when this optimum condition above the first valve is attained and, further, if the second valve is so placed and designed that it remains in the annulus gas column and open when the gas-injection pressure is reduced sufficiently at the surface to close the top valve, then gas injection into the second valve will commence.

To allow the second valve to remain open when the first is closed, the bellows-pressure charge of the second valve must be less than that of the first, and so on down the string. Gas injection into each valve is continued until optimum flow conditions are attained in the tubing from that particular valve depth to the surface.

Since optimum flow conditions are attained at each valve depth, the disadvantage of the differential-valve string is overcome; intermittent-type valves may be spaced at wider intervals than differential valves. However, the valve-stem area must be relatively small (usually $^3/_8$ in. or less); otherwise, the ratio of this area to the bellows area would be large, and valve opening would become highly dependent upon the pressure in the tubing. Thus the high gas rates required to gas-lift a large producer or to slug-lift a small well, for ex-

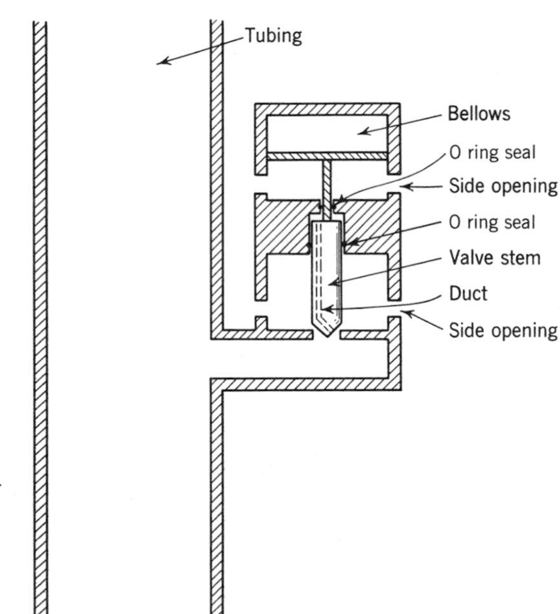

Fig. 7-9 Large-orifice intermittent-flow valve.

ample, cannot be passed through an intermittent-flow valve of the type just described.

For this reason the *large-orifice intermittent-flow valve* (sketched roughly in Fig. 7-9) was designed. To prevent the tubing pressure from having a controlling effect on the opening of the valve despite the large valve-stem area, the tubing pressure is ducted to a space, sealed off by O rings, where it can act over a considerable percentage of the upper valve-stem area in such a way as to nullify, to a large extent, the opening effect of the tubing pressure. No attempt is made to cancel completely this effect of tubing pressure, because otherwise the valve would lose its snap opening and closing property.

Both pressure-loaded valves described above suffer from the same disadvantage. Since the top valve operates at the highest pressure and the bottom valve at the lowest, the pressure required, and thus the compressor horsepower, is reduced on moving down the string. In an attempt to overcome this deficiency, the *constant-flow valve* (shown diagrammatically in Fig. 7-10) was designed. In this valve the injected gas has to pass through the chokes placed in the valve walls; hence, although the opening of the valve is controlled by the same considerations that apply to the intermittent-flow valve, the closing is somewhat different. As the pressure in the tubing drops, the pressure in the valve chamber also drops and the valve closes automatically when the pressure in the tubing has fallen to a certain level. Thus the bellows pressure charge may remain the same throughout a string of constant-flow valves, and the full available gas-injection pressure may be used down to the lowest valve in the string. However, because of the metering effect of the side chokes, the

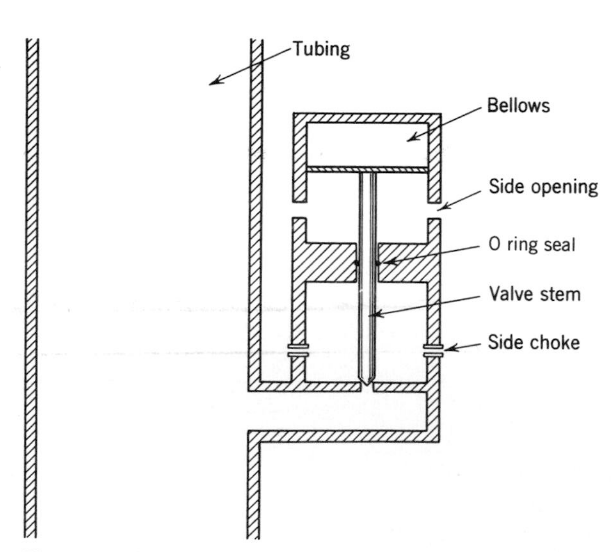

Fig. 7-10 Constant-flow valve.

constant-flow valve suffers from the main disadvantage of the differential valve, namely, that it cannot be used where high gas-injection rates are required.

7-6 EXAMPLE OF STRING DESIGN: PRESSURE-CHARGED BELLOWS VALVES

Suppose that it is desired to design a string of pressure-charged bellows valves for use in the well of Sec. 7-4. It will be supposed that the first valve has a bellows pressure of 500 psi and that the bellows pressure is reduced by 25 psi per valve on moving down the string.

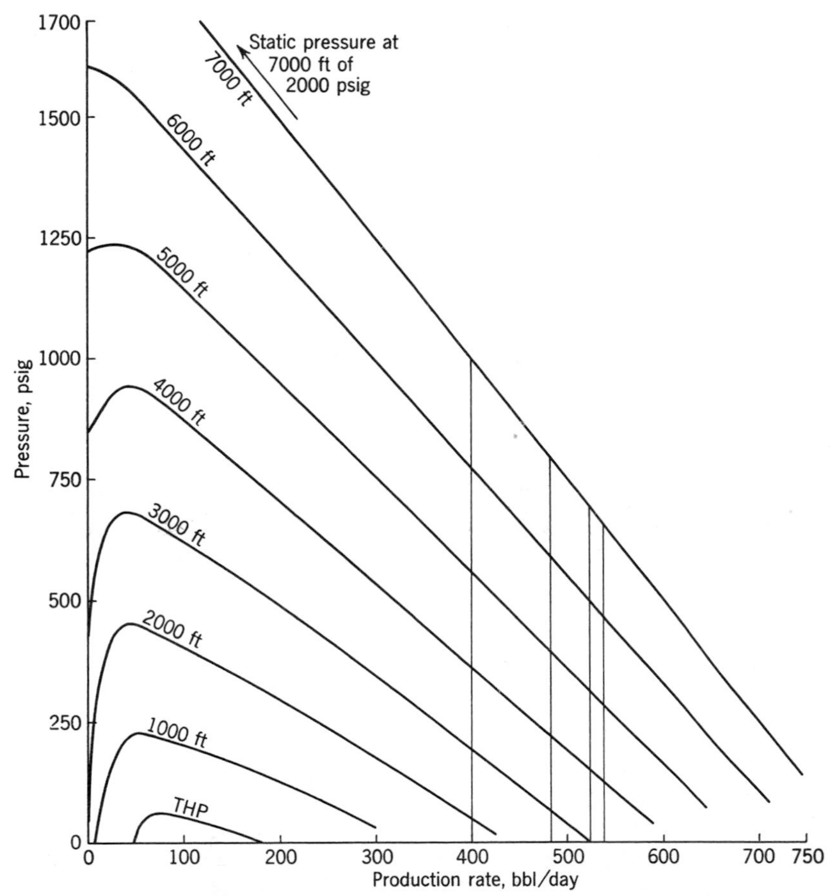

Fig. 7-11 Pressure-charged valve string design: pressure versus rate curves at different points in the tubing.

As in Sec. 7-4, either a 3½- or a 2⅞-in. tubing string would appear to be a reasonable choice. Suppose it is decided that a 2⅞-in. string is sufficient. The positioning of the top valve is carried out as for the differential-valve string; it will be placed at a depth of 3150 ft.

Plots are then made of the pressure versus rate curves in the tubing at various depths, using the natural formation GLR of 200 cu ft/bbl. The results are shown in Fig. 7-11, which repeats Fig. 7-3; the required information is listed in Table 7-1.

Now a plot is made of the pressure versus rate curve in the tubing at the depth of the first valve, using the natural formation GLR (Table 7-8, Fig. 7-12). Since the optimum GLR can be utilized with pressure-loaded valves for flow in the string above the valve, a plot of pressure versus rate at optimum conditions from 3150 ft with a THP of, say, 50 psi is now prepared (Table 7-9). The resulting curve is shown in Fig. 7-12; the point at which it intersects the pressure versus rate curve based on flow from 7000 to 3150 ft with a GLR of 200 cu ft/bbl gives the flow rate (400 bbl/day) and pressure (220 psi) in the tubing at 3150 ft when the optimum volume of gas is being injected into the top valve.

The closing pressure of the second valve is 475 psi, and the depth at which this valve should be placed can be found by making a cross plot of Fig. 7-11 at the rate of 400 bbl/day, obtaining a pressure versus depth curve at this rate. This cross plot is presented in Fig. 7-13 and shows that at a depth of 4600 ft the pressure of 475 psi is just sufficient to overcome the pressure in the fluid column. Continuing this process, Tables 7-8 and 7-9 and Figs. 7-12 and 7-13 are built up, and the valve string comes out as shown in Table 7-10.

It is apparent from Fig. 7-13 that only four valves are required to gas-lift

TABLE 7-8 Pressure-Charged Valve String Design: Pressure versus Rate Data at the Depths of the Various Valves, Based on the Pressure Loss in the Tubing between the Shoe and the Valve in Question

		Production Rate, bbl/day					
		0	50	100	200	400	600
p_{wf}, psi		2000	1875	1750	1500	1000	500
Equivalent depth, ft			7100	7300	6800	5400	3100
Equivalent depth	3150 ft, ft		3250	3450	2950	1550	
Pressure at	3150 ft, psi	500	720	660	500	220	
Equivalent depth	4600 ft, ft		4700	4900	4400	3000	700
Pressure at	4600 ft, psi	1060	1125	1025	840	480	100
Equivalent depth	5350 ft, ft		5450	5650	5150	3750	1450
Pressure at	5350 ft, psi	1360	1340	1250	1030	630	220
Equivalent depth	5650 ft, ft		5750	5950	5450	4050	1750
Pressure at	5650 ft, psi	1470	1430	1340	1120	700	270

TABLE 7-9 Pressure-Charged Valve String Design: Pressure versus Rate Data at the Depth of the Various Valves, Based on the Pressure Loss in the Tubing between the Valve in Question and the Surface

		Production Rate, bbl/day				
		50	**100**	**200**	**400**	**600**
Optimum GLR, mcf/bbl		8.8	6.3	4.3	3.25	2.4
Equivalent depth 50-psi THP, ft		2800	2000	2000	1100	1000
Equivalent depth	3150 ft, ft	5950	5150	5150	4250	4150
Pressure at	3150 ft, psi	120	150	170	220	240
Equivalent depth	4600 ft, ft	7400	6600	6600	5700	5600
Pressure at	4600 ft, psi	150	190	230	290	340
Equivalent depth	5350 ft, ft	8150	7350	7350	6450	6350
Pressure at	5350 ft, psi	175	220	270	340	390
Equivalent depth	5650 ft, ft	8450	7650	7650	6750	6650
Pressure at	5650 ft, psi	190	230	280	350	420

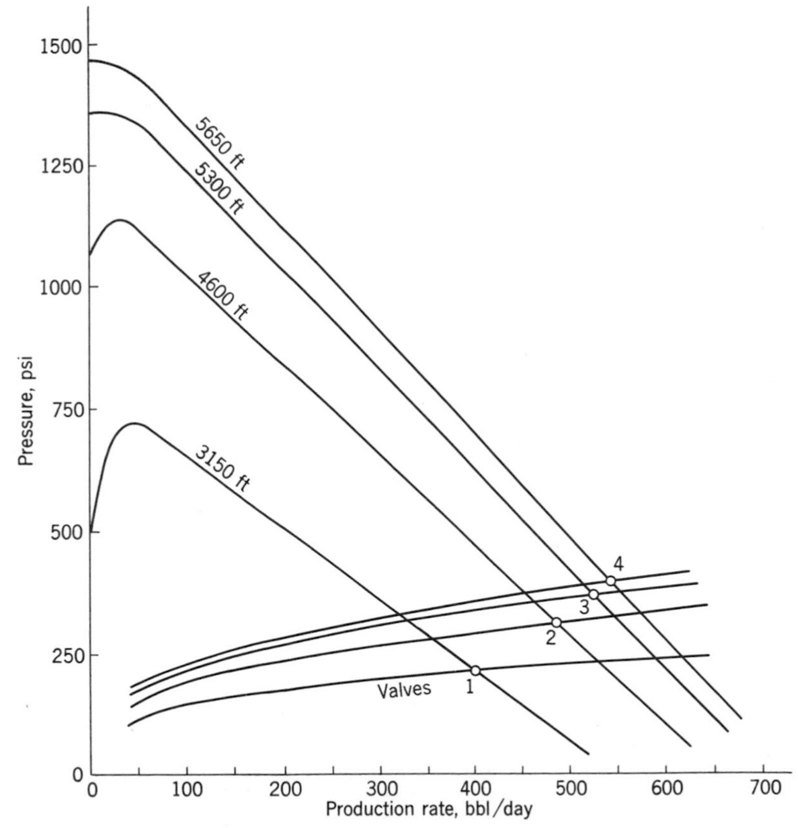

Fig. 7-12 Pressure-charged valve string design: flow rate attained when gas is passing through an intermediate valve in the string.

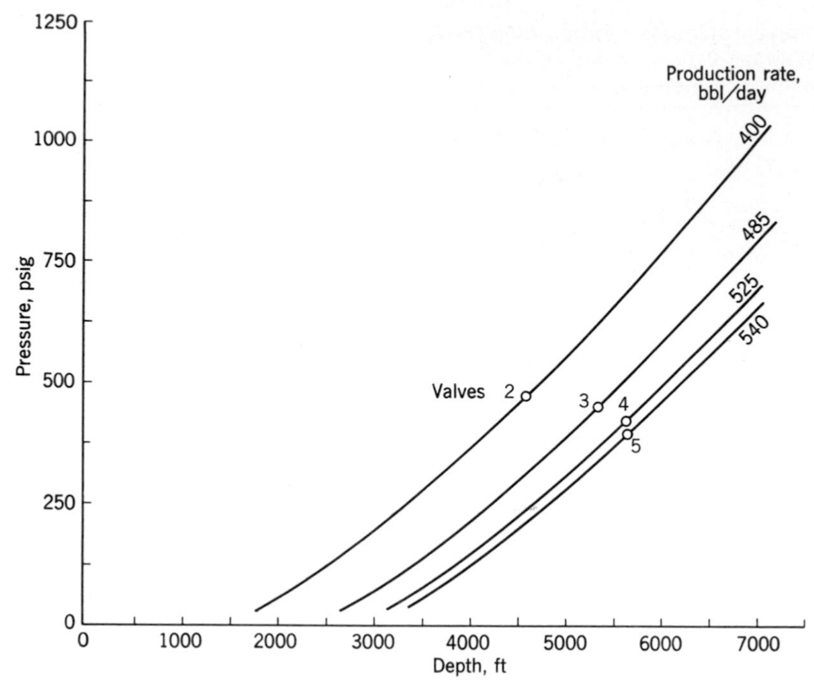

Fig. 7-13 Pressure-charged valve string design: determination of valve-setting depths.

the well at very close to the maximum rate possible through the tubing size under consideration and with the gas pressure available. A fifth valve is unnecessary, as it would have to be located so close to the fourth valve as to render it almost valueless.

Because the string design has been built around the use of optimum GLRs above each valve and because the flow below and above each valve has been made consistent (Fig. 7-12), no check is needed to determine whether the flowing pressures indicated in Table 7-10 can in fact be maintained; this point is assured by the method.

TABLE 7-10 Pressure-Charged Valve String Design: Details of Valve String

Valve	Depth, ft	Flow Rate at This Depth under Opt. GLR, bbl/day	Pressure in Tubing Opposite Valve under Opt. Conditions, psi
1	3150	400	220
2	4600	485	310
3	5350	525	370
4	5650	540	400

TABLE 7-11 Pressure-Charged Valve String Design: Daily Rate at Which Valves Must be Capable of Passing Gas

Valve	Optimum GLR, mcf/bbl	Injected GLR, mcf/bbl	Prod. Rate at This Stage, bbl/day	Daily Gas Vol. Injected through Valve, mcf
1	3.25	3.05	400	1220.0
2	2.88	2.68	485	1299.8
3	2.70	2.50	525	1312.5
4	2.63	2.43	540	1312.2

To determine the daily gas rate required at each valve, and hence the valve port sizes and the daily gas-injection rate needed, a plot of optimum GLR against production rate, as shown in Fig. 7-14, is useful; this plot is made from the data summarized in Table 7-9. From this, Table 7-11 can be prepared. For pressure-charged bellows valves, the upstream pressure is the opening pressure of the valve, and the downstream pressure is the pressure in the tubing (at the depth of the valve) at the optimum GLR. Thus, utilizing Eq. (7-7) and Fig. 7-2, the port sizes required in the various valves are found as shown in Table 7-12. It may be noted that the third and fourth valves should be of the large-orifice type (Sec. 7-5).

In summary, if a $2^7/_8$-in. string of pressure-charged bellows valves is to be

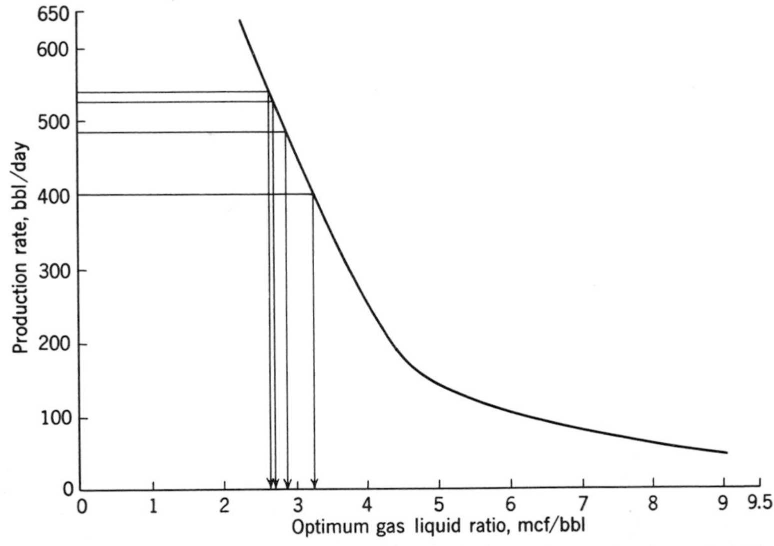

Fig. 7-14 Pressure-charged valve string design: determination of daily injection gas rate required at each valve.

TABLE 7-12 Pressure-Charged Valve String Design: Determination of Port Size Required in Each Valve

Valve	p_1, psia	p_2, psia	$\frac{p_2}{p_1}$	$F\left(\frac{p_2}{p_1}\right)$	A, sq in.	Port Diameter, in.	Nearest Larger 1/64 in.
1	515	235	0.456	0.210	0.0830	0.324	21
2	490	325	0.663	0.202	0.0965	0.350	23
3	465	385	0.828	0.166	0.1251	0.398	26
4	440	415	0.943	0.096	0.229	0.540	35

used, the maximum gas-lift rate that can be expected is 540 bbl/day, with four valves placed in the string as tabulated below.

Valve	Setting, psig	Port, in.	Depth, ft
1	500	21/64	3150
2	475	23/64	4600
3	450	26/64	5350
4	425	35/64	5650

Valve 4 would be the lowest injection point, and 1312 mcf/day of 425-psi gas would be required.

As with the example illustrating differential-valve string design in Sec. 7-4, it might, in practice, be advisable to place each valve two or three joints higher than the depth indicated in order to allow some safety margin. Further, the port sizes should, if anything, err on the large side.

It should also be mentioned that the valve string assumed in this example is an intermittent-flow or large-orifice intermittent-flow string. A string of constant-flow valves would require a somewhat different approach.

7-7 CONCLUDING COMMENTS

It must be emphasized that the methods used in the examples of Secs. 7-4 and 7-6, while illustrating the principles involved, require adaptation to take into account a number of factors other than those considered, among which may be mentioned design peculiarities of the various types of valves, the effects of temperature on pressure-charged valves, any limitation in gas supply, and the influence of fluid back pressure on the opening of pressure-charged valves.

This discussion of gas-lift string design has been based on the (implicit) assumption that the tubing was acting as the eductor string, but the methods considered and the valves illustrated may equally well be applied to wells in which the casing-tubing annulus is used as the eductor string, with gas injection taking place down the tubing; all that is required is a set of pressure-distribution curves for annular flow.

REFERENCES

1. Kirkpatrick, C. V.: *The Power of Gas,* 2d ed., Camco, Inc., Houston, Tex., 1954.
2. *Handbook of Gas Lift,* Garrett Oil Tools, Division of U.S. Industries, Inc., New York, 1959.
3. Binder, R. C.: *Fluid Mechanics,* 5th ed., Prentice-Hall, Inc., Englewood Cliffs, N.J., 1973.
4. Gilbert, W. E.: "Flowing and Gas-Lift Well Performance," *API Drill. Prod. Practice,* 1954, p. 126.

Liquid Production by Slugs

8

8-1 INTRODUCTION

If formation productivity and static pressure are such that steady flow cannot be maintained, some form of artificial lift is required. For a variety of reasons pumping may not be appropriate (for example, in offshore locations, in very deep or highly deviated holes) so that gas lifting becomes the preferred production method. But continuous gas lift may not be successful, perhaps because the gas compression requirement is too great to allow an economical operation, perhaps because a technically efficient gas-lift rate is too high (either for the good of the formation itself, or because of a maximum allowable rate for the well in question), or perhaps because the formation pressure and productivity are so low that continuous circulation of injected gas inhibits flow from the formation into the well bore.

Under such circumstances it is reasonable to turn to a form of intermittent gas lift, either a standard semiclosed or closed system or a modification of such a system in the form of plunger lift or chamber lift.

In this chapter the characteristics of such production are discussed, but any consideration of design of a string of gas-lift valves for such an installation is omitted. For such design, the reader is referred to Brown's *Gas Lift Theory and Practice* (Ref. 1), to the handbooks of the various manufacturers of gas-lift equipment, or to the general methods outlined in the preceding chapter.

Although the material included here is largely of an analytical nature, it is hoped that the conclusions contained in Sec. 8-4 (and in particular in Fig. 8-8) on optimum cycle frequency and efficiency will be of practical value and will contribute to a fuller understanding of those factors that should be taken into account in designing intermittent-lift systems.

8-2 THREE PRODUCING METHODS INVOLVING LIQUID SLUGS

Intermittent Gas Lift

Intermittent gas-lift installations are, typically, of two types, *semiclosed* (Fig. 8-1*a*) in which a packer but no standing valve is installed, and *closed* (Fig. 8-1*b*), which includes both a packer and a standing valve. In the former, production from the formation occurs throughout the cycle, albeit against a high back pressure in the tubing during the gas-injection phase. In the latter the formation is effectively closed in during gas injection; such an installation is used when there is a danger of losing injected gas to the formation. This could occur in medium- to high-productivity formations with low to medium static BHPs.

Intermittent production becomes less efficient, the larger the minimum back pressure and the longer the slug in the tubing. The magnitude of the minimum back pressure is controlled by the extent of the "liquid fallback" during the production phase, and this fallback is due in the main to injection gas channeling through the slug rather than driving the slug ahead of it. To gain some idea of the order of magnitude of the back pressure exerted by the fallback, it is of interest to note that an oil film of thickness 0.05 in. on 5000 ft of $2^7/_8$-in. tubing has a volume of about 2.32 bbl; 2.32 bbl of 30°API oil in

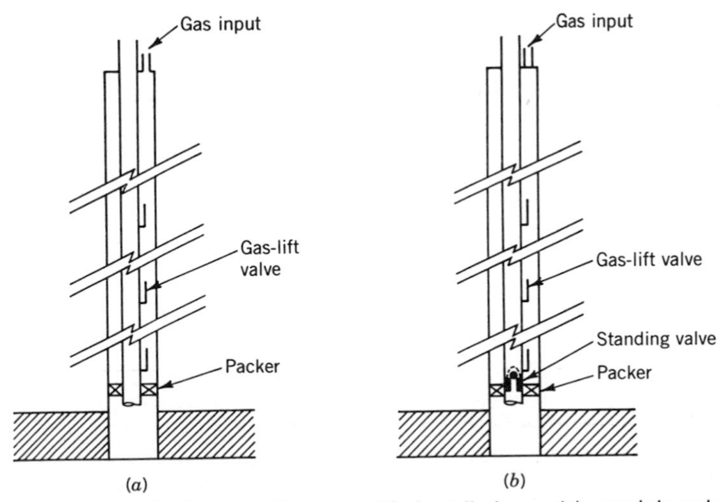

Fig. 8-1 Typical intermittent gas-lift installations: (*a*) semiclosed; (*b*) closed.

$2^7/_8$-in. tubing exerts a pressure of 150 psi. This calculation indicates that a back pressure against the formation of up to 400 or 500 psi might easily be exerted by the fallback liquid, and such a back pressure will evidently reduce the production rate to a value considerably below the potential theoretically obtainable from the formation (Sec. 8-4).

Evidently the degree of fallback could be reduced by reducing the excess pressure of the injected gas, but too great a reduction will result in slow slug movement up the tubing, and so in fewer cycles per day—itself a cause of inefficiency. To overcome this particular problem, a free-fall plunger may be installed.

The length of the slug, and consequently the back pressure exerted by the accumulation of a certain volume of liquid, may be reduced by using a large-diameter tubing as an accumulation chamber at the bottom of the hole. Alternatively the installation may be so designed that the casing itself is used as the chamber.

These two modifications to the straightforward intermittent gas-lift installation will now be considered.

Plunger Lift

Figure 8-2 illustrates a free-fall plunger installation in which no annulus packer is installed (*natural* plunger lift). A steel plunger containing a simple valve device is located in the tubing string, at the bottom of which is a seat containing an opening through which gas and liquid can pass into the tubing. When the plunger falls and lands against this seat, the valve located in the plunger is closed; hence, the tubing is closed at its lower end, and any production from the formation must pass into the annulus. The BHP consequently rises; as soon as it reaches a value greater than the sum of the pressures exerted by the plunger itself, by the oil and gas in the tubing above the plunger, and by the surface-trap pressure, the plunger starts to rise, and the oil above the plunger is lifted up the tubing and so produced.

At the upper end of the tubing a bumper is located above the flow-line offtake. When the plunger strikes this bumper, the valve in the plunger is opened, the pressure below is released to the flow line, and the plunger is free to drop back down the tubing. While the plunger is dropping, the well produces into the tubing against the back pressure created by the trap, the gas column in the tubing, and the steadily lengthening oil column (resulting from the formation's production) in the tubing. As soon as the plunger reaches bottom, the cycle is repeated, producing oil from the well by slugs.

If a well does not produce with a sufficiently high GLR for a naturally occurring plunger cycle to take place, gas must be intermitted into the casing-tubing annulus, the intermitting time being so arranged that gas injection commences at the time the plunger reaches the tubing shoe and continues until the plunger strikes the surface bumper, when it is discontinued. Even in those wells with a sufficiently high GLR to make gas injection unnecessary,

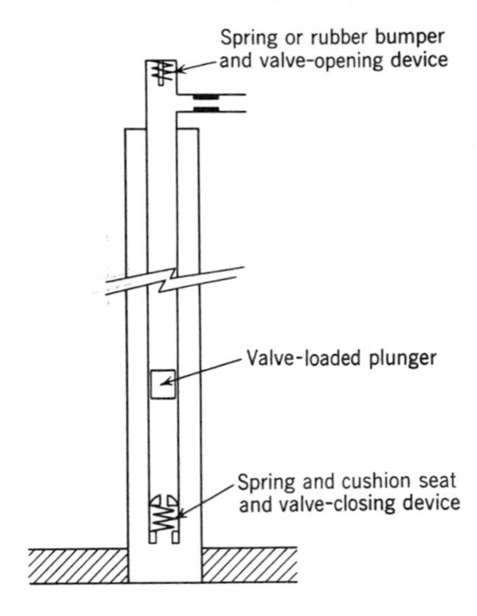

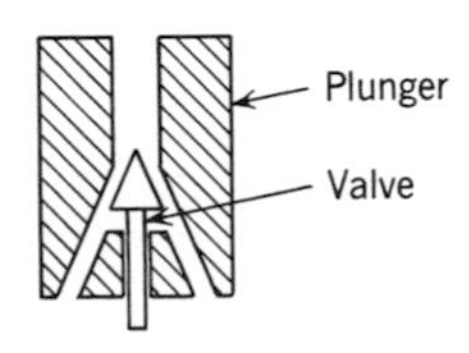

Fig. 8-2 Diagrammatic sketch of plunger-lift installation.

facilities for gas injection should be available in case the plunger fails to reach the surface for some reason, for instance, if the plunger is frozen by wax or sand in the tubing.

A plunger may also be installed in a well with an annulus packer, and in such a case is used to overcome problems of gas channeling.

Although plunger lift is evidently not designed for wells in which sand production is a problem, it can be used successfully in certain wells in which deposition of wax on the inner wall of the tubing might otherwise cause some difficulties. The continuous motion of the plunger up and down the tubing frees the wax and keeps the tubing wall clean.

Chamber Lift

The essential items of equipment required in this method of artificial lift are shown schematically in Fig. 8-3, and the part each plays may be illustrated by describing the chamber-lift cycle (Ref. 2).

1. When the control valve at the surface is closed, the well produces into the chamber via the standing-valve (SV) port, the equalizing valve permitting the fluid levels inside and outside the stinger to remain the same. As pro-

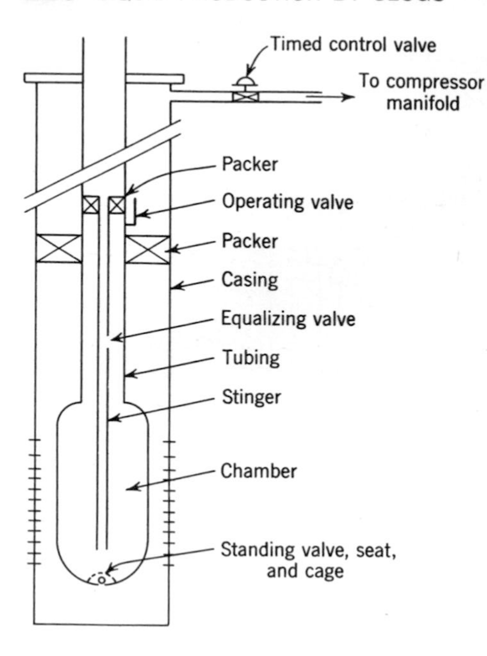

Fig. 8-3 Diagrammatic sketch of chamber-lift installation. (*After Nind, Ref. 2. Courtesy Can. Inst. Mining and Metallurgy.*)

duction accumulates inside the chamber, the back pressure on the formation increases, so that the formation production rate steadily decreases.

2. At a preselected time the control valve opens, and gas is injected into the casing-tubing annulus above the packer. The casing pressure rises and eventually reaches a level at which the operating valve opens. This permits gas to pass down into the stinger-tubing annulus. The equalizing valve and the SV both snap shut because of the high differential pressure. The gas drives the liquid in the chamber into the stinger and thence up into the tubing.
3. At a preselected time the control valve closes, and the liquid slug is forced to the surface by expansion of the gas in the casing-tubing annulus. The casing pressure drops, and the operating valve is adjusted so that it closes as soon as the slug reaches the surface, or immediately thereafter. The cycle is now complete.

8-3 LIQUID BUILDUP IN AN OPEN-ENDED VERTICAL CYLINDER

A formation with a PI of J bbl/(day)(psi) is producing into an open-ended vertical cylinder of cross-sectional area a sq ft, the back pressure against the upper end of the cylinder being P_1 psia. The vertical pressure gradient of the liquid in the cylinder is w psi/ft.

If there is a volume Q bbl of liquid in the cylinder at the time t (measured in days), then the length of the liquid column at that time is

$$h = \frac{5.614Q}{a} \qquad \text{ft} \tag{8-1}$$

so that the pressure exerted at the foot of the cylinder, that is, against the formation is

$$p = P_1 + 5.614 \frac{w}{a} Q \qquad \text{psi} \tag{8-2}$$

if it is assumed that the free gas bubbling up through the liquid has no lifting effect (Fig. 8-4). The pressure drawdown on the formation is, then,

$$p_s - \left(P_1 + 5.614 \frac{w}{a} Q\right) \qquad \text{psi}$$

and so the rate at which the formation is producing liquid into the cylinder is

$$q = J\left(p_s - P_1 - 5.614 \frac{w}{a} Q\right) \qquad \text{bbl/day} \tag{8-3}$$

Differentiating Eq. (8-3) with respect to time, remembering that

$$q = \frac{dQ}{dt}$$

gives

$$\frac{dq}{dt} = -5.614 \frac{w}{a} qJ$$

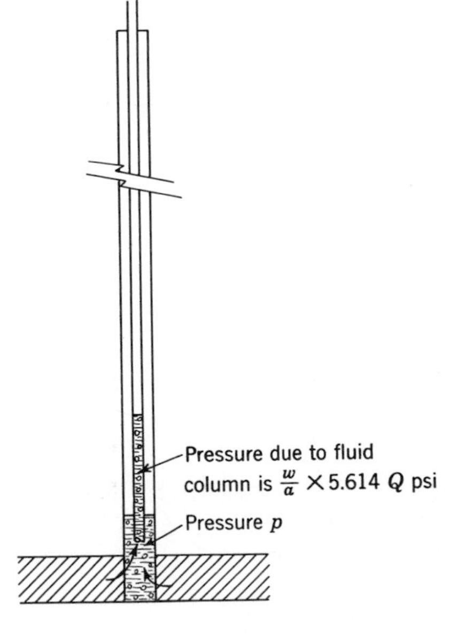

Fig. 8-4 Liquid buildup in open-ended tubing.

which on integration yields

$$\ln q = -5.614 \frac{Jwt}{a} + C \tag{8-4}$$

where C is a constant.

If Q_m bbl is the minimum volume of liquid in the cylinder, and if this volume occurs at time $t = 0$, then from Eq. (8-2) the minimum pressure against the formation is

$$p_m = P_1 + 5.614 \frac{w}{a} Q_m \tag{8-5}$$

while the (maximum) production rate q_m achieved by the formation (at time $t = 0$) is

$$q_m = J(p_s - p_m) \tag{8-6}$$

But from Eq. (8-4), putting $q = q_m$ at $t = 0$

$$\ln q_m = C$$

so that

$$C = \ln J\,(p_s - p_m)$$

Substituting this value of C back into Eq. (8-4) gives

$$q = J\,(p_s - p_m) \exp\left(-5.614 \frac{Jw}{a} t\right) \tag{8-7}$$

Using the fact that the increase $Q - Q_m$ in the volume of liquid in the cylinder is equal to $\int_0^t q\,dt$, it follows that

$$Q - Q_m = \frac{a}{5.614w} (p_s - p_m) \left[1 - \exp\left(-5.614 \frac{Jw}{a} t\right)\right] \tag{8-8}$$

If the system is such that the cylinder is purged n times a day and that the well is effectively closed in (by the shutting of an SV, say) for a fraction f of each day, then the producing time in each cycle is $(1 - f)/n$ of a day. From Eq. (8-8) the liquid production in each cycle is

$$\frac{a}{5.614w} (p_s - p_m) \left[1 - \exp\left(-5.614 \frac{Jw}{a} \frac{1-f}{n}\right)\right]$$

and the liquid production per day is

$$q = \frac{an}{5.614w} (p_s - p_m) \left[1 - \exp\left(-5.614 \frac{Jw}{a} \frac{1-f}{n}\right)\right]$$

$$= J\,(p_s - p_m)(1 - f) \frac{1 - \exp(-x)}{x} \tag{8-9}$$

where

$$x = 5.614 \frac{Jw}{a} \frac{1-f}{n} \tag{8-10}$$

The function $[1 - \exp(-x)]/x$ is plotted in Fig. 12-25 for x lying between 0.1 and 10.0. For values of x greater than 10.0, the function is close to $1/x$, while for values of x less than 0.1, $(1 - x/2)$ is a good approximation.

The efficiency of the lifting technique may be defined as the actual production rate divided by the potential Jp_s, so that from Eq. (8-9)

$$\text{Efficiency} = \left(1 - \frac{p_m}{p_s}\right)(1 - f)\frac{1 - \exp(-x)}{x} \tag{8-11}$$

It is apparent from this equation, taken in conjunction with Fig. 12-25 and Eq. (8-10), that production by slugs is at its best, from the standpoint of production efficiency, when

Static pressure p_s is high
Minimum back pressure p_m [Eq. (8-5)] is low
The nonproductive fraction of the day f is low
PI value J is low
Cross-sectional area of vertical cylinder (tubing, casing) is large, and
The number of cycles per day n is large

8-4 OPTIMUM CYCLE FREQUENCY AND EFFICIENCY

The general descriptions of intermittent gas lift, plunger lift (but not natural plunger lift—see Sec. 8-6), and chamber lift given in Sec. 8-2 illustrate the fact that during the gas-injection phase of the cycle little or no production takes place from the formation. It follows that the fraction of the day during which the well is effectively closed in, f, depends upon the number of cycles per day n. This raises the possibility that there is an optimum frequency, that is, some number of cycles per day that should lead to a maximum recovery rate under slug-lift production.

If T (expressed as a fraction of a day) is the gas-injection time per cycle, then

$$f = Tn$$

and Eq. (8-9) can be written

$$q = J(p_s - p_m)(1 - Tn)\frac{1 - \exp(-x)}{x} \tag{8-12}$$

where

$$x = 5.614\frac{Jw}{a}\frac{1 - Tn}{n} \tag{8-13}$$

Assuming that p_m (that is, the volume of fallback liquid) is insensitive to n—although not to an attempt to reduce T by increasing the gas-injection pressure in order to achieve higher slug velocities—then for any particular well $J(p_s - p_m)$ may be regarded as a constant. Further, substituting from Eq.

(8-13) in Eq. (8-12) gives

$$q = \frac{a(p_s - p_m)}{5.614w} n\{1 - \exp[-b(1/n - T)]\} \tag{8-14}$$

where

$$b = 5.614 \frac{Jw}{a} \tag{8-15}$$

so that the optimization of q, regarded as a function of n, resolves itself into a question of the optimization of

$$f(n) = n\{1 - \exp[-b(1/n - T)]\} \tag{8-16}$$

It is possible to study this expression theoretically, but first it will be illustrated by means of an example.

Example 8-1 Consideration is being given to a closed intermittent-gas-lift installation on Well 1387/BX. The installation would use $2^7/_8$-in. tubing hung just above the top perforations at 5872 ft. The static pressure in the formation is 1120 psig and the IPR is almost a straight line, the average PI being 0.8 bbl/(day)(psi). Similar installations in the same field confirm that an average upward slug velocity of 900 ft/min during the gas-injection stage is reasonable, and that the fallback to be anticipated at that velocity is 3 bbl. The back pressure on the tubing from the surface facilities is 100 psig and the pressure gradient exerted by the liquid from the formation is 0.31 psi/ft.

Determine the optimum number of cycles per day and also the sensitivity of that optimum number to variations in Jw/a and in the average upward slug velocity. Calculate the production rate to be expected on the intermittent-gas-lift installation.

From the data given, and referring to Eqs. (8-5), (8-14), and (8-15),

$$\begin{aligned} p_m &= 100 + 5.614 \times \frac{0.31}{0.0325} \times 3 \\ &= 100 + 161 \\ &= 261 \text{ psig} \end{aligned}$$

$$\begin{aligned} \frac{a(p_s - p_m)}{5.614w} &= \frac{0.0325 \times 859}{5.614 \times 0.31} \\ &= 16.04 \end{aligned}$$

$$\begin{aligned} b &= \frac{5.614 \times 0.8 \times 0.31}{0.0325} \\ &= 42.84 \end{aligned}$$

$$\begin{aligned} T &= 5872/(900 \times 1440) \\ &= 0.0045 \end{aligned}$$

Table 8-1 outlines the calculations for determining $f(n)$ of Eq. (8-16), and the results are plotted in Fig. 8-5. The optimum value of n is about 57 (that is, 57 cycles/day), and, from Eq. (8-14), the optimum value of the production rate is

$$q_{\text{opt}} = 24.5 \times 16.04 = 393 \text{ bbl/day}$$

The question of the sensitivity of $f(n)$ to changes in Jw/a and in the upward

TABLE 8-1 EXAMPLE 8-1: Determination of $f(n)$ for Various Values of n ($5.614Jw/a = 42.84$)

n	$1/n$	$1/n - T$	$\exp[-b(1/n - T)]$	$1 - \exp[-b(1/n - T)]$	$f(n)$
20	0.0500	0.0455	0.142	0.858	17.16
30	0.0333	0.0288	0.290	0.710	21.30
40	0.0250	0.0205	0.415	0.585	23.40
60	0.0167	0.0122	0.592	0.408	24.48
80	0.0125	0.0080	0.709	0.291	23.28
100	0.0100	0.0055	0.789	0.211	21.10

slug velocity is investigated in Tables 8-2 and 8-3 and the results are plotted in Figs. 8-6 and 8-7. Table 8-4 shows numerical results at the optimum condition for each of the cases investigated.

The following comparative points may be noted:

1. As the value of Jw/a increases—namely, as the PI of the well increases and/or the diameter of the tubing in the hole decreases—the optimum number of cycles per day also increases, although the sensitivity is not particularly high. In Table 8-4 it is assumed that the increase in Jw/a is entirely due to an increase in the PI and it is apparent that the efficiency of the lifting method (measured in terms of production rate as a percentage of potential) drops with increasing PI.
2. However, this decrease in lift efficiency with increasing PI can be offset by increasing the cross-sectional area of the accumulation chamber. Referring to the third case of Table 8-4, if the value of a were doubled, $5.614\,Jw/a$ would

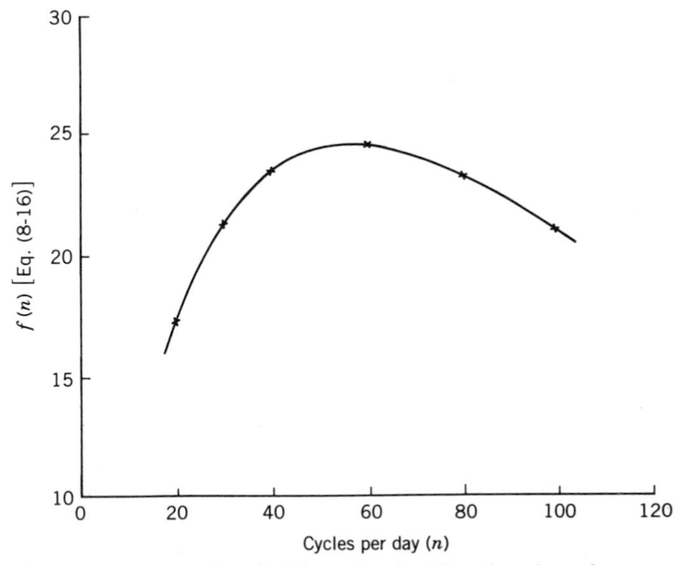

Fig. 8-5 Example 8-1: $f(n)$ from Eq. (8-16) as function of n.

TABLE 8-2 EXAMPLE 8-1: $f(n)$ as a Function of n for Values of $5.614Jw/a$ Equal to 21.42 and 85.68 ($T = 0.0045$)

		$\exp[-b(1/n - T)]$		$1 - \exp[-b(1/n - T)]$		$f(n)$	
n	$1/n - T$	21.42	85.68	21.42	85.68	21.42	85.68
20	0.0455	0.376	0.020	0.624	0.980	12.48	19.60
30	0.0288	0.538	0.084	0.462	0.916	13.86	27.48
40	0.0205	0.644	0.172	0.356	0.828	14.24	33.12
60	0.0122	0.770	0.351	0.230	0.649	13.80	38.94
80	0.0080	0.842	0.503	0.158	0.497	12.64	39.76
100	0.0055	0.889	0.624	0.111	0.376	11.10	37.60
120	0.0038	....	0.722	....	0.278	....	33.36

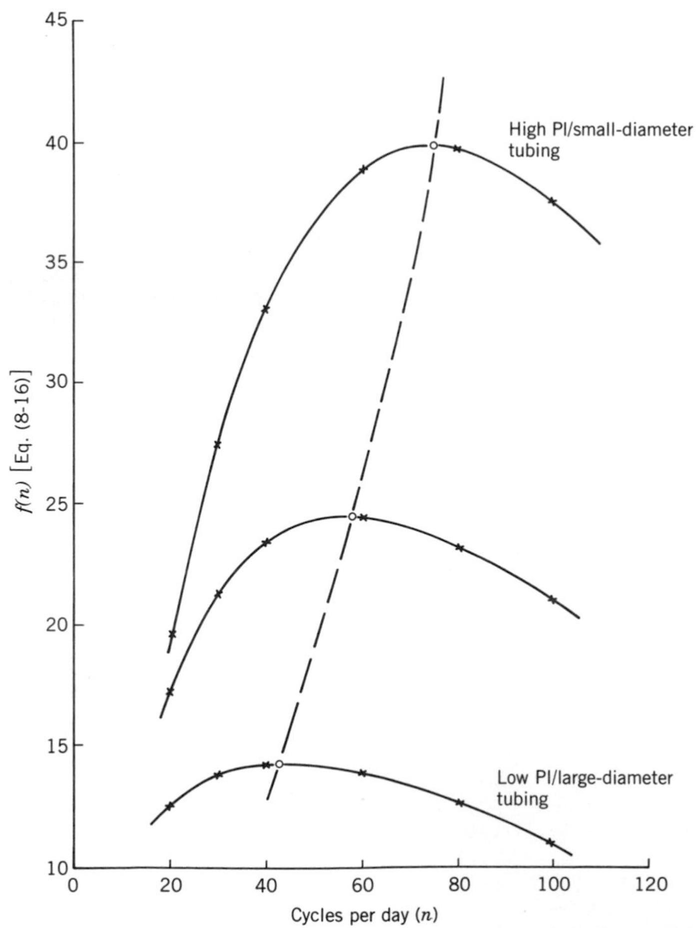

Fig. 8-6 Example 8-1: Effect of change in value of Jw/a on $f(n)$ [Eq. (8-16)].

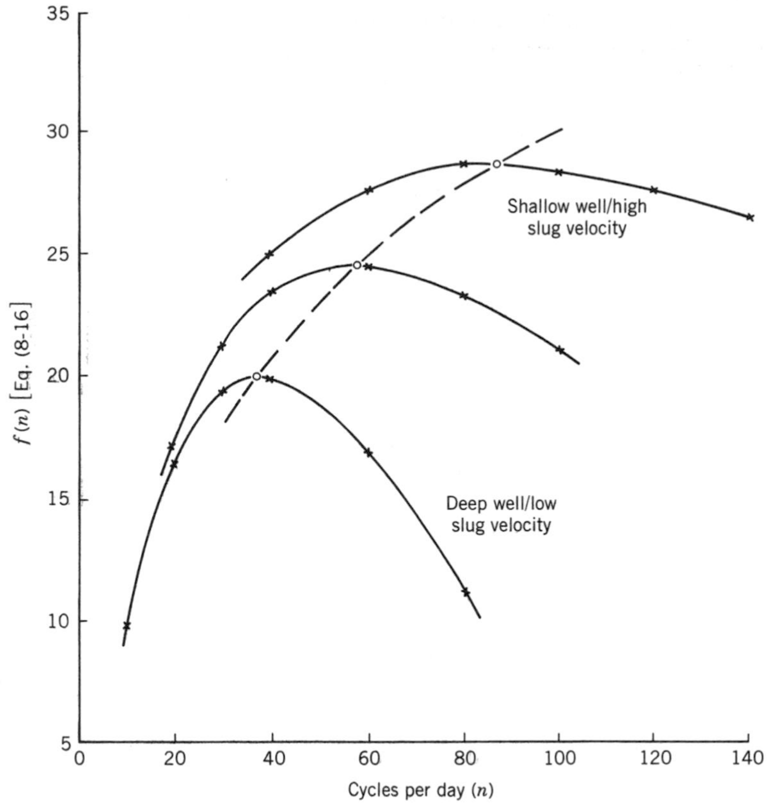

Fig. 8-7 Example 8-1: Effect of change in slug transit time on $f(n)$ [Eq. 8-16)].

TABLE 8-3 EXAMPLE 8-1: $f(n)$ as a Function of n for Values of T Equal to 0.00225 and 0.0090 ($5.614Jw/a = 42.84$)

	$1/n - T$		$\exp[-b(1/n - T)]$		$1 - \exp[-b(1/n - T)]$		$f(n)$	
n	0.00225	0.0090	0.00225	0.0090	0.00225	0.0090	0.00225	0.0090
10	. . . . .	0.0910	. . . . .	0.020	. . . . .	0.980	. . . . .	9.80
20	. . . . .	0.0410	. . . . .	0.172	. . . . .	0.828	. . . . .	16.56
30	. . . . .	0.0243	. . . . .	0.353	. . . . .	0.647	. . . . .	19.41
40	0.02275	0.0160	0.376	0.504	0.624	0.496	24.96	19.84
60	0.01445	0.0077	0.538	0.718	0.462	0.282	27.72	16.92
80	0.01025	0.0035	0.644	0.861	0.356	0.139	28.48	11.12
100	0.00775	. . . .	0.717	. . . .	0.283	. . . .	28.30	
120	0.00608	. . . .	0.770	. . . .	0.230	. . . .	27.60	
140	0.00489	. . . .	0.811	. . . .	0.189	. . . .	26.46	

TABLE 8-4 EXAMPLE 8-1: Optimum Cycles per Day, Optimum Slug Size, and Optimum Production Rate as Percentage of Potential (Assuming No Changes in p_s, p_m, a, w)

J, bbl/(day) (psi)	Transit Time for Slug, min	Opt. Cycles/Day	Opt. Value of $f(n)$	Opt. Slug Size, bbl/cycle	Opt. Rate as % of Potential
0.4	6½	43	14.4	5.4	51.6
0.8	3¼	87	28.8	5.3	51.6
0.8	6½	57	24.5	6.9	43.9
0.8	13	37	19.8	8.6	35.6
1.6	6½	75	39.8	8.5	35.7

be halved and would drop from 42.84 to 21.42; that is, it would become equal to the value of $5.614\,Jw/a$ used in the first case of Table 8-4. It follows that doubling the cross-sectional area of the accumulative chamber in the third case reduces the optimum number of cycles per day from 57 to 43.

Referring to Eq. (8-14) and Table 8-4, prior to doubling the value of a the optimum production rate was

$$q_1 = \frac{0.0325 \times 859}{5.614 \times 0.31} \times 24.5$$
$$= 393 \text{ bbl/day}$$

or 393/(0.8 × 1120), that is, 43.9 percent of potential.

Doubling the value of a results in an optimum production rate of

$$q_2 = \frac{0.0650 \times 859}{5.614 \times 0.31} \times 14.4$$
$$= 462 \text{ bbl/day}$$

or 462/(0.8 × 1120), that is, 51.6 percent of potential. The optimum slug size rises from 6.9 bbl/cycle to 10.8 bbl/cycle.

This is, of course, the reason for a chamber-lift installation, particularly in those wells with a good production potential.

3. The efficiency of the operation is sensitive to the number of cycles per day when the PI is high (or the tubing diameter small), but relatively insensitive for low-productivity wells.
4. As the transit time for the slug increases, the optimum cycle frequency decreases, and the effect is quite marked. Thus, deep wells should be operated at fewer cycles per day than shallow wells.
5. The efficiency of the operation is sensitive to the number of cycles per day for deep wells (or wells in which the average slug velocity up the tubing is low), but relatively insensitive for shallow wells (or operations in which the slug velocity is high, but note the danger of gas channeling in such circumstances).

Returning now to Eq. (8-16), the optimum may be found by the standard techniques of differential calculus. Carrying out the necessary differentiation, it is found that n_{opt} is given by

$$\exp\left(\frac{b}{n_{opt}}\right) = \left(1 + \frac{b}{n_{opt}}\right)\exp(bT) \tag{8-17}$$

and the maximum production rate is then

$$q_{opt} = \frac{a(p_s - p_m)}{5.614w}\frac{b}{1 + b/n_{opt}} \tag{8-18}$$

Figure 8-8 is based on Eq. (8-17) and gives values for n_{opt} for various transit times (vertical axis) with b ($=5.614Jw/a$) as the parameter. From a knowledge of n_{opt}, the anticipated production rate may be determined by substitution into Eq. (8-18).

Example 8-2 Well 231 West Block is intermitting at 460 bbl/day from 7421 ft on $3^1/_2$-in. tubing (internal cross-sectional area of 0.049 sq ft). The intermitting frequency is 40 cycles/day, and the gas-injection time required to bring the slug to the surface on each cycle is 8 min. The static pressure of the formation is

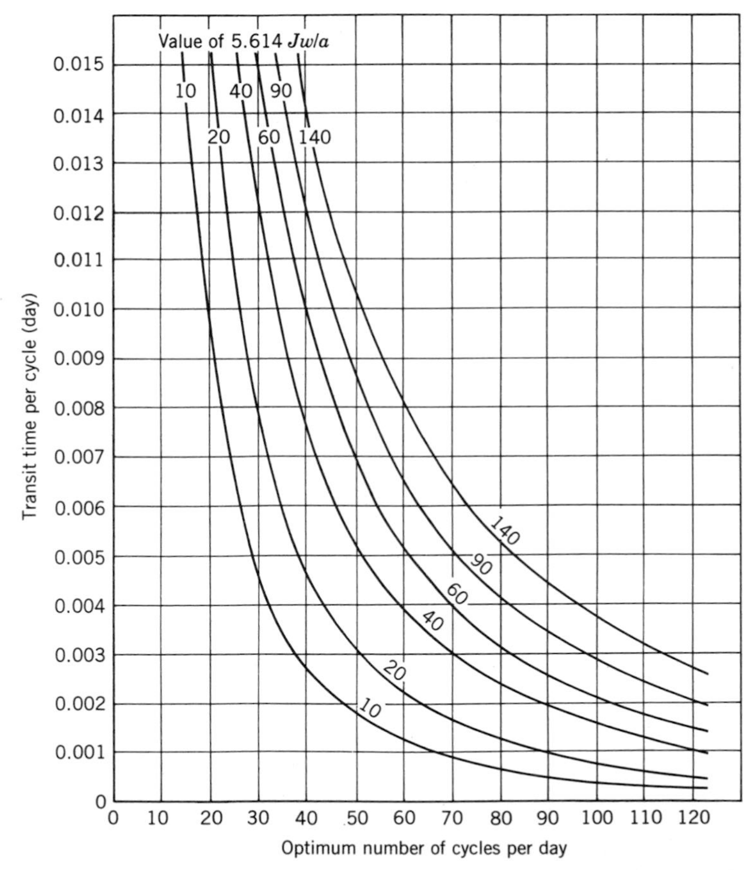

Fig. 8-8 Optimum number of cycles per day as function of slug transit time and of Jw/a.

1340 psig and the PI of the well is 1.3 bbl/(day)(psi). Could the production rate from the well be substantially improved by changing the number of cycles per day?

Using a figure of 0.32 psi/ft for w, Eq. (8-15) leads to

$$b = \frac{5.614 \times 1.3 \times 0.32}{0.049}$$

$$= 47.7$$

Using

$$n = 40$$
$$b = 47.7$$
$$T = 8 \text{ min} = 0.0056 \text{ day}$$

in Eq. (8-16) gives the value of $f(n)$ as 24.16. Reference to Fig. 8-8 shows that 51 cycles/day is an optimum for the b and T values for the well, and with n equal to 51, $f(n)$ is 24.89. Assuming that p_m would remain unchanged at the higher number of cycles per day, the optimum production rate (on 51 cycles) would be

$$\frac{24.89}{24.16} \times 460 \qquad \text{or} \qquad 474 \text{ bbl/day}$$

which is a marginal increase only.

It is of interest to note that the fallback may be calculated from the data given. From Eq. (8-14), the value of $p_s - p_m$ for the actual operations is determined. This is

$$\frac{5.614 \times 0.32 \times 460}{0.049 \times 24.16} \qquad \text{or} \qquad 700 \text{ psi}$$

But $p_s = 1340$ psig, so that $p_m = 640$ psig. If the line pressure is 100 psig, this implies a fallback pressure of 540 psig, which is equivalent to a column of liquid of length 1690 ft in the tubing, or a fallback volume of 14.7 bbl.

Example 8-3 A well producing from 14,020 ft is making 120 bbl/day on intermittent gas lift, 40 cycles/day, through $2^7/_8$-in. tubing (internal cross-sectional area 0.0325 sq ft). The gas-injection time per cycle is 20 min. It is estimated that the static pressure of the producing horizon is 1450 psig and that the PI is 0.3 bbl/(day)(psi). Could the production rate from the well be substantially improved by changing the number of cycles per day?

Using a figure of 0.32 psi/ft for w, calculations similar to those shown in Example 8-2 give

$$b = 16.6$$
$$T = 0.0139$$
$$f(n) = 6.72 \text{ as the well is being operated [Eq. (8-16)]}$$
$$\text{Opt. } n = 19 \text{ (Fig. 8-8)}$$
$$\text{Opt. } [f(n)] = 9.03 \text{ Eq. [(8-16)]}$$
$$\text{Opt. } q = \frac{9.03}{6.72} \times 120 = 161 \text{ bbl/day}$$

so that an increase of 41 bbl/day (34.2 percent) in production rate is possible by decreasing the number of cycles per day to the optimum of 19.

Again, it is of interest to calculate the value of p_m (as in the preceding example) from the operating value of 6.72 for $f(n)$.

$$p_s - p_m = 987 \text{ psi}$$
$$p_m = 1450 - 987 = 463 \text{ psig}$$

If the (minimum) line pressure is 100 psig, the fallback pressure must be 363 psig, so that the length of the fallback liquid column in the tubing is 1133 ft, equivalent to 6.6 bbl of fallback liquid.

8-5 GLR AS FUNCTION OF SLUG SIZE

Let the minimum THP, which will occur during the accumulation of the slug at the bottom of the hole, be P_1 psi and let the maximum THP, which will occur while the liquid is actually being produced at the surface, be P_2 psi. Let the length of the tubing be D ft and its cross-sectional area be a sq ft, and let the length of the accumulation chamber (of cross-sectional area $\bar{a}$ sq ft) at the foot of the tubing be $\bar{D}$ ft.

The volume of each liquid slug as it reaches the surface is q/n bbl, that is, 5.614 q/n cu ft. This slug, together with the THP during production, exerts a back pressure of amount

$$P_2 + \frac{5.614q}{an} w \qquad \text{psi}$$

on the gas that is driving it to the surface.

On the other hand, when the slug is at the bottom of the hole and just before it starts to move up the tubing, its volume is $5.614(q/n + Q_m)$ cu ft when Q_m bbl represents the fallback at each cycle. Hence the pressure immediately below the slug is

$$p_g = P_1 + \frac{5.614w}{a}\left(\frac{q}{n} + Q_m\right) \qquad \text{psi} \tag{8-19}$$

and the gas-injection pressure must be slightly in excess of this.

It has been found (see, for example, Uren, Ref. 3, page 212; Beeson et al., Ref. 4; and Garrett Oil Tools, Ref. 5, page 307) that a reasonable rising velocity for the liquid slug in the tubing is of the order of 1000 ft/min and that in order to achieve this rate, the gas pressure exerted on the slug from below should exceed the back pressure due to the slug itself by an amount roughly equal to 33 to 40 percent of the pressure due to the slug. In this analysis a figure of 33 percent will be used. This implies a gas pressure behind the slug, as it reaches the surface, of amount

$$p_g = P_2 + \frac{4}{3} \times \frac{5.614w}{a}\frac{q}{n}$$
$$= P_2 + \frac{7.48wq}{an} \tag{8-20}$$

Assuming no change in gas-injection pressure as the slug moves to the surface, Eqs. (8-19) and (8-20) permit P_2 to be estimated if Q_m is known.

If V cu ft is the volume occupied by gas in the well when the liquid slug has just reached the surface, then this volume expressed at standard conditions is

$$\frac{1}{14.7}\left(P_2 + \frac{7.48qw}{an}\right)V \qquad \text{scf}$$

neglecting temperature variations and supercompressibility. Immediately after the liquid slug has been produced, gas production takes place until the THP has dropped to P_1. Thus the volume of gas in the well after the slug's production is

$$\frac{1}{14.7}P_1V \qquad \text{scf}$$

The volume of gas produced with q/n bbl of liquid is therefore

$$\frac{1}{14.7}\left(P_2 - P_1 + \frac{7.48qw}{an}\right)V \qquad \text{scf}$$

and the producing GLR is

$$R = \left[(P_2 - P_1)\frac{n}{14.7q} + 0.51\frac{w}{a}\right]V \qquad \text{cu ft/bbl} \tag{8-21}$$

In intermittent-gas-lift installations, the volume V is aD if the valves are set to close as soon as the pressure starts to bleed down after the slug has been produced. In chamber lift, the volume is $(aD + \bar{a}D)$, while in natural (that is, without packer) plunger lift, V is $(a + A)D$, where A sq ft is the cross-sectional area of the casing-tubing annulus.

Example 8-4 Using the data of Example 8-1 determine the volume of gas that would be needed daily to intermit the well at the optimum number (57) of cycles per day. Also determine the value of the THP during production of a slug at the surface.

Taking the latter part of the question first, Eqs. (8-19) and (8-20) give

$$P_2 = P_1 + \frac{5.614wq}{an} + \frac{5.614wQ_m}{a} - \frac{7.48wq}{an}$$

since in this example $\bar{a} = a$ (that is, no special accumulation chamber). Inserting the given values

$$\begin{aligned} P_2 &= 100 + 161 - \frac{1.866 \times 0.31}{0.0325} \times \frac{393}{57} \\ &= 100 + 161 - 123 \\ &= 138 \text{ psig} \end{aligned}$$

In Eq. (8-21)

$$R = \left[38 \times \frac{57}{14.7 \times 393} + 0.51 \times \frac{0.31}{0.0325}\right] \times 5872 \times 0.0325 \text{ cu ft/bbl}$$
$$= [0.375 + 4.865] \times 5872 \times 0.0325 \text{ cu ft/bbl}$$
$$= 1000 \text{ cu ft/bbl}$$

so the gas that must be circulated to lift 393 bbl/day is 393,000 cu ft/day.

Typical curves of GLR per 1000 ft of lift as a function of liquid production per cycle (q/n) are shown in Fig. 8-9. Values used in preparing these curves are as follows (the expression $aD + \overline{aD}$ has been approximated by aD in presenting the chamber-lift curve):

Size of tubing	$2^7/_8$-in. OD	(a = 0.0325 sq ft[1])
Size of casing	$4^1/_2$-in.	(A = 0.0448 sq ft[1])
	$5^1/_2$-in.	(A = 0.0848 sq ft[1])
	7-in.	(A = 0.1748 sq ft[1])
Trap pressure (P_1)		50 psig
Maximum THP (P_2)		100 psig
Liquid specific gravity		1.0 (w = 0.433)

Figure 8-9 and Eq. (8-21) illustrate several points:

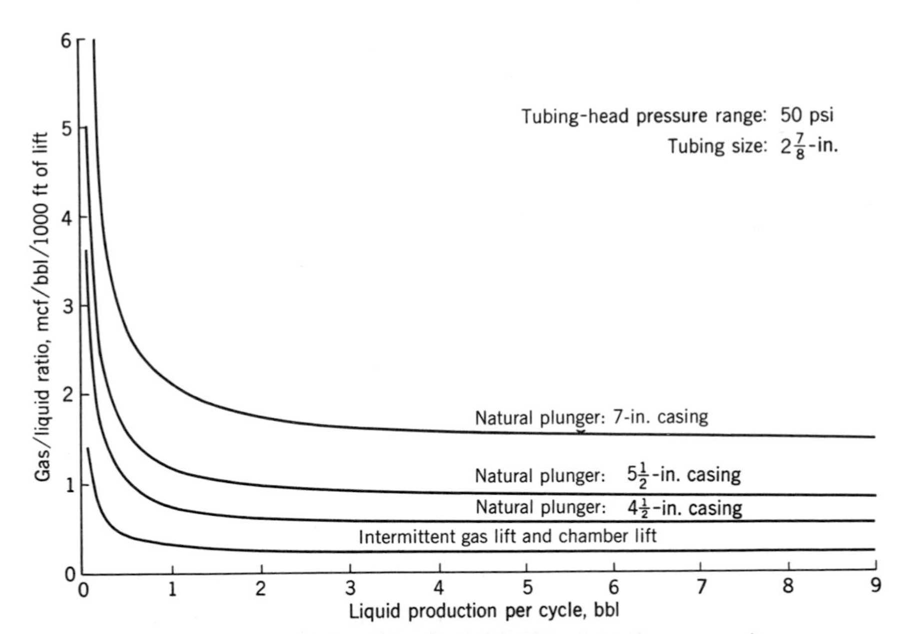

Fig. 8-9 Intermittent lift: GLR as function of liquid production per cycle.

[1] These are average figures; the precise values will depend upon the weight of the tubing and casing.

1. The required GLR falls as the liquid production per cycle increases. In other words, for a given daily liquid rate, gas requirements decrease as the cycle frequency n decreases.
2. On the other hand, it must be remembered that Fig. 8-9 was prepared for a THP range of 50 psi. The form of Eq. (8-21) makes it obvious that the bigger the THP range, the higher the GLR; high fallback will lead to a higher THP range.
3. Natural plunger lift has the higher gas requirements and becomes less and less efficient with regard to gas usage as the annulus size increases (that is, as the casing diameter increases).
4. As tubing diameter is increased, the GLR requirements for intermittent gas lift and chamber lift are also increased. This follows from Eq. (8-21) on substituting aD for V, so that the equation becomes

$$R = (P_2 - P_1)\frac{n}{14.7q}aD + 0.51wD$$

For natural plunger lift, V is equal to $(A + a)D$ so that Eq. (8-21) becomes

$$R = (P_2 - P_1)\frac{n}{14.7q}(A + a)D + 0.51w\left(\frac{A}{a} + 1\right)D$$

As the tubing area is increased, the annulus area is decreased to approximately the same extent, so that $A + a$ remains roughly constant, whereas $A/a + 1$ decreases. Thus for natural plunger lift, the larger the tubing size, the lower the GLR requirement.
5. Even in $4^1/_2$-in. casing with $2^7/_8$-in. tubing, where the annulus cross-sectional area is less than $6^1/_2$ sq in., the minimum GLR requirement for natural plunger lift is about 500 cu ft/bbl with a THP variation of 50 psi. Since these are, to all intents and purposes, minimum requirements, it may be said that natural plunger lift needs a GLR of at least 500 cu ft/bbl per 1000 ft of lift.

8-6 NATURAL PLUNGER LIFT

Although not widely used because of the restricted range of conditions under which such an installation is suitable and because it requires a completion in which the casing-tubing annulus is not packed off, there are circumstances in which this production method is efficient and cheap to operate. Moreover, the analysis is somewhat different from that of Sec. 8-4 because the formation may be expected to be producing throughout the cycle; that is, f of Eq. (8-9) is equal to zero.

Let v be the average velocity (ft/min) of the plunger in the tubing on its upward journey; the average velocity of fall is then of the order of $1.5v$ (Ref.

4).[2] The number of cycles per day n is given by the equation

$$n\left(\frac{D}{v}+\frac{D}{1.5v}\right)=1440$$

or

$$n=\frac{864v}{D} \tag{8-22}$$

The plunger will start its upward journey from bottom (Fig. 8-2) as soon as the pressure in the annulus exceeds the pressure due to the plunger itself, the liquid above the plunger, and the THP. Moreover, the pressure behind the plunger will increase steadily as the plunger moves up the hole because of the continued production from the formation. As an average—see the discussion of Sec. 8-5—it will be assumed that the pressure below the plunger exceeds the downward pressure by an amount equal to 33 percent of the pressure exerted by the liquid slug itself. Moreover, under a plunger-lift operation the fallback should be very small. It follows that, in place of Eqs. (8-19), (8-20), or (8-5), the (minimum) pressure on the upstroke may be taken as

$$P_1+\frac{7.48qw}{an}$$

while the time for the upstroke is $D/1440v$, expressed as a fraction of a day.

Substituting in Eq. (8-8), the liquid production into the tubing while the plunger is on the upstroke is

$$\frac{a}{5.614w}\left(p_s-P_1-\frac{7.48qw}{an}\right)\left[1-\exp\left(-5.614\frac{Jw}{a}\frac{D}{1440v}\right)\right]$$

To a reasonable degree of approximation, provided x is less than 0.25, $[1-\exp(-x)]$ may be replaced by x. Using typical values for a plunger-lift installation, say $w=0.33$, $a=0.0325$, $D=4000$, $v=1000$, the value of $5.614JwD/1440av$ is about $0.16J$, so that provided the PI is not greater than 1.5 bbl/(day)(psi), the liquid production into the tubing on the upstroke may be approximated by

$$\frac{JD}{1440v}\left(p_s-P_1-\frac{7.48qw}{an}\right)$$

At the start of the downstroke this volume of liquid is present in the tubing, so, from Eq. (8-5), the minimum back pressure (at the start of the downstroke) is

$$P_1+5.614\frac{w}{a}\frac{JD}{1440v}\left(p_s-P_1-\frac{7.48qw}{an}\right)$$

[2] Since the plunger falls freely, the average downward velocity may be expected to be greater than the average upward velocity.

This expression, used in Eq. (8-8), leads to a value for the liquid production into the tubing on the downstroke. The approximation for $[1 - \exp(-x)]$ may be applied, and the values for the liquid production on the upstroke and that on the downstroke added to give q/n, the production per cycle. Carrying out these steps, it is found that, provided $5.614JwD/1440av$ is less than 0.25,

$$\frac{q}{n} = \frac{JD}{1440v}\left(p_s - P_1 - \frac{7.48qw}{an} + \frac{p_s}{1.5} - \frac{P_1}{1.5}\right)$$

to a reasonable degree of accuracy. Applying Eq. (8-22), it follows that

$$q = \frac{864}{1440}J\left(\frac{2.5}{1.5}p_s - \frac{2.5}{1.5}p_1 - \frac{7.48qw}{an}\right)$$

Rearrangement of this equation leads to

$$q = \frac{J(p_s - P_1)}{1 + KJ} \tag{8-24}$$

where

$$K = \frac{4.5w}{an} \tag{8-25}$$

If p_c is the average CHP and if the weight of the static gas column is neglected,

$$q = J(p_s - p_c)$$

or

$$p_s = p_c + \frac{q}{J}$$

which, substituted in the right-hand side of Eq. (8-24), gives

$$Kq = p_c - P_1 \tag{8-26}$$

that is, the daily liquid capacity of natural plunger lift is proportional to the difference between the average CHP and the trap pressure. This pressure difference is known as the *net operating pressure* p_0.

The factor K, as shown by Eq. (8-25), depends on the tubing size and on the number of cycles per day [or, from Eq. (8-22), on the tubing depth if a constant average plunger velocity is used].

Substituting from Eq. (8-22) in Eq. (8-25) and using a value of v of 1000 ft/min and a value of w of 0.433 psi/ft (water),

$$K = \frac{0.00226D}{a} \quad \text{psi/(bbl)(day)} \tag{8-27}$$

where the depth D is measured in thousands of feet. For example, using 6000 ft of $2\frac{7}{8}$-in. tubing, K is equal to 0.42 psi/(bbl)(day).

The potential q' of the well is Jp_s, so the natural-plunger-lift production rate, expressed as a percentage of the potential, is, from Eq. (8-24),

$$\frac{p_s - P_1}{p_s(1 + KJ)} \times 100$$

TABLE 8-5 Natural-Plunger-Lift Analysis: Theoretical Values of *K* for Various Tubing Sizes

Tubing Size, in.	Av. Internal Area, sq ft	*K* Divided by Depth, 1000 ft
1.9	0.0141	0.160
2⅜	0.0217	0.104
2⅞	0.0325	0.070
3½	0.0489	0.047
4	0.0687	0.033
4½	0.0853	0.027

If P_1 is regarded as negligible compared with p_s, the natural plunger-lift production rate, expressed as a percentage of the potential, is

$$\frac{100}{1 + KJ}$$

This expression may be used in conjunction with Eq. (8-27) to prepare a chart for estimating the effectiveness of natural plunger-lift applications (see Tables 8-5 and 8-6 and Fig. 8-10). However, although the picture presented in Fig. 8-10 is probably qualitatively valid, the many assumptions introduced into the theory make quantitative answers suspect. Particular note should be taken of the following points:

1. An upward plunger velocity of less than 1000 ft/min (and/or a downward velocity of less than 1500 ft/min) decreases the number of cycles per day from that assumed in Eq. (8-27). This increases the value of *K* corresponding to a certain tubing size and depth and reduces the slopes of the

TABLE 8-6 Natural-Plunger-Lift Analysis: Theoretical Determination of Lift Efficiency [Values of 100/(1 + *KJ*)]

	J, bbl/(day)(psi)			
K, psi/(bbl)(day)	0.1	0.25	0.5	1.0
0.1	99.0	97.6	95.2	90.9
0.2	98.0	95.2	90.9	83.3
0.3	97.1	93.0	87.0	77.0
0.4	96.2	90.9	83.3	71.4
0.5	95.2	88.9	80.0	66.7
0.6	94.3	87.0	77.0	62.5
0.7	93.5	85.1	74.1	58.8
0.8	92.6	83.3	71.4	55.6
0.9	91.7	81.6	69.0	52.6
1.0	90.9	80.0	66.7	50.0

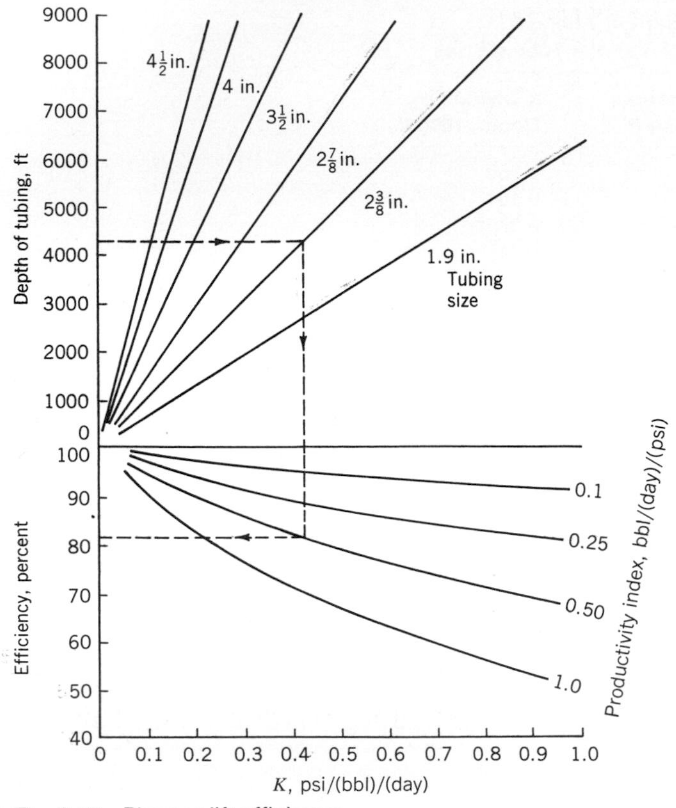

Fig. 8-10 Plunger-lift efficiency.

lines in the upper part of Fig. 8-10 and therefore (and obviously) reduces the efficiency of the plunger lift.

2. The trap pressure P_1 may not be negligible compared with the static BHP p_s. If P_1 were 20 percent of p_s, the natural-plunger-lift production rate expressed as a percentage of the potential would be

$$\frac{80}{1 + KJ}$$

and the natural-plunger-lift production rates would be 80 percent of those indicated by Table 8-6 and Fig. 8-10.

Thus, on each of two very real counts the curves of Fig. 8-10 tend to overestimate the effectiveness of natural plunger lift.

Conclusions relating to natural plunger lift as a producing mechanism are:

1. Its efficiency decreases with depth (Fig. 8-10).
2. Its efficiency decreases with increasing well PI (Fig. 8-10). It is suited to conditions in which the PI is below, say, 0.5 bbl/(day)(psi).

3. Its efficiency increases as the tubing size is increased (Fig. 8-10); moreover, increasing the tubing size decreases the GLR requirement (Sec. 8-5).
4. The minimum gas requirement is some 500 cu ft/bbl per 1000 ft of lift, and this increases as the casing size increases (Sec. 8-5).
5. It is not a final depletion mechanism. As the static BHP drops, the trap pressure becomes a larger and larger percentage of this static pressure, thus reducing the efficiency [Eq. (8-24)].

REFERENCES

1. Brown, Kermit E.: *Gas Lift Theory and Practice,* The Petroleum Publishing Co., Tulsa, Okla., 1973.
2. Nind, T. E. W.: "A Study of Chamber Lift," *Trans. Can. Inst. Mining and Metallurgy,* **63:**310 (1960).
3. Uren, L. C.: *Petroleum Production Engineering: Oil Field Exploitation,* 3d ed., McGraw-Hill Book Company, Inc., New York, 1953.
4. Beeson, Carrol M., Donald G. Knox, and John H. Stoddard: "Plunger-Lift Correlation Equations and Nomographs," *30th Annual Fall Meeting, Petroleum Branch,* AIME, Dallas, Tex., Paper 501-G, 1955.
5. *Handbook of Gas Lift,* Garrett Oil Tools, Division of U.S. Industries, Inc., New York, 1959.

Sucker Rod Pumping

9

9-1 INTRODUCTION

In this chapter and the two that follow some aspects of sucker rod pumping and of the instruments used to analyze pumping well behavior are discussed. Many different pumping systems are in current use, for example, conventional sucker rod pumping, long-stroke pumping, and hydraulic, centrifugal, and sonic pumping, but no attempt is made to deal here with any other method than conventional sucker rod pumping. This method, so common and mechanically so simple, has many features of interest, and there remain several questions still not satisfactorily resolved, such as, for example, the determination of the pump-setting depth and size of plunger that will lead to the maximum pumping rate in a particular well. Another interesting problem is that of gas anchors (bottom-hole separators); yet another is the interpretation of the results obtained from devices for determining the depth of the fluid level in the annulus during pumping.

The need for a greater understanding of the pumping process and for the ability to engineer pump installations to a close degree of tolerance is being accentuated by the trend to pump from greater and greater depths. Problems that are relatively minor when the pump is set at 2000 ft may become extremely important when the pump is at 10,000 ft.

In the present chapter a critical survey is made of the mechanical system and of the results that may be expected from the operation of such a system. Chapter 10 contains a summary of the principal instruments used to analyze pumping-well performance, while Chap. 11 addresses some of the questions to which the answers are still outstanding.

9-2 THE PUMPING CYCLE

In its simplest form the pump consists of a working barrel or liner suspended on the tubing; the plunger is moved up and down inside this barrel by the sucker rod string, which consists of a series of screwed steel rods attached at the surface to the pumping unit. The unit and prime mover at the surface supply the oscillating motion to the sucker rod string and so to the pump. At the bottom of the working barrel is a stationary ball-and-seat valve—the *standing valve* (SV)—while a second ball-and-seat valve—the *traveling valve* (TV)—is located in the plunger.

The principal features of the pumping cycle are illustrated in Fig. 9-1, and the four sketches included in this figure refer to the situations described below.

Plunger Moving Down, Near the Bottom of the Stroke

Fluid is moving up through the open TV while the weight of the fluid column in the tubing is supported on the SV, which is consequently closed (if the flowing BHP were greater than the fluid column weight, the SV would be

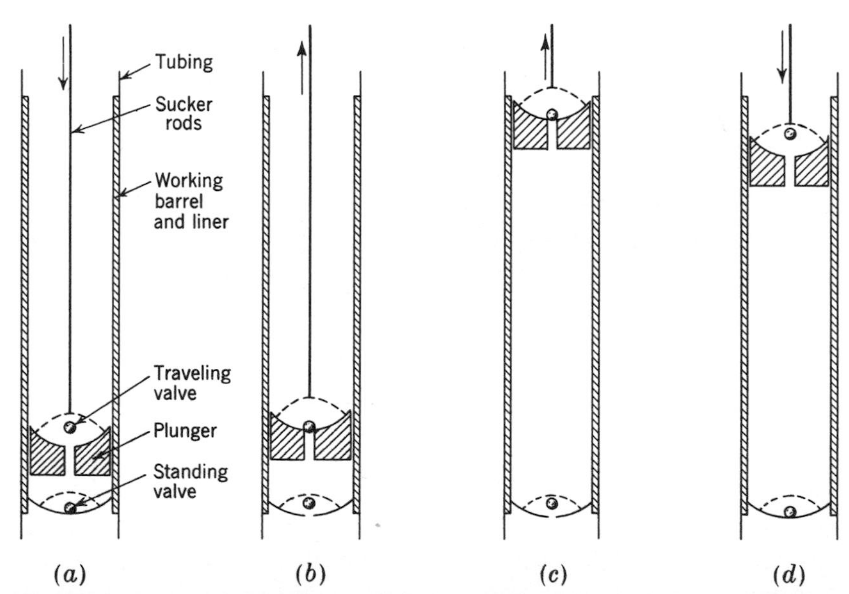

Fig. 9-1 The pumping cycle: (*a*) plunger moving down, near bottom of stroke; (*b*) plunger moving up, near bottom of stroke; (*c*) plunger moving up, near top of stroke; (*d*) plunger moving down, near top of stroke.

open even when the plunger was near the bottom of its stroke, and the well would be flowing, or possibly agitating; see Sec. 11-3).

Plunger Moving Up, Near the Bottom of the Stroke

The TV is now closed; consequently, the load due to the fluid column has been transferred from the tubing to the rod string. The SV (shown as open in Fig. 9-1*b*) opens as soon as the pressure below it exceeds that above; the position on the upstroke at which this occurs depends on the pump *spacing,* that is, on the volume included between the SV and TV at the bottom of the stroke, and on the percentage of free gas in the fluid trapped in this volume.

Plunger Moving Up, Near the Top of the Stroke

If there is any pumping production at all from the well, the SV must be open by this time, permitting the formation to produce into the tubing. The TV is closed.

Plunger Moving Down, Near the Top of the Stroke

The standing valve is closed by the increased pressure resulting from the compression of the fluids in the volume between the SV and TV. The TV is shown as open in Fig. 9-1*d*, but the point of the downstroke at which it opens depends on the percentage of free gas in the trapped fluids, since the pressure below the valve must exceed that above (that is, the pressure due to the fluids in the tubing above the plunger) before the TV will open.

9-3 SURFACE AND SUBSURFACE EQUIPMENT

A typical arrangement of surface equipment is illustrated in Fig. 9-2. The rotary motion of the *crank arm* is converted into an oscillatory motion by means of the *walking beam.* The *horse's head* and *hanger cable* arrangement is used to ensure that the pull on the sucker rod string is vertical at all times, so that no bending moment is applied to that part of the string above the stuffing box. The *polished rod* and *stuffing box* combination is used to maintain a good liquid seal at the surface.

Conventional beam-pumping units of the type shown in Fig. 9-2 are available in a wide range of sizes, with stroke lengths varying from 12 to almost 200 in. The stroke length for any particular unit is variable within limits, about six different lengths being possible as a rule; these are achieved by varying the position of the *pitman* connection in the crank arm. Walking beam ratings, expressed in maximum allowable polished rod loads (PRLs), vary from some 3000 to 35,000 lb. Other types of units, for example, *air-balanced units* and

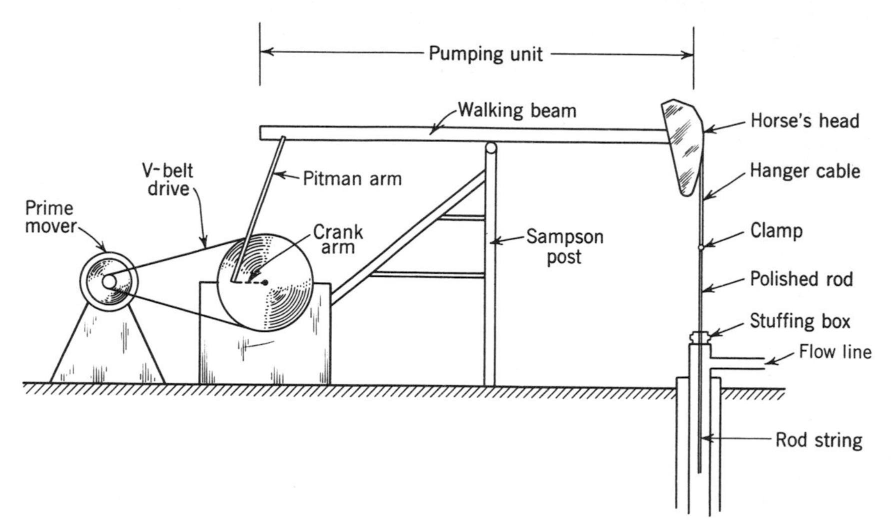

Fig. 9-2 Typical surface installation (diagrammatic).

long-stroke hydraulic units, are also available, each having ranges of application for which it is particularly advantageous (see Sec. 9-4 for mention of some of the outstanding features and special applications of the air-balanced unit).

Counterbalance (Sec. 9-5) for conventional beam units is accomplished by placing weights directly on the beam in the smaller units, or by attaching weights to the rotating crank arm or by a combination of the two methods in the larger units. In more recent designs, the rotary counterbalance can be adjusted by shifting the position of the weight on the crank arm by a jackscrew or rack and pinion.

The two basic types of *prime mover* are electric motors and internal combustion engines. The modern tendency is to equip each well with its own motor, although some multiple-power systems are still in use; in these, power is developed at a central plant and transmitted to the wells by means of reciprocating shackle lines. The main advantages of electric motors over gas engines lie in lower initial and maintenance costs, dependable all-weather service, and the ease with which they can be fitted into an automatic system; on the other hand, gas engines have the advantages of more flexible speed control, operation over a wider range of load conditions, and inexpensive fuel (frequently casinghead gas).

The main constituent of all *sucker rods* is iron, which constitutes over 90 percent of the entire composition of the rod. However, iron in its pure state is soft and weak, so various other elements are added to improve the strength, hardness, and corrosion resistance of the rods. Among these alloying elements special mention may be made of carbon (for increasing strength, hardness, and susceptibility to heat treatment), manganese or silicon (to cut

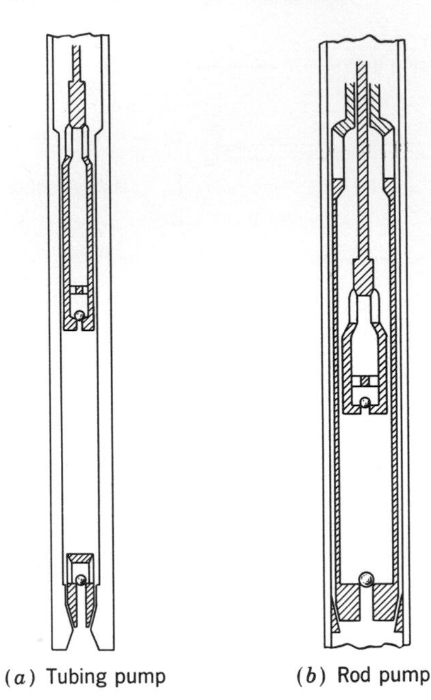

(*a*) Tubing pump (*b*) Rod pump

Fig. 9-3 Typical subsurface pumps.

down on the formation of iron oxides, which weaken the alloy), nickel (to combat corrosive conditions), molybdenum (for increasing strength), and copper (for resistance to atmospheric and other corrosive environments). In general, it is found that maximum polished rod stresses should not be permitted to rise above about 30,000 psi, this value decreasing as the corrosive properties of the fluid to be pumped increase. This figure of 30,000 psi is based on laboratory tests involving a so-called *infinite life* of 10 million stress reversals, but it should be noted that at a fairly typical pumping speed of 15 strokes/min, 10 million stress reversals are achieved in about 15 months. Sucker rods are available in various sizes, the standards being $^{5}/_{8}$-, $^{3}/_{4}$-, $^{7}/_{8}$-, 1-, and $1^{1}/_{8}$-in. diameter (Ref. 1).

Subsurface pumps are of two main types (see Fig. 9-3), although there are many variations (Ref. 2). The basic designs are the *tubing pump* and the *rod pump;* the advantage of the latter is that the entire pump assembly, including liner and SV, is run on the rod string; replacement and repair then constitute a relatively inexpensive operation since it is not necessary to pull the tubing. The disadvantage of the rod pump compared with the tubing pump is that the plunger diameter must be smaller (for a given tubing size), which cuts down on the pump's capacity. Plunger diameters may vary between $^{5}/_{8}$ in. and $4^{3}/_{4}$ in., the plunger area varying from 0.307 to 17.721 sq in.

Typical sucker rod *pumping speeds* vary from 4 to 40 strokes/min, depending on a variety of well and fluid properties.

9-4 MOTION OF THE POLISHED ROD

In Fig. 9-4 two typical arrangements of the pitman arm relative to the sampson post and polished rod are illustrated, corresponding to the conventional and the air-balanced units. Since the crank may be assumed to rotate at a constant angular velocity, the point of connection between the pitman arm and the crank generates simple harmonic motion in the vertical direction, so that the motion of the horse's head, and of the polished rod, is a modification of simple harmonic. The variations from true harmonic motion are due to the geometry of the lever system and differ from unit to unit. Referring to Fig. 9-4, the following two points should be noted in particular:

1. In the *conventional-type unit* the acceleration at the bottom of the stroke is somewhat greater than true simple harmonic acceleration, whereas it is less at the top of the stroke. Herein lies one of the major drawbacks of the conventional unit, namely, that at the bottom of the stroke, just at the time the traveling valve is closing and the fluid load is being transferred to the rods, the acceleration force on the rods is at its maximum. These two factors combine to create a maximum stress on the rod system that is one

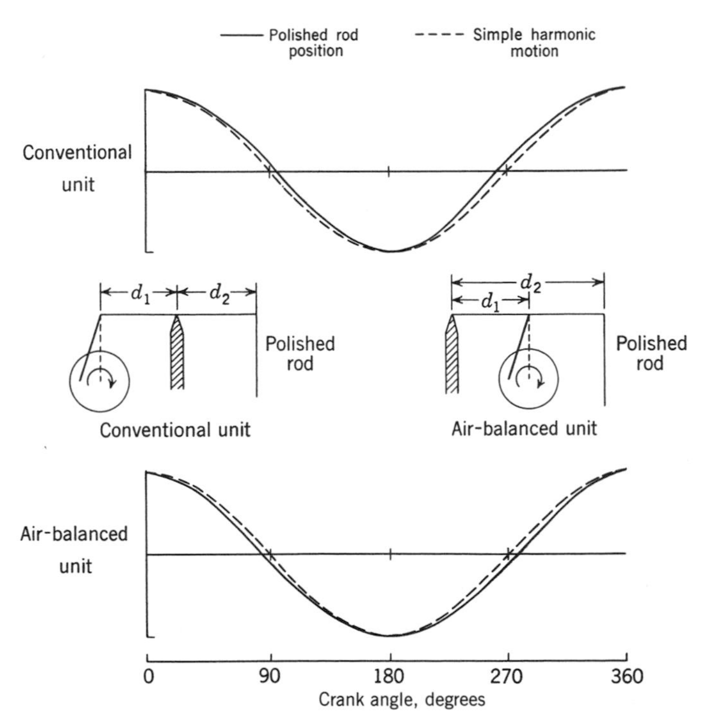

Fig. 9-4 Polished rod motions: conventional and air-balanced units.

of the limiting factors in installation design, as will be seen below (Sec. 9-5).

2. The *air-balanced unit* lever system causes the maximum acceleration to occur at the top of the stroke, whereas the acceleration at the bottom of the stroke is less than that under simple harmonic motion. Thus a lower maximum stress is set up in the rod system by using an air-balanced rather than a conventional-type unit, other factors being equal.

An approximate analysis will now be presented that assumes that the center of the crank is vertically beneath the point of connection between the pitman arm and the walking beam and that this point of connection moves in a vertical straight line rather than in the arc of a circle, which is the case in practice. Referring to Fig. 9-5 for definition of the various points and angles, it is seen that

$$(AB)^2 = (OA)^2 + (OB)^2 - 2(OA)(OB)\cos AOB \tag{9-1}$$

where AB = length of pitman arm, say h
OA = length of crank arm, say c
OB = distance from crank center O to pitman arm–walking beam connection B

If x denotes the distance of B below its top position C and if time is measured from the instant at which the crank arm and pitman arm are in the vertical position with the crank arm vertically upward, then Eq. (9-1) gives

$$h^2 = c^2 + (h + c - x)^2 - 2c(h + c - x)\cos \omega t$$

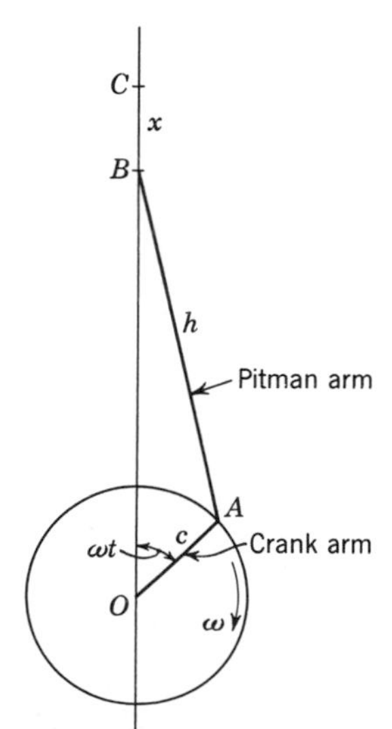

Fig. 9-5 Approximation to motion of point of connection between pitman arm and walking beam.

where ω is the angular velocity of the crank. This equation reduces to

$$x^2 - 2x[h + c(1 - \cos \omega t)] + 2c(h + c)(1 - \cos \omega t) = 0$$

so that

$$x = h + c(1 - \cos \omega t) \pm [c^2 \cos^2 \omega t + (h^2 - c^2)]^{1/2}$$

When ωt is zero, x is also zero (Fig. 9-5), which means that the negative root sign must be taken, so that finally

$$x = h + c(1 - \cos \omega t) - [c^2 \cos^2 \omega t + (h^2 - c^2)]^{1/2} \tag{9-2}$$

To find the acceleration of point B, Eq. (9-2) must be twice differentiated with respect to time. Carrying out these steps, it is found that the maximum acceleration occurs when ωt is equal to zero (or an even multiple of π radians) and that this maximum value is

$$\omega^2 c \left(1 + \frac{c}{h}\right)$$

It also appears that the minimum value of the acceleration (or the maximum value in the direction OC, as opposed to the direction CO) occurs when ωt is π radians or any odd multiple of π radians. In this position the acceleration is

$$\omega^2 c \left(1 - \frac{c}{h}\right)$$

If N is the number of pumping strokes per minute,

$$\omega = \frac{2\pi N}{60} \qquad \text{rad/sec}$$

so that the maximum downward acceleration of point B, which occurs when the crank arm is vertically upward, is

$$\frac{cN^2}{91}\left(1 + \frac{c}{h}\right) \qquad \text{ft/sec}^2$$

or

$$\frac{cN^2}{2940}\left(1 + \frac{c}{h}\right) \times g$$

Similarly, the maximum upward acceleration of point B, which occurs when the crank arm is vertically downward, is

$$\frac{cN^2}{2940}\left(1 - \frac{c}{h}\right) \times g$$

It follows that in a conventional pumping unit the maximum (upward) acceleration of the horse's head occurs at the bottom of the stroke and is equal to

$$\frac{d_2}{d_1}\frac{cN^2}{2940}\left(1 + \frac{c}{h}\right) \times g$$

where d_1 and d_2 are the distances shown in Fig. 9-4*a*. But $2cd_2/d_1$ is equal to the polished rod stroke length S, so if S is measured in inches,

$$c\frac{d_2}{d_1} = \frac{S}{24} \quad \text{ft}$$

Substituting this value, the maximum polished rod acceleration occurs at the bottom of the stroke in the conventional unit and is equal to

$$\frac{SN^2}{70{,}500}\left(1 + \frac{c}{h}\right) \quad g$$

or, maximum acceleration (at bottom of stroke) equals

$$\frac{SN^2M}{70{,}500} \times g \tag{9-3}$$

where M is the so-called *machinery factor* (Ref. 3) defined by

$$M = 1 + \frac{c}{h} \tag{9-4}$$

Similarly, maximum downward acceleration (at top of stroke in the conventional unit) equals

$$\frac{SN^2}{70{,}500}\left(1 - \frac{c}{h}\right) \times g \tag{9-5}$$

In the air-balanced unit, because of the arrangement of the levers, the acceleration defined by Eq. (9-5) occurs at the *bottom* of the stroke, whereas that defined by Eq. (9-4) occurs at the *top*. (With the lever system in the air-balanced unit, the polished rod is at the top of its stroke when the crank arm is vertically upward; see Fig. 9-4*b*.)

9-5 APPROXIMATE ANALYSIS OF SUCKER ROD PUMPING: SINGLE ROD SIZE

In the approximate analysis that follows no allowance will be made for the reflected loads that are continuously transmitted up and down the sucker rod string and are superimposed as additive or subtractive loads on those due to rod and fluid weights and to the acceleration of the mass of the system. With this simplification, the PRL during that portion of the upstroke during which the TV is closed (see Sec. 9-2) is made up of the following terms:

Pressure exerted by fluid in tubing multiplied by net plunger area[1]
plus weight of plunger
plus weight of rods

[1] Net plunger area is the gross plunger area less the sucker rod cross-sectional area.

plus acceleration term
plus friction term
less upthrust from below on plunger

It has been found in practice that no force attributable to fluid acceleration is required (Refs. 3, 4), so the acceleration term involves only acceleration of the rods. The reasons for this probably lie in the relatively high compressibility of the fluid pumped and the relatively low speed of propagation of sound in the fluid compared with that in the sucker rods.

In order to size the rod string correctly (the analysis of this section is limited to the case of a single rod size) and to determine the limiting factors on pumping speed, length of stroke, plunger cross-sectional area, and so on, it is necessary to determine the maximum PRL, remembering that the maximum stress in the polished rod must be kept lower than a certain practical limit which depends on the rods themselves and also on whether or not the operating conditions are corrosive (Sec. 9-3). In deriving the expression given below, the friction term and the weight of the plunger are neglected, in addition to the influence of reflected stresses. These approximations tend to make the value of the maximum PRL as calculated less than the actual maximum. To compensate for this, certain simplifying assumptions that act in the opposite direction are also made. These are, first, that there is zero thrust on the plunger from below (that is, that the pressure in the well at the pump intake is zero—that the well is *pumped off*) and, second, that the TV closes at the instant at which the acceleration term reaches its maximum value; in practice these two things may not occur precisely simultaneously, so that the acceleration term may be somewhat less than the maximum—and decreasing—at the time that the fluid load is applied. With these simplifications, then, the maximum PRL is made up of the following terms:

Pressure exerted by fluid in tubing multiplied by net plunger area
plus weight of rods
plus maximum acceleration term (rods only)

In symbols, for the conventional unit,[2]

$$\text{Maximum PRL} = \frac{62.4\rho D(A_p - A_r)}{144} + \frac{w_s D A_r}{144} + \left(\frac{w_s D A_r}{144}\,\frac{SN^2M}{70{,}500}\right) \tag{9-6}$$

where ρ = fluid specific gravity
A_p = gross plunger area, sq in.
A_r = sucker rod cross-sectional area, sq in.
D = length of sucker rod string, ft
w_s = density of steel, lb/cu ft

If W_r is the weight of the rod string in air, so that

[2] For the air-balanced unit, factor M should be replaced by $1 - c/h$ (see Sec. 9-4).

$$W_r = \frac{w_s D A_r}{144} \tag{9-7}$$

then $$\frac{62.4\rho D A_r}{144} = \frac{62.4\rho W_r}{w_s} = 0.1275\rho W_r \quad \text{approximately}$$

taking the density of steel to be 490 lb/cu ft. This term is frequently expressed by the symbol W_{rb} and is called the *buoyancy force* on the rods.

The term $62.4\rho A_r D/144$ appears as a subtractive term in the expression for the maximum PRL so, to be on the conservative side in the design of installations, this term should be taken to have its lowest practical value. A crude with an API gravity of 50° has sp gr 0.78, and in this case 0.1275ρ is equal to 0.1. Writing W_f for the fluid load on the full plunger area A_p, so that

$$W_f = \frac{62.4\rho D A_p}{144} \tag{9-8}$$

Eq. (9-6) reduces to

$$\text{Maximum PRL} = W_f + (0.9 + F_1)W_r \tag{9-9}$$

where F_1 is the maximum acceleration of the rod string at the bottom of the stroke; that is

$$F_1 = \frac{SN^2(1 + c/h)}{70{,}500} \quad \text{for conventional units}$$

and $$F_1 = \frac{SN^2(1 - c/h)}{70{,}500} \quad \text{for air-balanced units} \tag{9-10}$$

The minimum PRL occurs while the TV is open so that the fluid column weight is carried by the tubing and not by the rods. Theoretically, the minimum load occurs if the TV opens at the top of the stroke, as in this position the downward acceleration of the rod string is also at a maximum. When the TV is open, there is a net buoyancy effect on the rod string of amount $62.4\rho A_r D/144$ lb, or $0.1\ W_r$ lb as above, so that the net weight of the rod string when the TV is open is $0.9W_r$. Neglecting the weight of the plunger and neglecting also the friction term (which will be acting to reduce the minimum PRL), it is seen that

$$\begin{aligned}\text{Minimum PRL} &= 0.9W_r - F_2 W_r \\ &= (0.9 - F_2)W_r\end{aligned} \tag{9-11}$$

where F_2 is the maximum acceleration of the rod string at the top of the stroke; that is,

$$F_2 = \frac{SN^2(1 - c/h)}{70{,}500} \quad \text{for conventional units}$$

and $$F_2 = \frac{SN^2(1 + c/h)}{70{,}500} \quad \text{for air-balanced units} \tag{9-12}$$

To cut down the power requirements of the prime mover, a counterbalance load is used on the walking beam (or rotary crank); a good first approximation to the effective counterbalance load C will be the average PRL. Thus as a first approximation, from Eqs. (9-9) and (9-11),

$$C = {}^1\!/_2 W_f + 0.9W_r + {}^1\!/_2(F_1 - F_2)W_r$$
$$= {}^1\!/_2 W_f + W_r\left(0.9 \pm \frac{cSN^2}{70{,}500h}\right) \tag{9-13}$$

The peak torque value exerted is usually calculated on the most severe possible assumption that the peak (polished rod less counterbalance) load occurs when the effective crank length is also a maximum, that is, when the crank arm is roughly horizontal (Fig. 9-4). With this assumption, the peak torque is

$$T = c[C - (0.9 - F_2)W_r]\frac{d_2}{d_1}$$
$$= {}^1\!/_2 S[C - (0.9 - F_2)W_r] \quad \text{in.-lb}$$
$$= {}^1\!/_2 S[{}^1\!/_2 W_f + {}^1\!/_2(F_1 + F_2)W_r] \quad \text{in.-lb}$$
$$= {}^1\!/_4 S\left(W_f + \frac{2SN^2W_r}{70{,}500}\right) \quad \text{in.-lb} \tag{9-14}$$

There is a limiting relationship between stroke length and cycles per minute that is frequently of considerable practical value. As was shown in Sec. 9-4, the maximum value of the downward acceleration (which occurs at the top of the stroke) is equal to $SN^2\ (1 \pm c/h)/70{,}500$ times the acceleration due to gravity, where the minus sign refers to conventional units and the plus sign to air-balanced units. If this maximum acceleration exceeds unity, the downward acceleration of the hanger is greater than the free-fall acceleration of the rods at the top of the stroke; this leads to severe pounding when the polished rod shoulder once again falls onto the hanger and consequently to frequent rod failures. It is often found that in any one field it is possible to establish an upper limit for permissible values of the maximum downward acceleration, values in excess of this limit leading to excessive sucker rod breakages. This limit is usually about 0.5, but if it is written as L (L will vary from field to field), then

$$\frac{SN^2}{70{,}500}\left(1 \mp \frac{c}{h}\right) \leq L \tag{9-15}$$

so that, for a given stroke length S (in inches), the maximum allowable pumping speed is given by

$$N = \left[\frac{70{,}500L}{S(1 \mp c/h)}\right]^{1/2} \tag{9-16}$$

In the special case when L is equal to 0.5, Eq. (9-16) reduces to

$$N = \frac{188}{[S(1 \mp c/h)]^{1/2}} \tag{9-17}$$

In Eqs. (9-15) to (9-17), the minus sign should be applied in calculations referring to conventional pumping units and the plus sign in calculations referring to air-balanced units.

It should be noted that, neglecting rod and tubing stretch and overtravel (that is, assuming that the plunger motion is equal to the polished rod motion), the existence of the restriction implied by Eq. (9-15) means that for a given plunger size, maximum plunger displacement rate (that is, maximum swept volume per day) is obtained by using the longest stroke length available, even at the expense of accepting fewer strokes per minute.[3] The reason is that the displacement rate is proportional to SN, which from Eq. (9-16) is equal to

$$\left(\frac{70{,}500L}{1 \mp c/h}\right)^{1/2} S^{1/2}$$

and so increases as the stroke length S increases.

Because of the elastic properties of the rod string, the motion of the plunger does not coincide with that of the polished rod. Two major sources of the difference are *stretch* and *overtravel.*

Stretch is caused by the periodic transfer of the fluid load from the SV to the TV and back again, and it consists therefore of two components, namely, *rod stretch* and *tubing stretch.* The former is caused by the weight of the fluid column in the tubing coming on to the rods at the bottom of the stroke when the TV closes; this load is released from the rods at the top of the stroke when the TV opens (Fig. 9-6). It is apparent that the plunger stroke will be less than the polished rod stroke S by an amount equal to the rod stretch. The magnitude of the rod stretch is

$$S_r = \frac{W_f D}{A_r E} \qquad \text{ft} \qquad (9\text{-}18)$$

where W_f = weight of fluid, lb
D = length of rod string, ft
A_r = cross-sectional area, sq in.
E = modulus of elasticity for steel, which is 30×10^6 psi (approximately)

Tubing stretch may be expressed by an equation similar to Eq. (9-18), with the cross-sectional area of steel in the tubing A_t replacing the cross-sectional area of the rods A_r. But, because the tubing cross-sectional area is large compared with that of the rods, so that the tubing stretch is relatively small, and because the tubing is frequently anchored, the tubing-stretch term is often neglected. The reason for using a tubing anchor is that any periodic stretch and contraction in the tubing causes wear at all points of contact between the casing and the tubing, which, if allowed to continue, eventually seriously weakens the casing.

[3] The same result holds true when rod stretch and overtravel are taken into account, as can be shown by an analysis of Eq. (9-21).

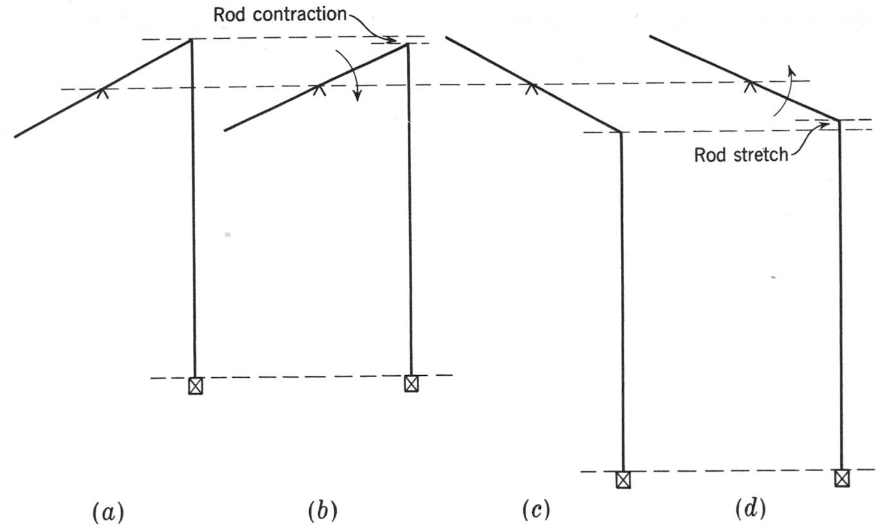

Fig. 9-6 Rod stretch: (*a*) top of stroke, TV closed; (*b*) top of stroke, TV open; (*c*) bottom of stroke, TV open; (*d*) bottom of stroke, TV closed.

Plunger overtravel at the bottom of the stroke is a result of the upward acceleration imposed on the downward-moving sucker rod system, which may, for the purposes of this argument, be regarded as an ideal elastic system. An approximation to the extent of the overtravel may be obtained by considering a sucker rod string being accelerated vertically upward at a rate of F times the acceleration due to gravity. The vertical force required to supply this acceleration is W_rF, where W_r is the weight of the rods, and the magnitude of the rod stretch due to this force is

$$S_o = \frac{W_rFD}{A_rE} \quad \text{ft}$$

But the maximum acceleration is $SN^2(1 \pm c/h)/70{,}500$, so that

$$S_o = \frac{W_rD}{A_rE}\frac{SN^2(1 \pm c/h)}{70{,}500} \quad \text{ft} \tag{9-19}$$

the plus sign applying to conventional units and the minus sign to air-balanced units. Restricting the argument to conventional units for simplicity, and writing M for $(1 + c/h)$,

$$\begin{aligned} S_o &= \frac{W_rD}{A_rE}\frac{SN^2M}{70{,}500} \quad \text{ft} \\ &= 490\frac{A_r}{144}D\frac{D}{A_r \times 30 \times 10^6}\frac{SN^2M}{70{,}500}12 \quad \text{in.} \\ &= 1.93 \times 10^{-11}D^2SN^2M \quad \text{in.} \end{aligned} \tag{9-20}$$

which is the familiar Coberly expression for the overtravel (Ref. 3). It should be noted that this expression makes no allowance for any possible overtravel that might occur at the top of the stroke. Experience indicates that the overtravel (or rod buckling) at the top of the stroke is very limited in extent, being damped out almost entirely by the oil column in the tubing.

Combining Eqs. (9-18) and (9-19), it is seen that the plunger stroke is approximately

$$S_p = S - S_r + S_o$$

or

$$S_p = S - \frac{12D}{A_r E}\left(W_f - \frac{SN^2M}{70{,}500} W_r\right) \quad \text{in.} \tag{9-21}$$

If pumping is being carried out at the maximum permissible speed [Eq. (9-16)], the plunger stroke becomes

$$S_p = S - \frac{12D}{A_r E}\left(W_f - \frac{1 + c/h}{1 - c/h} L\, W_r\right) \quad \text{in.} \tag{9-22}$$

For an air-balanced unit, the factor $(1 + c/h)/(1 - c/h)$ in this formula should be replaced by its reciprocal.

The necessity for including a factor M equal to $1 + c/h$ in the formulas pertaining to conventional pumping units was pointed up when actual pump strokes were measured by Gilbert using the pump dynagraph (Ref. 4) (Secs. 10-2 and 10-3).

Many other pumping formulas have been suggested in the literature, and the various formulas lead to results that differ to a fairly wide degree, especially in the deeper wells (Refs. 1, 5). The great problem, of course, is to find a simple approximation that will describe reasonably accurately the extremely complex situation created by the act of sucker rod pumping. In the rough formulas presented in this section, no reference has been made to reflected stresses, which will in reality play an important, and possibly in certain circumstances a decisive, role in pumping dynamics. Modern deep-well pumping is creating many situations in which the usual approximations are not sufficiently accurate to describe the various motions and stresses with any degree of reliability. The reader is referred to the work by Gibbs (Refs. 6, 7) for more detailed analysis.

Example 9-1 A conventional-type unit has the following characteristics:

Maximum PRL	25,000 lb
Maximum polished rod stroke	72 in.
Peak torque of gearbox	256,000 in.-lb
Maximum counterbalance effect	15,000 lb
Crank/pitman ratio	0.25

The unit is to be used for pumping 25°API gravity crude from 3120 ft. Plunger sizes of $2^1/_2$-, $2^3/_4$-, 3-, and $3^1/_4$-in. diameter are available. It is desired to maximize the volume swept daily by the plunger within the limitations that $^7/_8$-in. rods be used, that the peak stress be less than 30,000 psi, and that the maximum downward acceleration be less than 0.4 times the acceleration due to gravity.

Let S be the polished rod stroke in inches and N be the number of strokes per day. The limitation that the maximum downward acceleration shall be less than 0.4 times the acceleration due to gravity implies [Eq. (9-15)] that

$$\frac{SN^2}{70{,}500}(1 - 0.25) \leq 0.4$$

or

$$SN^2 \leq 37{,}600 \qquad (9\text{-}23)$$

The limitation that the peak stress in the rod string shall be less than 30,000 psi implies [Eq. (9-9)] that

$$W_f + (0.9 + F_1)W_r \leq 30{,}000\ A_r \qquad (9\text{-}24)$$

But

$$W_f = \frac{62.4\rho D A_p}{144}$$

so that (since 25°API crude has sp gr 0.9042) if d is the diameter of the plunger,

$$W_f = 62.4 \times 0.9042 \times \frac{\pi}{4} d^2 \times \frac{3120}{144}$$

$$= 960\ d^2 \quad \text{lb}$$

Also,

$$F_1 = \frac{SN^2(1 + c/h)}{70{,}500}$$

$$= \frac{1.25}{70{,}500} SN^2$$

$$= 1.773 \times 10^{-5}\ SN^2$$

and

$$W_r = w_s \frac{A_r}{144} D$$

$$= 490 \times \frac{\pi}{4} (7/8)^2 \times \frac{3120}{144}$$

$$= 6380 \text{ lb}$$

Finally,

$$30{,}000 A_r = 30{,}000 \frac{\pi}{4} (7/8)^2$$

$$= 18{,}040 \text{ lb}$$

Thus Eq. (9-24) becomes

$$960d^2 + (0.9 + 1.773 \times 10^{-5} SN^2) \times 6380 \leq 18{,}040 \qquad (9\text{-}25)$$

Combining the results of Eqs. (9-23) and (9-25) gives the values shown in Table 9-1. From Eq. (9-21) the plunger stroke length S_p is equal to

$$S - \frac{12 \times 3120}{(\pi/4)(7/8)^2 \times 30 \times 10^6}(960d^2 - 1.773 \times 10^{-5} SN^2 \times 6380) \quad \text{in.}$$

or

$$S - 2.07 \times 10^{-3}(960d^2 - 0.113 SN^2) \quad \text{in.}$$

TABLE 9-1 EXAMPLE 9-1: Determination of Maximum Allowable Values of SN^2, Various Plunger Sizes

Plunger Diameter, in.	Maximum SN^2, Eq. (9-25)	Maximum SN^2, Eqs. (9-25), (9-23)
2½	55,800	37,600
2¾	45,100	37,600
3	32,700	32,700
3¼	19,700	19,700

The greatest swept volume attainable with any given plunger size is obtained when S is equal to the maximum stroke length available—72 in. in this case. Thus the greatest swept volume for any one plunger is

$$\frac{A_p N}{144 \times 12}[72 - 2.07 \times 10^{-3}(960d^2 - 0.113SN^2)] \qquad \text{cu ft/min}$$

Values of this expression for various plunger sizes are listed in Table 9-2, from which it is seen that the maximum swept volume is obtained with a 3-in. plunger, the unit operating on a 72-in. stroke at 21.3 strokes/min.

Finally, it is necessary to check that this operating condition is within the peak PRL, counterbalance, and peak torque limitations of the unit. For the conditions outlined,

$$F_1 = \frac{SN^2(1 + c/h)}{70{,}500} \qquad \text{[from Eq. (9-10)]}$$
$$= 0.58$$

Thus, from Eq. (9-9),

$$\text{Maximum PRL} = W_f + (0.9 + F_1)W_r$$
$$= 8640 + 1.48 \times 6380$$
$$= 18{,}080 \text{ lb}$$

which is within the limit. From Eq. (9-13),

TABLE 9-2 EXAMPLE 9-1: Determination of Plunger Swept Volume, Various Plunger Sizes

Plunger Diameter, in.	Maximum Plunger Stroke Length, in.	No. of Strokes/Min at 72-in. Stroke	Swept Volume, $\frac{A_p}{144}\frac{S_p N}{12}$ cu ft/min
2½	68.4	22.8	4.43
2¾	65.8	22.8	5.15
3	61.8	21.3	5.39
3¼	55.6	16.5	4.41

$$C = {}^1/_2 W_f + 0.9W_r + \frac{cSN^2}{70{,}500h} W_r$$
$$= 4320 + 6380\,(0.9 + 0.116)$$
$$= 10{,}800 \text{ lb}$$

which is within the limit. From Eq. (9-14),

$$T = {}^1/_4 S \left(W_f + \frac{2SN^2}{70{,}500} W_r\right)$$
$$= {}^{72}/_4(8640 + 5920)$$
$$= 262{,}000 \text{ in.-lb}$$

which is somewhat greater than the rated torque of the gearbox. However, most pumping-unit gearboxes can be safely run with a continuous overload, so that this value of 262,000 in.-lb is probably within the capacity of the gearbox.

9-6 THE USE OF TAPERED ROD STRINGS IN SUCKER ROD PUMPING

For deep-well pumping applications it may be found that a single sucker rod size is impractical because a polished rod stress of such a magnitude may be exerted by the weight of the rods themselves that only a very small diameter plunger can be tolerated. A method of overcoming this difficulty is to taper the rod string, placing large-sized rods at the top of the hole but reducing the size (and so the weight) by stages down the hole. Most commonly, either two or three sizes of rod are used, and typical combinations of rod sizes in tapered strings are as follows:

$^3/_4$- and $^5/_8$-in. rods
$^7/_8$- and $^3/_4$-in. rods
1- and $^7/_8$-in. rods
$^7/_8$-, $^3/_4$-, and $^5/_8$-in. rods
1-, $^7/_8$-, and $^3/_4$-in. rods

The design of such strings to take into account the various dynamic stresses becomes very complex, and for that reason tapered strings are frequently calculated on static loads alone.

Since the design of such a string involves no new principles, it will not be discussed further in this book beyond a mention of the fact that one of two possible criteria is generally employed: either the stress in the top rod of each rod size should be the same throughout the string, or the stress in the top rod of the smallest (deepest) set of rods should be the highest (say, 30,000 psi) and the stress should progressively decrease in the top rods of the higher sets of rods. The reason for employing the second method is that it is, in general, preferable that any rod breaks occur near the bottom of the string; otherwise, the weight of the heavier rods at the top falling against the lighter rods below might buckle the latter.

9-7 RATE OF SLIPPAGE OF OIL PAST PUMP PLUNGERS

In relation to plunger fits, a bulletin (Ref. 2) issued by Oil Well Supply, Dallas, Texas, has this to say:

> Plunger fits: Metal-to-metal plungers are commonly available with plunger-to-barrel clearances on the diameter of −0.001″, −0.002″, −0.003″, −0.004″, and even −0.005″. Such fits are referred to as minus one (−1), minus two (−2), etc.; meaning that the plunger is 0.001″, 0.002″, etc., smaller than the nominal inside diameter of the barrel. In selecting a plunger, it is very important to consider the viscosity of the oil to be pumped. A plunger with a loose fit may operate with acceptable efficiency in a well producing a highly viscous oil and yet, when installed at the same depth, fail to deliver any oil to the surface in a well producing oil of low viscosity.
> (a) Low viscosity oil (centipoise of 1 to 20) can be pumped with a metal-to-metal plunger fit of −0.001″.
> (b) High viscosity oil (centipoise of over 400) will probably carry sand in suspension; and in most cases where a loose fit metal-to-metal plunger of −0.005″ is used, the sand particles will slip past the plunger rather than sticking or galling it.

In a progress report on an experimental study of the performance of sucker rod pumps conducted jointly by the Mid-Continent District Topical Committee on Production Practice and the University of Oklahoma, Stearns (Ref. 8) made use of an empirical equation that had been presented by Davis (Ref. 9) to determine the pressure gradient under annular flow of a viscous liquid. Using conventional oil field units, this equation may be written

$$q_s = \frac{K}{\mu} \frac{(d_2 - d_1)^{2.9}(d_2 + d_1)}{d_2^{0.1}} \frac{\Delta P}{L} \tag{9-26}$$

where q_s = slippage rate through the annulus between the plunger and the liner, bbl/day
K = constant
d_1 = plunger diameter, in.
d_2 = linear diameter, in.
ΔP = differential pressure across the plunger, psi
L = length of the plunger, in.

This equation assumes viscous-flow conditions in the annular space between pump plunger and liner, which seems reasonable. Stearns reported that the average value of the constant K observed in the slippage tests was 4.17×10^6, the range of variation being from 2.77×10^6 to 6.36×10^6; these results are based on 28 test runs.

Families of curves, taken from Stearns's paper (Ref. 8), giving the slippage in bbl/day as a function of the plunger fit and the effective pump depth for various liner diameters are shown in Fig. 9-7. They are based on a plunger

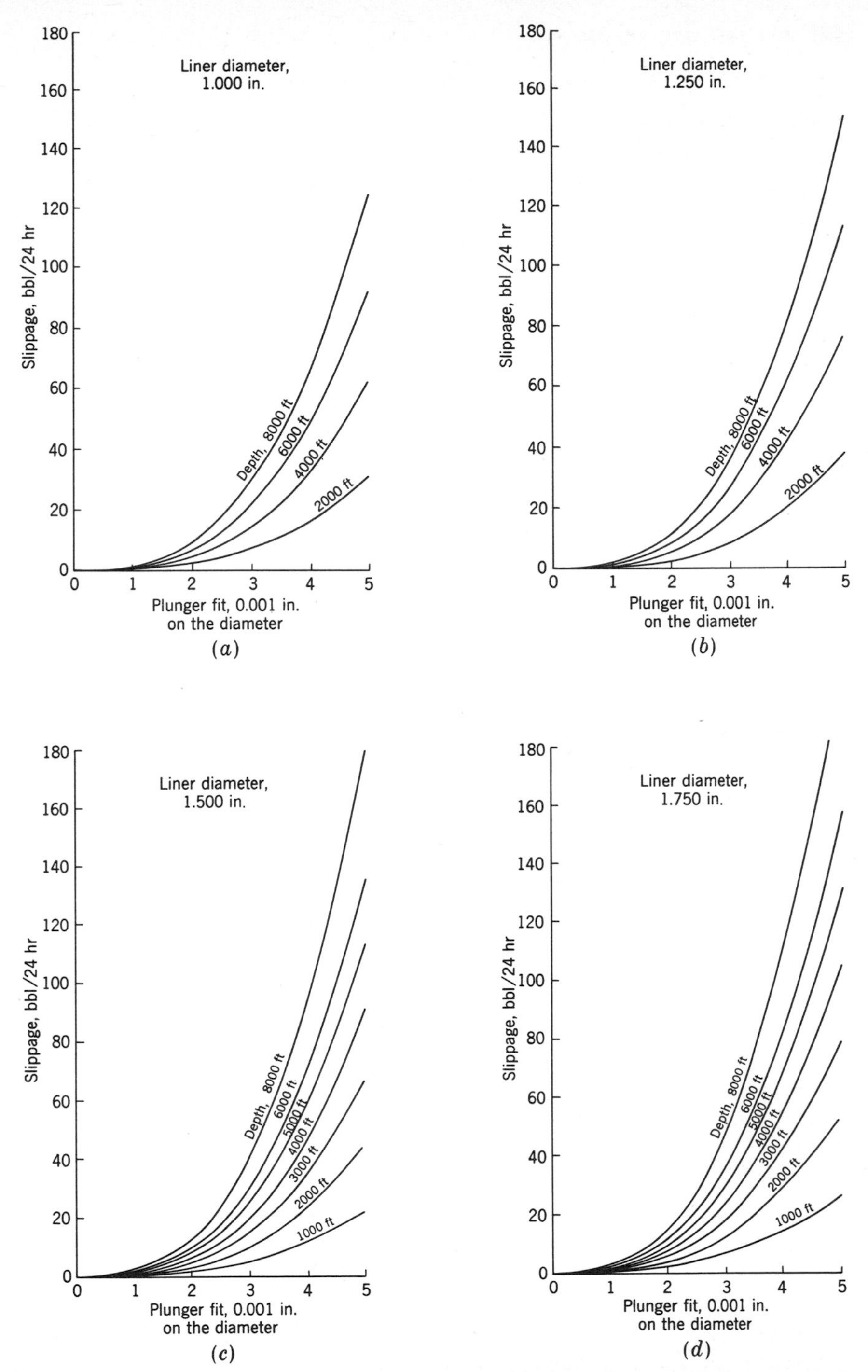

Fig. 9-7 Slippage past plunger. (*After Stearns, Ref. 8. Courtesy* API Drill. Prod. Practice.) (*a*) Liner diameter, 1.000 in.; (*b*) liner diameter, 1.250 in.; (*c*) liner diameter, 1.500 in.; (*d*) liner diameter, 1.750 in.; (*e*) liner diameter, 2.000 in.; (*f*) liner diameter, 2.250 in.

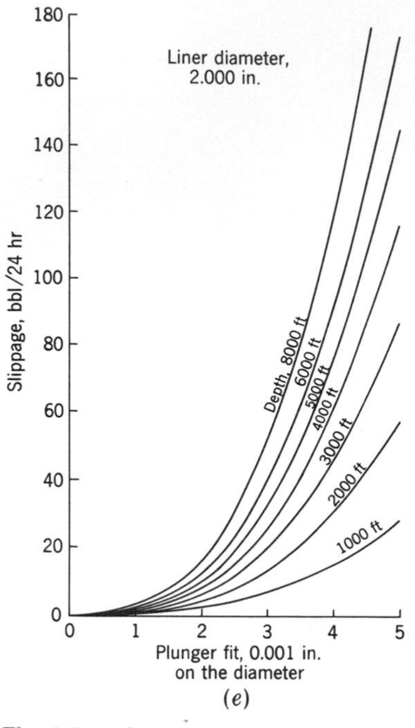

(e)

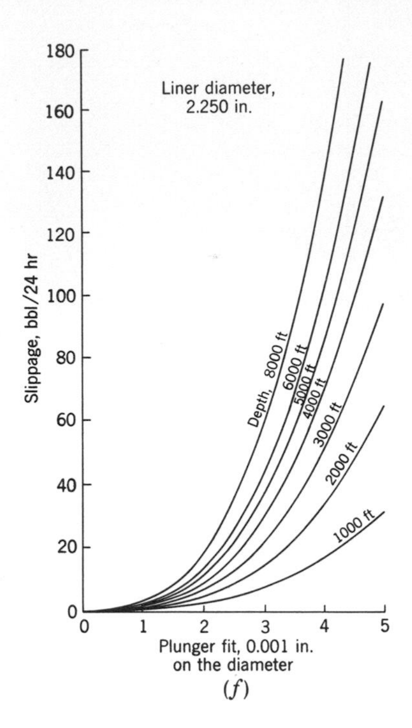

(f)

Fig. 9-7 (*Continued*)

length of 48 in., an oil viscosity of 1.0 cSt,[4] an average K value of 4.17×10^6, and an oil density of unity (so that the effective pump depth is a direct measure of the differential pressure across the plunger). When these curves are used, it should be remembered that there is no slippage while the TV is open, so that for a steadily pumping well the daily volume of liquid that slips past the plunger is only one-half the value indicated on the ordinates of Fig. 9-7.

Example 9-2 Estimate the slippage in a pumping well, given the following data:

Depth to fluid level in casing, ft	4500
Diameter of plunger, in.	$2^1/_4$
Plunger fit, in.	0.003
Length of plunger, in.	60
Viscosity of oil, cSt	1.75

The effective plunger depth is 4500 ft (as a rough approximation; see, however, Sec. 10-8). So on turning to Fig. 9-7*f*, the slippage rate is seen to be

$$33.5 \times \frac{1}{1.75} \times \frac{48}{60} = 15.3 \text{ bbl/day}$$

and the volume of oil slipping past the plunger per day is half of this, or 7.65 bbl.

If, in this example, the stroke of the plunger were 50 in. and if the unit were

[4] Viscosity in centistokes equals viscosity in centipoises divided by fluid density.

pumping at 10 strokes/min, the total daily volume swept by the plunger would be 295 bbl/day. Thus if the TV failed to open, slippage would amount to 5.2 percent of the plunger displacement. The effect of slippage on the so-called *gas-locked pump* phenomenon will be discussed in Sec. 10-3.

REFERENCES

1. *Sucker Rod Handbook,* Bethlehem Steel Company, Bethlehem, Pa., 1958.
2. *Selection and Application of Subsurface Pumps,* Oil Well Supply, Division of United States Steel Corporation, Dallas, Tex., 1957.
3. Coberly, C. J.: "Problems in Modern Deep-Well Pumping," *Oil Gas J.,* May 12, 1938; May 19, 1938.
4. Gilbert, W. E.: "An Oil-Well Pump Dynagraph," *API Drill. Prod. Practice,* 1936, p. 94.
5. Rieniets, R. W.: "Plunger Travel on Oil-Well Pumps," *API Drill. Prod. Practice,* 1937, p. 159.
6. Gibbs, S. G.: "Predicting the Behavior of Sucker-Rod Pumping Systems," *J. Petrol. Technol.,* **15**(7):769 (1963).
7. Gibbs, S. G.: "Computing Gearbox Torque and Motor Loading for Beam Pumping Units with Consideration of Inertia Effects," *J. Petrol. Technol.* **27**(9):1153 (1975).
8. Stearns, G. M.: "An Evaluation of the Rate of Slippage Past Oil-Well Pump Plungers," *API Drill. Prod. Practice,* 1944, p. 25.
9. Davis, Elmer S.: "Heat Transfer and Pressure Drop in Annuli," *Trans. Am. Soc. Mech. Engrs.,* **65**(7):755 (1943).

Pumping Well Instruments 10

10-1 INTRODUCTION

In this chapter the operation and interpretation of some of the commoner instruments used in the analysis of pumping well behavior are described. In particular, reference is made to the polished rod (surface) dynamometer, to fluid-level surveying instruments, and to permanently installed BHP recorders. A discussion of the bottom-hole, or pump, dynagraph is included, not because it is a widely used operational tool, but because a knowledge of the results obtained with it is essential to the correct interpretation of surface dynamometer cards. A section is devoted to the nature of the fluid in the casing-tubing annulus when the well is being pumped with the annulus open and to the conclusions that may be drawn from measurements of the depth to the free fluid level in the annulus.

Finally, seven methods of obtaining values for the flowing BHP (intake pressure) of a pumping well are described. Although each method has drawbacks, it is probable that at least one is applicable to each and every pumping well, thus ensuring that the IPR may be found.

10-2 THE PUMP DYNAGRAPH: GENERAL DESCRIPTION

The pump dynagraph, which is illustrated diagrammatically in Fig. 10-1, was designed by Gilbert (Ref. 1) to record the plunger stroke as well as the

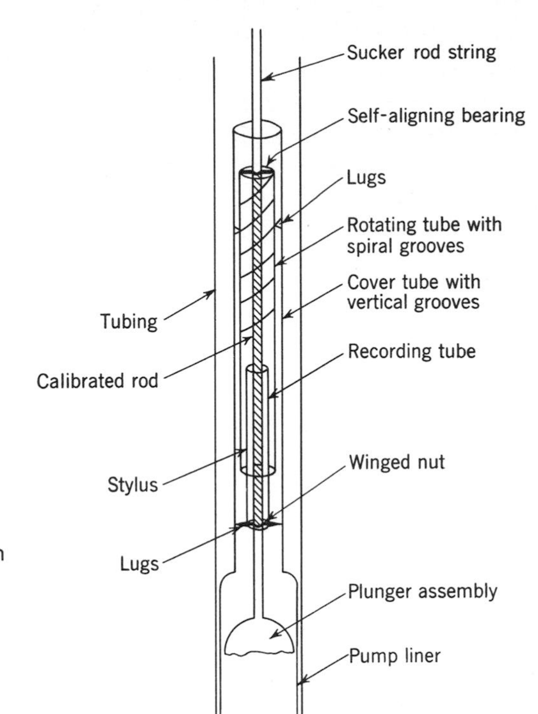

Fig. 10-1 Diagrammatic sketch of the pump dynagraph.

loads carried by the plunger during the pumping cycle. The instrument is for use with a rod-type pump (Fig. 9-3*b*) and is located immediately above the pull rod. Relative motion between the cover tube, which is attached to the pump liner and hence anchored to the tubing, and the calibrated rod, which is an integral part of the sucker rod string, is recorded as a horizontal line on the recording tube. This is achieved by having the recording tube mounted on a winged nut threaded onto the calibrated rod and prevented from rotating by means of two lugs, attached to the winged nut, which run in vertical grooves in the cover tube. The stylus is mounted on a third tube, which is free to rotate and is connected by a self-aligning bearing to the upper end of the calibrated rod. Lugs attached to the cover tube run in spiral grooves cut in the outer surface of the rotating tube. Consequently, vertical motion of the plunger assembly relative to the liner results in rotation of this third tube, and the stylus cuts a horizontal line on the recording tube.

Any change in plunger loading causes a change in length of the section of calibrated rod between the winged nut supporting the recording tube and the self-aligning bearing supporting the rotating tube, so that a vertical line is cut on the recording tube by the stylus.

When the pump is in operation, the stylus traces a series of cards, one on top of the other. In order to obtain a new series of cards, the polished rod at the wellhead is rotated. This rotation is transmitted to the plunger in a few pump strokes. Since the recording tube is prevented from rotating by the winged-nut lugs that run in the cover-tube grooves, the rotation of the sucker rod string causes the winged nut to travel—upward or downward depending on the direction of rotation—on the threaded calibrated rod. Upon completion of a series of tests, the recording tube (which is 36 in. long) is removed, and it forms part of the permanent test record.

It is important to note that, although the bottom-hole dynagraph records the plunger stroke and the variations in plunger loading, no zero line is obtained; hence, quantitative interpretation of the cards becomes somewhat speculative unless a pressure element is run simultaneously (see Sec. 10-10).

10-3 THE PUMP DYNAGRAPH: INTERPRETATION OF RESULTS

An ideal pump dynagraph card may be expected to be rectangular, as shown in Fig. 10-2*a*, in which the arrow indicates the direction of the upstroke. The higher load during the upstroke is due to the static load of the fluid column in the tubing. The assumption of a constant load during the upstroke supposes that there is no inertial loading due to the acceleration imparted to the fluid column, and test results (Ref. 1) have shown that this is indeed true. This particular question was discussed to some extent in Sec. 9-5, and it seems probable that the reason no fluid-acceleration loads are imposed lies in the relatively high compressibility of the fluid and in the long fluid column that is usual in oil wells.

The ideal card depicted in Fig. 10-2*a* supposes instantaneous valve action

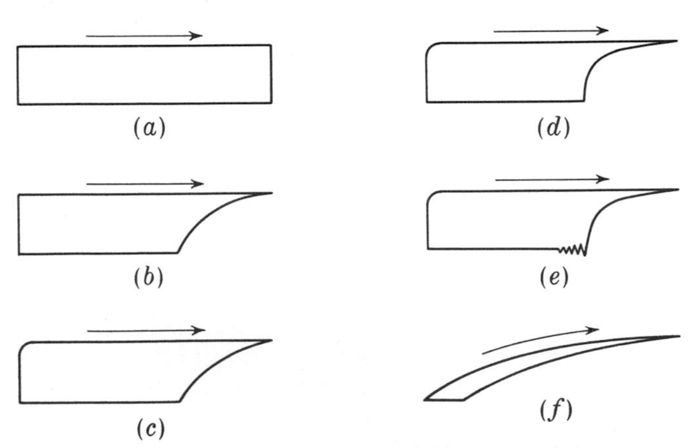

Fig. 10-2 Pump dynagraph cards: (*a*) ideal card; (*b*) gas compression on downstroke; (*c*) gas expansion on upstroke; (*d*) fluid pound; (*e*) vibration due to fluid pound; (*f*) gas lock.

at the top and the bottom of the stroke. In general, some free gas is drawn into the pump on the upstroke, so a period of gas compression occurs on the downstroke before the TV opens (Sec. 9-2). This results in a gradual decrease in load in the rod string above the plunger, and a card like that illustrated in Fig. 10-2*b* is obtained. In some circumstances, particularly if the clearance volume at the bottom of the downstroke is large, there may be a relatively gradual increase in load on the upstroke due to the expansion of any free gas trapped in this clearance volume (but see Sec. 11-2, footnote 1), to gas coming out of solution as the pressure is reduced, to liquid compressibility, and to the choking effect of the standing valve. The result may be some curvature of the load-increase line as shown in Fig. 10-2*c*, but this effect will, in general, be small.

Another possible occurrence is *fluid pound.* In this case the well is almost or completely pumped off, and the plunger displacement rate is higher than the formation's potential liquid rate. As a result, there is a volume of low-pressure gas in the pump chamber at the top of each stroke. As the plunger moves down, this gas is compressed, but insufficient pressure is built up below the TV to overcome the static load of the fluid in the tubing before the plunger strikes the relatively gas-free liquid in the lower part of the chamber. The result is a rapid falloff in stress in the rod string, and the suddenly imposed shock may be deleterious to the mechanical system. Such a fluid pound condition, illustrated in Fig. 10-2*d*, is frequently hard to differentiate from that of gas compression on the downstroke; indeed, the line of demarcation between the two is not clearly defined but is really only one of degree. In some instances the imposed shock load resulting from fluid pound may show up in the pump card[1] (Fig. 10-2*e*).

If the pump is operating at a very low volumetric efficiency, that is, if almost all the pump stroke is being lost in gas compression and expansion, a card of the type shown in Fig. 10-2*f* may result. In the limit, when no liquid at all is being pumped, no valve action takes place and the area within the card may all but disappear; in such a case, the pump is said to be *gas-locked.* Such a condition is only a temporary one, as the continual leakage of oil past the plunger (Sec. 9-7) gradually increases the fluid volume in the pump barrel, with the result that the maximum pressure at the bottom of the downstroke rises until eventually there is sufficient pressure to open the TV and pumping action recommences.

There are many other aspects of considerable interest in pump dynagraph interpretation (Ref. 1), but the few items mentioned in this section are sufficient for an outline to be given of some of the principles behind the interpretation of polished rod (or surface) dynamometer cards.

[1] This shock load may be felt in the polished rod at the surface, but it should not be taken as a sure indication of fluid pound as it sometimes occurs when considerable volumes of gas are being pumped.

10-4 THE SURFACE DYNAMOMETER: GENERAL DESCRIPTION

The use of the pump dynagraph involves the pulling of rods and pump from the hole both for installation of the instrument and for recovery of the recording tube; moreover, the dynagraph described in Sec. 10-2 cannot be used in a well equipped with a tubing pump. It follows that this type of instrument is a research rather than an operational tool, but once an understanding of the pump card has been developed, it is possible to apply this knowledge to the interpretation of surface dynamometer cards that are relatively simple to take but often difficult to analyze.

The surface, or polished rod, dynamometer is a tool that records the motion of, and the load history in, the polished rod during the pumping cycle. While this rod is forced, by the pumping-unit action, to follow a regular time versus position pattern, the load picture is severely distorted because the sucker rod string separates the polished rod from the point at which the load variations are occurring regularly during each stroke, namely, the plunger.

Before a general description of the instrument is given, it should be mentioned that surface cards have three principal uses: to obtain information to be used as a basis for making changes in pumping equipment and, in particular, to determine load, torque, and horsepower requirements (see Sec. 9-5); to improve operating conditions by adjustments in pumping speed and stroke length; and to check well conditions, after installation of the equipment, to prevent or to diagnose various operating problems.

No attempt will be made in this section or in the two that follow to give an exhaustive analysis of dynamometer card interpretation; all that will in fact be undertaken is a discussion of some of the more important features. Comprehensive studies of dynamometer cards are available from many sources (Refs. 2–9).

As to the instrument itself, mechanical, hydraulic, and electric models are available. One of the commoner mechanical instruments is the ring dynamometer; this is installed between the hanger bar and the polished rod clamp (Fig. 9-2) in such a way that the ring may be made to carry the entire well load. The deflection in the ring is proportional to the imposed load, and this deflection is amplified and transmitted to the recording arm by a series of levers. A stylus on the recording arm traces a record of the imposed loads on a waxed paper card located on a drum. The loads are obtained in terms of polished rod displacement by having the drum oscillate back and forth to reflect the polished rod motion. One way of achieving this is to attach the drum to a spring-loaded pulley from which a string is taken to some stationary object, such as the wellhead. On the upstroke, the string rotates the pulley, and hence the drum, against the spring; on the downstroke, the drum's rotation is reversed by the action of the spring. A zero line, or baseline corresponding to zero load, can be traced by placing the stylus in contact with the waxed paper after installing the dynamometer on the well but with the PRL carried by a clamp placed between the hanger bar and the ring.

The diagnostic techniques of surface cards using computer analysis (Refs. 6–9) require a strip-chart record of polished rod displacement and load plotted against time (Ref. 10). The strain-gauge load cell of the dynamometer is mounted, as above, between the hanger bar and the clamp, and the electrical signal from the cell is fed into one side of a dual-channel recorder. The polished rod displacement is measured by a potentiometer coupled to a constant-torque retractable pulley on which a cable is wound, the free end of the cable being attached to the polished rod. The output signal from the potentiometer is fed into the second channel of the recorder.

10-5 QUALITATIVE INTERPRETATION OF SURFACE DYNAMOMETER CARDS

Maximum and minimum PRLs may be read directly from the surface card with the use of the instrument calibration and the baseline, so that the load, torque, and horsepower requirements of the surface installation may be readily computed. Analysis of the qualitative aspects of the surface card is, however, more complex, and some of the complicating features are mentioned below.

Rod Stretch and Contraction

This phenomenon was discussed in Sec. 9-5 and evidently affects the surface dynamometer card. At the bottom of the stroke the tension in the rod string increases steadily as the polished rod is moved upward by the action of the pumping unit and as the TV closes. When this tension is sufficient to lift the fluid column supported by the plunger, the plunger itself begins its upward journey. The increased tension in the rods on the upstroke results in rod stretch, as illustrated in Fig. 9-6. At the top of the stroke, the tension is gradually relieved in the rod string as the horse's head moves downward and the rods contract somewhat.

These effects modify the vertical load-increase and load-release portions of the ideal pump card (Fig. 10-2*a*), and the ideal surface card will be as shown in Fig. 10-3*a*.

Acceleration Forces

The main effect of the rod acceleration forces, discussed in Sec. 9-5, is to rotate the card clockwise, that is, to raise the PRLs at the bottom of the stroke and to lower them at the top of the stroke. Referring to Fig. 10-3*b*, at the bottom of the stroke (point *A*), the PRL needed to accelerate the rod string

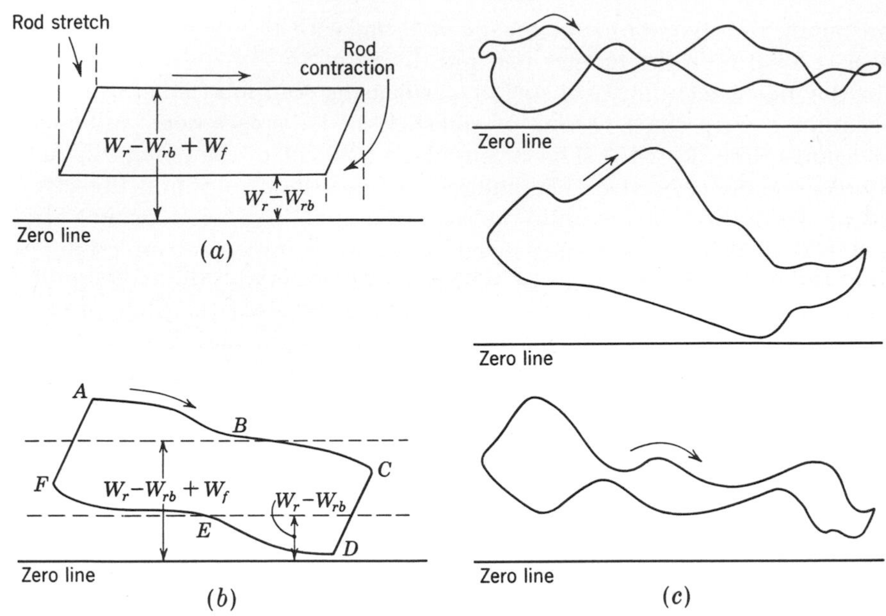

Fig. 10-3 Surface dynamometer cards: (*a*) ideal card, allowing for rod stretch and contraction; (*b*) ideal card, allowing for acceleration loads; (*c*) three typical cards.

upward is a maximum. In the middle of the stroke (point B) it becomes zero, and during the latter half of the upstroke the rods are being decelerated, the maximum deceleration being experienced by the rods at the top of the upstroke (point C). The picture on the downstroke is similar, the maximum acceleration load again being required at the bottom of the stroke (point F).

It will be noted that the values of the weight of the rods in the fluid, namely $W_r - W_{rb}$, and of this weight plus the weight of the fluid on the full plunger area, namely $W_r - W_{rb} + W_f$, are no longer immediately obtainable from the surface dynamometer card. For this reason SV and TV checks are sometimes made on a pumping well. These are discussed briefly in Sec. 10-10, where it is shown that the pressure at the pump intake (that is, at the SV) may be determined from these checks, provided that conditions are favorable.

Rod Vibration

A serious complicating factor in the interpretation of surface dynamometer cards is introduced by the damped[2] harmonics that occur in the string because of sudden changes in rod load caused by *pickup* (the closing of the TV) and by

[2] The damping is due to friction—of the rod string in the oil, of the rod string against the tubing wall, and of the plunger in the barrel.

fluid pound. Since the velocity of propagation of the stress wave through the sucker rod string is finite—of the order of 15,000 ft/sec—there is a time lag between the inception of a stress train at the plunger and its arrival at the surface. Moreover, since the lower end of the string is a free end, there is a stress reversal each time the wave is reflected from this end. This leads to stress highs and lows, damping out with time, in the polished rod, and these are recorded by the surface dynamometer. The time interval between successive highs and lows of any one train is a constant (equal in seconds to twice the length of the rod string in feet divided by 15,000), but because the dynamometer card is recorded against an abscissa of equal intervals of polished rod displacement and because the polished rod velocity varies throughout the stroke, the stress peaks and troughs do not appear on the card at regular intervals. In fact, the polished rod is moving slowly near the top and bottom dead centers, so the stress fluctuations will show up close together at either end of the card. They will be more widely spaced in the center since the rods have their maximum velocity half way through the stroke.

A method of analyzing the load fluctuations exhibited in surface dynamometer cards was suggested by Clark, Dangberg, and Kartzke (Ref. 11), and various theoretical studies of the problem were made (Refs. 12, 13), culminating in the work of Gibbs (Refs. 6, 9), Gibbs and Neely (Ref. 7), and Eickmeier (Ref. 8). These methods are discussed briefly in the next section. Gilbert (Ref. 1) has given some interesting comparisons of pump cards and surface cards on the same well.

Some typical surface dynamometer cards showing the effects of rod vibration are illustrated in Fig. 10-3*c*.

10-6 COMPUTER ANALYSIS OF DYNAMOMETER CARDS

Figure 10-4 (Ref. 7) shows part of a typical chart from a strain-gage-type dynamometer (Sec. 10-4), while curve *a* of Fig. 10-5 reproduces the data in a load versus displacement diagram. The latter is the easier to interpret visually but, as Gibbs and Neely point out (Ref. 7), such visual interpretation is dependent on personal judgment and experience. The analysis utilizing a digital computer enables qualitative and quantitative results to be obtained from which the personal element has been largely removed. Eickmeier (Ref. 8) makes the important point that not every pumping well in a field—nor every dynamometer card run—requires computer analysis; experience in the particular conditions encountered will ensure that visual interpretation of the standard load versus displacement dynamometer card is reasonably accurate in that it is based on some fully analyzed results.

The method is well illustrated by an example given by Gibbs and Neely. Quoting from their paper (Ref. 7):

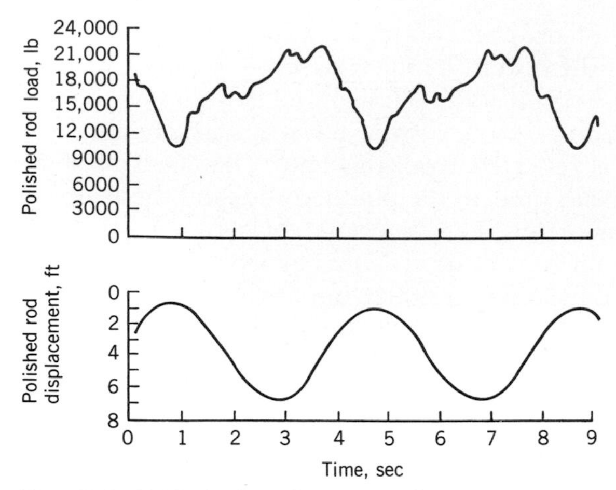

Fig. 10-4 Typical chart from a strain-gauge-type dynamometer. (*From Gibbs and Neely, Ref. 7. Courtesy AIME.*)

To illustrate the technique, the polished-rod data of [Fig. 10-4] are analyzed. These data are shown in the time-history form as measured by a special dynamometer and pertain to an 8,525-ft well having a three-taper rod string operated with a 74-in. conventional unit at 15.4 strokes/min. Measurement of polished rod data in time-history form is most convenient in view of the need for expressing these data in Fourier series as functions of time. . . . Ordinates from these curves for one complete cycle are read into the computer along with the pertinent rod design data, pumping speed, damping factor, etc., to yield the set of dynagraph cards shown in [Fig. 10-5]. Considerable diagnostic information can be gained from analysis of these cards. Shown are the calcu-

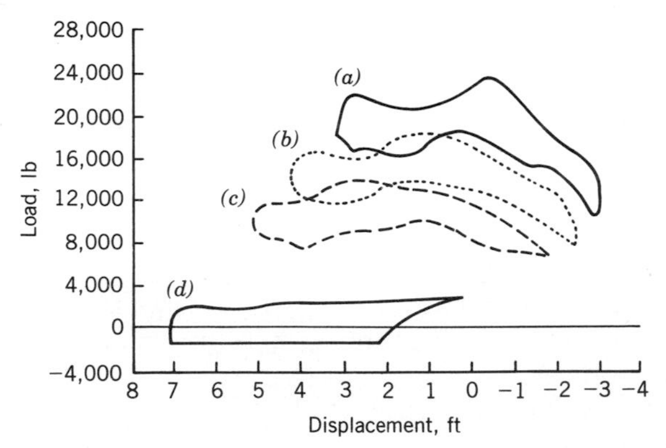

Fig. 10-5 Down-hole cards derived from surface dynamometer card. (*From Gibbs and Neely, Ref. 7. Courtesy AIME.*)

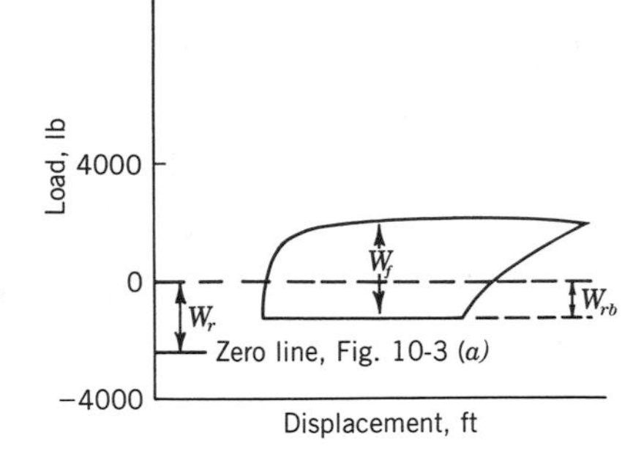

Fig. 10-6 Bottom-hole card derived from surface dynamometer card: compare with Fig. 10-3*a*.

lated dynagraph cards at the surface [(*a*)], at junction points in the combination rod string [(*b*) and (*c*)] and at the pump [(*d*)]. Peak stresses of 28,800 psi in the 1-in. rods, 29,600 psi in the 7/8-in. rods and 30,300 psi in the 3/4-in. rods are indicated. . . . The pump dynagraph card indicates a gross pump stroke of 7.1 ft, a net liquid stroke of 4.6 ft and a fluid load [W_f] of 3,200 lb. . . . The shape of the pump card indicates some down-hole gas compression. The shape also indicates that the tubing anchor is holding properly. A liquid displacement rate of 200 bbl/day is calculated which, when compared with measured production of 184 bbl/day indicates no serious tubing or flowing leak.

In Fig. 10-5 the buoyant weight of the rod string is shown as a negative load; see Fig. 10-6, which should be compared with Fig. 10-3*a*. The reason for the difference in the location of the zero line is that Fig. 10-3*a* is a representation of an idealized surface card, whereas curve *d* of Fig. 10-5 is a bottom-hole card, that is, the card that might result from the use of a pump dynagraph (Secs. 10-2 and 10-3). During the downstroke of the plunger the fluid load is on the SV, which is closed, and the only force acting on the rod string is the rod buoyancy.

An estimate of the flowing BHP may also be made from this form of analysis of dynamometer cards (Sec. 10-10).

10-7 PUMPING WITH ANNULUS CLOSED AND WITH ANNULUS OPEN

For varying reasons, many wells are pumped with the casing-tubing annulus closed at the surface. This is acceptable practice as long as there is little or no free gas in the produced fluids at the intake to the pump (for instance, when the flowing BHP is greater than or close to the bubble-point pressure, or when water cuts are very high), but it is an inefficient method otherwise. The reason is that all the free gas present in the liquid stream at the tubing shoe is forced to enter the pump barrel, and this gas reduces the volumetric efficiency of the pump (Secs. 9-2 and 11-2). Evidently, an annulus fluid-level survey[3] run on such a well will show the casing to be full of gas down to the pump intake (unless the volume of free gas is so low at BHPs that no gas is able to move up into the annulus) so that the well gives the false appearance of being pumped off. Although the determination of the flowing BHP, and so of the IPR, is greatly simplified when the annulus is full of gas, this is but small compensation for the high price paid for a low pump volumetric efficiency, and hence a low offtake rate, especially when the production rate can usually be significantly increased by the simple expedient of opening the casing-tubing annulus at the surface. It may even prove possible to increase the rate still further by the installation of a gas anchor to divert free gas into the annulus (Sec. 11-2).

The effect of opening the annulus is to permit a liquid column to form therein, and the free gas bubbling through this liquid creates a foamy mixture. The presence of this mixture, or *annulus fluid,* complicates the interpretation of the results of annulus fluid-level surveys; a discussion of this aspect of the matter will be the subject of the following section, and the surveying instrument itself is described in Sec. 10-9.

10-8 NATURE OF THE ANNULUS FLUID: PUMPING WITH ANNULUS OPEN

If the pump is set at a depth D ft, if the CHP is p_c psig (assuming that the gas from the annulus is bled off through a choked line), and if $\bar{\rho}$ is the average specific gravity of the annulus fluid, then the pressure p_{wf} (in psig) at the pump intake is equal to p_c plus the pressure due to a gas column H ft long plus the pressure due to the annulus fluid column of length $(D - H)$ ft plus the friction loss resulting from the gas movement up the annulus, where H is the depth in feet of the top of the annulus fluid column below the casinghead. Neglecting the pressures due to the gas column and the friction loss, H can be expressed in the following way:

$$H = D - \frac{p_{wf} - p_c}{0.433\bar{\rho}} \qquad (10\text{-}1)$$

If J is the PI of the formation (for the purposes of the present argument a straight-line IPR will be assumed), then

$$p_{wf} = p_s - \frac{q}{J}$$

where p_s is the static pressure of the formation and q is the gross production

[3] See Sec. 10-9.

rate. Using this relationship, the expression for H becomes

$$H = D - \frac{p_s - p_c}{0.433\bar{\rho}} + \frac{q}{0.433\bar{\rho}J} \tag{10-2}$$

If $\bar{\rho}$ is regarded as a constant, Eq. (10-2) states that the relationship between H and q is linear, as illustrated in Fig. 10-7. In other words, if it is reasonable to expect the average density of the annulus fluid to be independent of production rate, then it is reasonable to expect the annulus fluid level to drop steadily as the production rate is increased. But is this supposition—that the average density of the annulus fluid is independent of the production rate—a realistic one?

To answer this question, suppose the well is to be pumped at a very low rate. Then, since the daily (surface) gas rate from the formation is equal to the formation GLR multiplied by the liquid rate, it follows that the daily free-gas rate (at the pump intake, and hence expressed at pump intake pressure and temperature) is also low. Only a portion of this free gas will be diverted up the annulus, so there will be relatively little lightening of the annulus fluid and the annulus fluid gradient will approach the oil gradient. Now suppose the same well is to be pumped at a high rate. Not only will the daily free-gas rate at the pump intake be high because of the high liquid production rate and the low intake pressure, but there will also be additional gas out of solution and the formation itself may be producing at a higher free-gas/liquid ratio because of the excessive drawdown in the neighborhood of the well bore (Sec. 3-3). Hence, it is highly probable that a considerable daily volume of free gas is di-

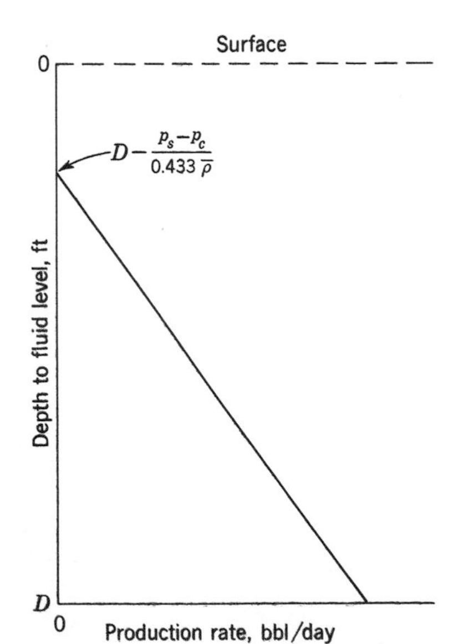

Fig. 10-7 Depth of annulus fluid level versus production rate: annulus fluid assumed to be of constant density.

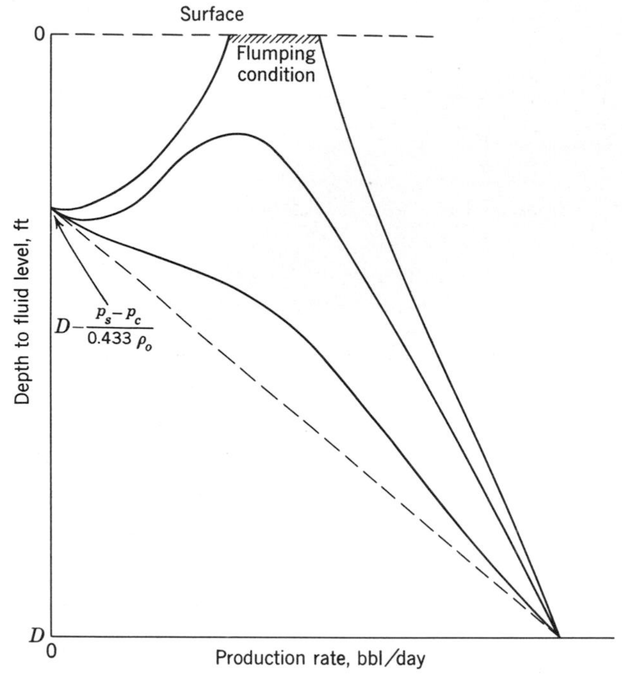

Fig. 10-8 Typical examples of variation in depth to annulus fluid level with production rate.

verted up the annulus,[4] thus greatly reducing the overall density of the annulus fluid.

It follows that $\bar{\rho}$ in Eq. (10-2) will, in general, decrease as the liquid flow rate increases, and the way in which the depth to the fluid level H varies as a function of q will depend on the interplay between the various terms in the equation. Reverting to Eq. (10-1), it is evident that H will increase with decreasing inflow pressure (that is, the fluid level will fall with increasing liquid rate) if $\bar{\rho}$ decreases more slowly than $p_{wf} - p_c$. On the other hand, if $\bar{\rho}$ decreases more rapidly than $p_{wf} - p_c$, the annulus fluid will rise up the hole as the production rate is increased. In certain circumstances it is possible that H may decrease to zero, in which case the fluid level is at the surface for a certain range of values in the inflow pressure and the well is flumping (Sec. 11-4).

Bearing in mind the variation in the average specific gravity of the annulus fluid with rate, only three definite statements can be made concerning the way in which the depth to the fluid level varies as the liquid offtake rate is changed. First, when the well is not in production, the depth to the fluid level will be

[4] Indeed, this must be so, for otherwise the free gas would pass into the pump and reduce the pumping volumetric efficiency to such an extent that the high liquid production rate postulated would not be attainable.

$$H_s = \frac{p_s - p_c}{0.433\rho_o} \tag{10-3}$$

where p_s is the static pressure and ρ_o is the specific gravity of the oil. Second, when the well is pumped off (that is, the formation is producing at its potential), the fluid level must be at the pump intake. Third, since $\bar{\rho}$ decreases with increasing production rate, the actual value of H is always less than—or equal to, in special circumstances—the value of H calculated on the assumption that the annulus fluid has an oil gradient. Some examples of the way in which the depth to the annulus fluid level may vary as a function of the production rate are shown in Fig. 10-8.

It should be emphasized here that if the results of fluid-level surveys are interpreted on the basis of an assumed straight-line relationship between H and q, such surveys will frequently appear to be self-contradictory; a correct understanding of the nature of the annulus fluid when a well is pumped with the casing open will usually resolve the apparent anomalies.

The variation in the average specific gravity of the annulus fluid with rate makes the problem of determining the pump intake pressure from a knowledge of the length of the fluid column in the annulus a complex one, to which no general solution has been given to date. This problem is discussed to some extent in Sec. 10-10.

10-9 LOCATION OF ANNULUS FLUID LEVELS

Turning briefly to the instruments that are used in fluid-level surveys, the general principle involved is to generate an energy wave at the surface and to record the time between the emission of this wave and the return of its reflection to the surface. The energy is commonly supplied by discharging a blank cartridge, but in those wells operating with a reasonably high CHP (greater than 75 psi, say), the use of a steel bottle may sometimes prove more convenient. This bottle may be made out of a foot or two of 5-in. casing, or something similar, and is equipped with a quick-release valve for filling and a bleed valve. The bottle is attached to the casing with both valves closed. The quick-release valve is opened, which permits the casing gas to expand suddenly into the bottle, thus setting up a wave train in the annulus. If a series of readings is required from a particular well, then the bleed valve simply has to be opened between each test.

One point of practical importance in running fluid-level surveys is to ensure that all the tubing joints in the hole are of approximately the same length, that is, all 20-ft joints or all 30-ft joints. The reason for this is that, instead of using the velocity of sound in the annulus in conjunction with the time difference between the emission of the wave train and the receipt of the first reflection, it is usual to determine the depth to the fluid level by secondary reflections from the tubing joints. The difficulty with the approach uti-

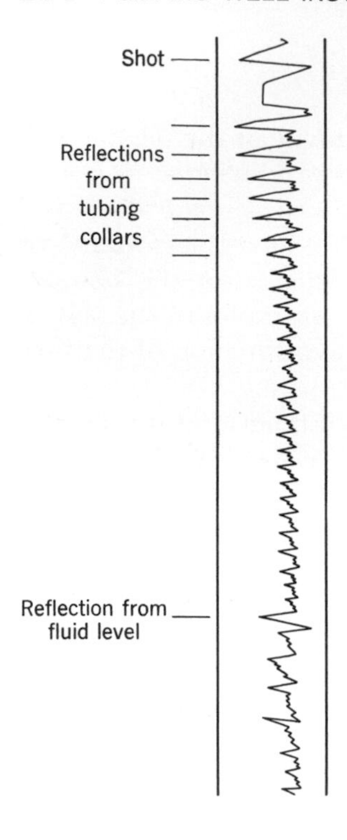

Fig. 10-9 Record resulting from fluid-level survey.

lizing the velocity of sound is that this velocity depends on the composition, temperature, and pressure of the gas and is usually not known precisely. On the other hand, secondary reflections take place from the tubing joints, so that the record of a fluid-level survey appears somewhat as shown in Fig. 10-9. If it is known that each tubing joint is approximately 20 ft long, for instance, then a depth scale can be immediately established and the depth of the fluid level may be read directly from the record.

10-10 DETERMINATION OF FLOWING BHPs: PUMPING WELLS

Pressure Element Run on a Wireline

In certain wells equipped with a sufficiently large casing and sufficiently small tubing, it is possible to run a pressure bomb down the annulus on a wireline. Such an operation is always hazardous to some extent because the line tends to become wrapped around the tubing string and considerable care must be taken not to break the line when pulling out. This problem is accentuated if the oil is viscous, because if the crude has a high viscosity, the bomb will move downward very slowly through the oil; if the wireline is paid out too rapidly on running it into the hole, it may become slack at times. Where such annulus bomb runs are carried out, a special offset flange at the surface is frequently used in an attempt to ensure that the tubing is not centered in the casing.

In many wells the opportunity to run down the annulus with a pressure element does not arise, even if the operator were willing to accept the consequent risk of losing a certain percentage of instruments, because the casing-tubing annulus is simply not large enough to accommodate the tool, and some other method of determining the flowing BHP is required.

Permanently Installed Pressure Elements

In certain key pumping wells, problem wells, or special test wells, it may be considered advisable to install pressure elements below the pump so that accurate pressure readings may be obtained whenever required. Such permanent installations require a conducting cable to be run from the surface to the element, and this cable is strapped to the outside of the tubing. One type of permanently installed pressure device operates on the principle of a stressed wire, or linear spring, vibrating in an electric field. One end of the wire is connected to a diaphragm, which is deflected by the well pressure. These deflections result in changes in the tension in the wire and so in changes in its natural vibration frequencies. These changes are measured electrically and transmitted to the surface. A second type of device utilizes a Bourdon tube that positions a code wheel. The position of this wheel is scanned by means of a contact, the resulting information being transmitted to the surface via the cable.

Although such permanently installed elements give reliable and readily obtainable results, the cost would be high if they were used in every pumping well in a field or area; in addition, the running and pulling of the cable with the tubing string adds to the length of any work-over job.

Walker's Method

The major assumption made in this method is that the average specific gravity of the fluid in the annulus is a constant for a particular set of producing conditions; that is, if the intake pressure at the pump remains constant so that the formation's producing rate does not alter, if all the liquid produced by the formation is pumped through the tubing but some of the gas is diverted up the annulus, and if the displacement rate of the pump remains unaltered so that the daily rate of annulus gas is also unchanged, then the average specific gravity of the annulus fluid is independent of the CHP applied.

Whether or not this assumption is a reasonable one can be inferred from the validity of the results obtained from the method based upon it. Since results do indeed appear to be satisfactory, the assumption may be said to be valid for engineering purposes.

The method consists of recording the depth to the top of the fluid column in the annulus when the well is pumping at a steady rate, the CHP being controlled to within 1 or 2 psi by a back-pressure regulator. The back pressure on the casing is then changed, and the well is permitted to restabilize at the same liquid production rate (that is, no alteration is made in the pump stroke or speed). When the condition is established, the depth to the annulus fluid level is again recorded and a value of the flowing BHP, which must have been the same in the two tests because of the unchanged production rate, is calculated as follows:

Reference to Fig. 10-10*a* shows that if H_1 is the depth to the annulus fluid level in feet when the CHP is p_{c1} (in psig) and if $\bar{\rho}$ is the average specific gravity of the fluid in the annulus, then

$$p_{wf} = p_{c1} + \text{pressure due to gas column} + 0.433(D - H_1)\bar{\rho}$$

Similarly, from Fig. 10-10*b*,

$$p_{wf} = p_{c2} + \text{pressure due to gas column} + 0.433(D - H_2)\bar{\rho}$$

using the assumption that the average specific gravity of the annulus fluid is the same in the two tests. The pressure due to the gas column may be neglected or it may be approximated, using, for example, Eq. (4-17); p_{wf} and $\bar{\rho}$ may then be estimated.

The major drawback to Walker's method is the operational time required to establish stability after the CHP has been changed. The quickest procedure is that suggested in Fig. 10-10, namely, to adjust the pressure regulator to a fairly high level and then to leave the well for, say, 1 week. At the end of this time the well is gauged and a fluid-level survey made. The regulator is then adjusted to a lower level. It has been found in practice that with this method the well will almost always come back to stabilized conditions within 24 hr. If,

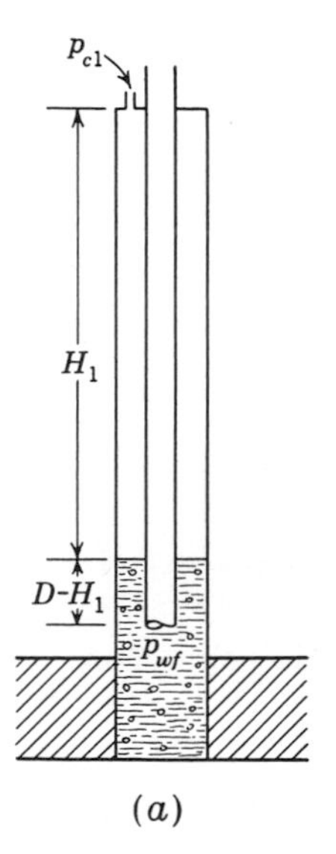

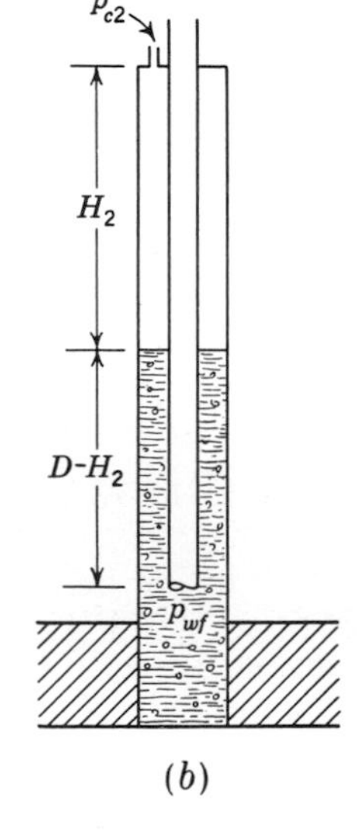

Fig. 10-10 Determination of pump intake pressure: Walker's method.

however, the lower CHP is used first, it may require considerable time before the well stabilizes at the higher.

Example 10-1

Depth to pump	4130 ft
Depth to fluid level when CHP is 120 psig	3015 ft
Depth to fluid level when CHP is 50 psig	2440 ft

Using Eq. (4-18), the pressure at the top of the fluid column in the case of the higher CHP is found to be, approximately,[5] 127 psig and the pressure at the top of the fluid column in the case of the lower CHP is 52 psig.

Thus the two equations from which the intake pressure p_{wf} and the average fluid column specific gravity $\bar{\rho}$ may be found are

$$p_{wf} = 127 + 0.433\bar{\rho}(4130 - 3015)$$

and

$$p_{wf} = 52 + 0.433\bar{\rho}(4130 - 2440)$$

From these,

$$\bar{\rho} = 0.301$$

$$p_{wf} = 272 \text{ psig}$$

Gas Blow-Round Method

This method requires the use of a back-pressure regulator and a polished rod dynamometer, and it has the drawback of being time-consuming, particularly in wells in which the annulus gas rate is fairly low. The method consists of imposing an ever-increasing back pressure against the casing until the annulus gas begins to blow around the foot of the tubing. The incidence of this condition can be noted on the surface dynamometer card, the area of which suddenly decreases. The back pressure on the casing is reduced slightly so that the well is pumping with the annulus fluid level close to the tubing shoe. The well is then permitted to pump steadily for a few hours to regain an equilibrium condition (during this period a few dynamometer checks should be taken to ensure that gas is not blowing around), and a gauge is taken. The back pressure on the annulus is gradually increased until the dynamometer card again collapses; the CHP is then backed off a few pounds and then increased until once again the polished rod card shows gas to be blowing around. In this way the value of the CHP when the annulus is filled with gas is established. A value for the pump intake pressure is obtained by adding to this CHP an allowance for the pressure due to the gas column.

Agnew's Method: SV and TV Checks

This method for determining the intake pressure of pumping wells by utilizing the dynamometer was proposed by Agnew (Ref. 14). In the discussion of maximum PRLs in Sec. 9-5, the upthrust exerted on the bottom of the

[5] Strictly speaking, Eq. (4-17) is applicable only to static gas columns.

plunger when the standing valve is open was neglected. This was reasonable in view of the purpose of that discussion, which was to establish a maximum possible PRL. But the force that was neglected was equal to the product of the formation inflow pressure and the cross-sectional area of the plunger (assuming the pressure loss across the SV port is small) and so cannot be ignored in any application in which the determination of the inflow pressure is the aim.

If W_f = weight of the fluid on the full plunger area, lb
W_r = weight of the rods in air, lb
W_{rb} = buoyancy force on the rods, lb
A_r = cross-sectional area of the (bottom) rods, sq in.
A_p = cross-sectional area of the plunger, sq in.
p_t = THP, psig
p_{wf} = flowing BHP of the well, psig

it follows that when the pumping unit is stopped (gradually, in order to eliminate acceleration loads) sufficiently near the top of a stroke to ensure that the SV is open and the TV is closed, the load recorded by a dynamometer attached to the polished rod is

$$\text{TV load} = W_f + W_r - W_{rb} - p_{wf}A_p + p_t(A_p - A_r) \tag{10-4}$$

This is known as the *TV check.*

Similarly, if the pumping unit is stopped (gradually) on the downstroke and sufficiently near the bottom of the stroke to ensure that the SV is closed and the TV is open, the PRL is simply the weight of the rods suspended in the fluid, or

$$\text{SV load} = W_r - W_{rb} \tag{10-5}$$

This is known as the *SV check.*

Using Eq. (10-5), Eq. (10-4) may be written

$$\text{TV load} = \text{SV load} + W_f - p_{wf}A_p + p_t(A_p - A_r)$$

from which it follows that

$$p_{wf}A_p = W_f - (\text{TV load} - \text{SV load}) + p_t(A_p - A_r) \tag{10-6}$$

But
$$W_f = 0.433\bar{\rho}DA_p$$
$$= \frac{A_p}{A_r} W_{rb}$$

since
$$W_{rb} = 0.433\bar{\rho}DA_r$$

Hence, from Eq. (10-5)

$$W_f = \frac{A_p}{A_r}(W_r - \text{SV load})$$

which, when substituted into Eq. (10-6), gives

$$p_{wf}A_p = \frac{A_p}{A_r}(W_r - \text{SV load}) - (\text{TV load} - \text{SV load}) + p_t(A_p - A_r)$$

$$p_{wf} = \frac{W_r - \text{SV load}}{A_r} - \frac{\text{TV load} - \text{SV load}}{A_p} + p_t\left(1 - \frac{A_r}{A_p}\right) \qquad (10\text{-}7)$$

from which p_{wf} may be calculated.

Example 10-2 A pumping well is completed with 3000 ft of 7/8-in. rods and a 1 3/4-in. plunger. An SV check is made and a load of 5700 lb recorded; a TV check gives a load of 6500 lb. The THP is 60 psig. Estimate the inflow pressure, that is, the pressure immediately below the SV.

$$A_r = \frac{\pi}{4}\left(\frac{7}{8}\right)^2 = 0.601 \text{ sq in.}$$

$$A_p = \frac{\pi}{4}\left(\frac{7}{4}\right)^2 = 2.405 \text{ sq in.}$$

Using a figure of 490 lb/cu ft for the density of steel,

$$W_r = 490 \times 3000 \times \frac{0.601}{144}$$

$$= 6135 \text{ lb}$$

$$W_r - \text{SV load} = 6135 - 5700 = 435 \text{ lb}$$

$$\text{TV load} - \text{SV load} = 6500 - 5700 = 800 \text{ lb}$$

$$A_r/A_p = 0.25$$

Substitution of these values in Eq. (10-7) gives

$$p_{wf} = 724 - 333 + 45$$

$$= 436 \text{ psig}$$

One practical limitation of this method is imposed by friction losses, which, if large, will evidently lead to a considerable error in the calculated value of p_{wf}. Such friction might be due to paraffin accumulation, galling of the pump by sand or scale, or tubing hung in compression, to mention some of the possibilities. Impulse loads created when stopping the pumping unit may also give rise to errors. Evidently, best results will be obtained by this method if the test is taken shortly after a new or repaired pump is installed in the hole and if great care is taken to stop the unit gently.

Method Involving Calculation of Tubing Gas Rate

This method depends on a calculation of the rate at which gas may be expected to be produced through the tubing. This rate is determined as a function of the flowing BHP, so that if the actual tubing gas rate is measured, the flowing pressure can be inferred. One major drawback of the method is that the pump displacement rate must be estimated; as was seen in Sec. 9-5, this is a lengthy calculation that cannot, in general, be regarded as giving very accu-

TABLE 10-1 EXAMPLE 10-3: Assumed PVT Data

Pressure, psia	B_o, bbl/bbl	B_g, bbl/scf	R_s, scf/bbl
2500	1.250	0.0015	450
2200	1.225	0.0018	412
1900	1.203	0.0023	377
1600	1.180	0.0030	340
1300	1.156	0.0039	303
1000	1.133	0.0050	268
700	1.108	0.0065	231
400	1.080	0.0085	165
200	1.045	0.0110	97

rate results. Another feature that may prove difficult to overcome in certain circumstances is that it is the tubing gas rate and not the total gas rate that is required. If the casing of the well is hooked into the flow line downstream of the wellhead, which is quite a common arrangement, then this bypass line must be closed and the casing gas—probably—vented to the atmosphere while the tubing gas rate is being measured.[6]

Suppose that D = pump displacement rate, bbl/day
q_o = oil production rate, bbl/day
q_w = water production rate, bbl/day
B_o = oil formation volume factor, bbl/bbl
B_g = gas formation volume factor, bbl/cu ft
R_s = gas solubility factor, cu ft/bbl

Of the total daily pump displacement, a volume q_w is occupied by water and a volume $B_o q_o$ by oil. Thus the volume occupied by free gas is

$$(D - q_w - B_o q_o) \qquad \text{bbl}$$

This gas volume is measured at the pump intake pressure (no pressure loss across the SV being assumed) and the bottom-hole temperature. To convert to gas at standard conditions, it is necessary to divide by the gas formation volume factor. In addition to the free gas present at the pump intake, some gas will come out of solution as the oil is produced up the well and to the stock tanks. By the definition of gas solubility, this will amount to $q_o R_s$ scf of gas daily.

It follows that

$$\text{Tubing gas rate} = \frac{1}{B_g}(D - q_w - B_o q_o) + R_s q_o \qquad \text{(10-8)}$$

[6] If this is the procedure adopted, great care must be taken not to alter the CHP during the operation, since any such surface pressure change will lead to a change in pressure at the foot of the tubing and thus in the production from the well.

Example 10-3 Suppose that a well is pumping at 330 bbl/day oil and 210 bbl/day water, that the tubing GOR is 300 scf/bbl, that the pump displacement rate is 750 bbl/day, and that the PVT data for the oil are as shown in Table 10-1. Estimate the flowing BHP.

The tubing gas rates at various assumed pressures are calculated from Eq. (10-8) (Table 10-2) and are plotted as a function of pressure in Fig. 10-11. The actual tubing gas rate is 330 × 300 scf/day, or 99 mcf/day; entering this value on Fig. 10-11 it is seen that the flowing BHP is 665 psia.

It is apparent from Fig. 10-11 that a 10 percent reduction in the measured gas rate would reduce the estimated pressure from 665 to 550 psia, whereas a 10 percent increase would raise the estimated pressure to 785 psia. This example is no exception to the rule; in general, the estimated pressure is sensitive to the value assigned to the tubing gas rate. In view of the relative inaccuracy of field gas measurements, this sensitivity is, of course, a drawback to the method. Other possible sources of error are possible pressure losses across the SV, oil slippage past the plunger (which will reduce the volume of

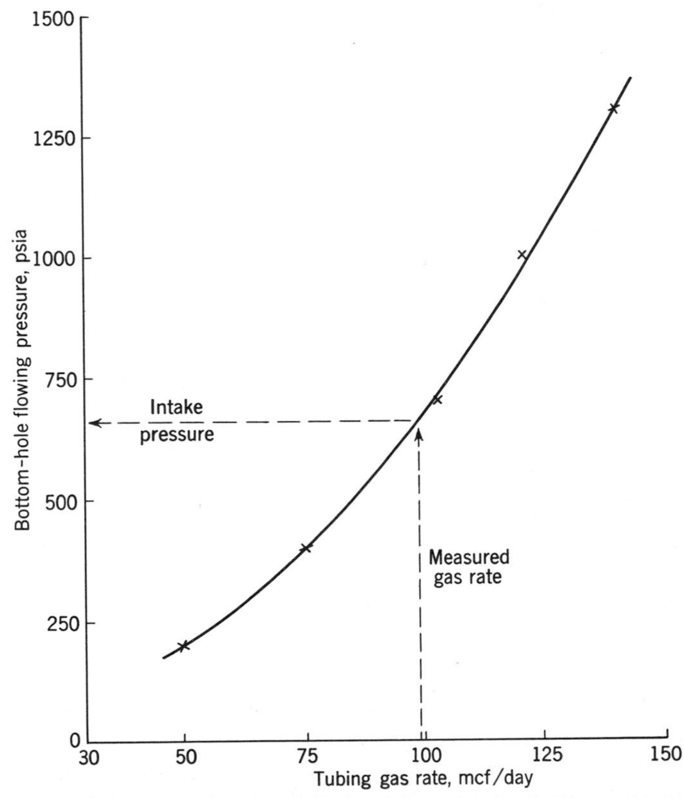

Fig. 10-11 Example 10-3: Determination of pump intake pressure using measured tubing gas rate.

TABLE 10-2 EXAMPLE 10-3: Calculation of Tubing Gas Rates at Various Assumed Intake Pressures

Assumed p_{wf}, psia	$B_o q_o$, bbl/day	$B_o q_o + q_w$, bbl/day	$D - B_o q_o - q_w$, bbl/day	$\frac{D - B_o q_o - q_w}{B_g}$, scf/day	$R_s q_o$, scf/day	Daily Tubing Gas, mcf/day
200	345	555	195	17,700	32,000	49.7
400	356	566	184	21,600	54,500	76.1
700	366	576	174	26,800	76,200	103.0
1000	374	584	166	33,200	88,400	121.6
1300	381	591	159	40,800	100,000	140.8

free gas in the pump—a correction may be made for this on the basis of Fig. 9-7), and the influence of temperature and gas supercompressibility effects.

From Computer Analysis of Dynamometer Cards

If p_p is the pressure immediately above the pump due to the combination of the fluid head in the tubing and the THP, then the pump load carried by the plunger during the upstroke is

$$p_p\,(A_p - A_r)$$

where A_p is the pump area and A_r is the area of the lowest set of rods. Offsetting this load is the force beneath the plunger resulting from the formation producing pressure, namely p_{wf} (or an intake pressure p_i that may be derived from p_{wf} if the pump is located some distance above the perforations). This upward force is $p_{wf}\,A_p$.

Thus the net plunger load is

$$p_p\,(A_p - A_r) - p_{wf}\,A_p$$

so that

$$p_{wf} = p_p\left(1 - \frac{A_r}{A_p}\right) - \frac{\text{net plunger load}}{A_p} \tag{10-9}$$

The value of the net plunger load is obtained from the analysis of the dynamometer card (Sec. 10-6, Fig. 10-5), while A_r and A_p are known for any particular installation. The value to be assigned to p_p is more difficult to determine. Given some knowledge of the GLR in the tubing, an approximation to p_p may be obtained from gradient considerations, that is, the assignment of a value for the average fluid density in the tubing, or from a two-phase flow pressure traverse. As Gibbs and Neely state (Ref. 7), an upper limit to p_{wf} can be calculated on the assumption of a pure liquid gradient in the tubing. Eickmeier (Ref. 8) makes the point that reasonable estimates of the average fluid density in the tubing may result from comparisons with dynamometer card analyses on wells of similar depth that are known to be pumping off.

Example 10-4 As an illustration of the calculation using Eq. (10-9) reference may once again be made to the example of Gibbs and Neely, Sec. 10-6 and Fig. 10-5. From the pump card of Fig. 10-5, a net plunger load (that is, predicted load on the plunger during the upstroke) of 2650 lb is obtained. If the plunger diameter is 1.5 in., then $A_p = 1.77$ sq in. The bottom set of rods are of ¾-in. diameter, so that $A_r = 0.44$ sq in., and $1 - A_r/A_p = 0.75$. Assuming 2⅜-in. tubing and a THP of the order of 50 psi, the pressure-distribution curve of Fig. 4-18 for 200 bbl/day and 0.1 mcf/bbl gives a value of about 2500 psi for p_p (at 8525 ft). It is evident that this determination of p_p is subject to a large degree of error. Inserting the values in Eq. (10-9).

$$\begin{aligned} p_{wf} &= 2500 \times 0.75 - 2650/1.77 \\ &= 1875 - 1500 \text{ or } 375 \text{ psig} \end{aligned}$$

The fact that a relatively small answer results from taking the difference between two much larger numbers, at least one of which is suspect, reduces the degree of reliance that may be placed upon the answer.

Use of Fluid-Level-Locating Instruments

Enough has already been said in Secs. 10-8 and 10-9 to indicate that the determination of the inflow pressure from a measurement of the length of the annulus fluid column is not at all straightforward, particularly in those cases in which this fluid column is several hundred feet long. One way of overcoming the difficulty is to use Walker's method described above.

It appears probable that a wholehearted effort at predicting the average annulus-fluid specific gravity in any particular set of circumstances should be successful, for this average specific gravity must depend on, among other things, the annulus gas rate, the pressure at the free-fluid level [that is, through Eq. (4-18), the CHP], and the inflow pressure itself. It will also presumably depend on the cross-sectional area of the annulus and on the oil gravity. However, to date no satisfactory correlation enabling the prediction of average specific gravity of the annulus fluid has appeared in the literature; until such a correlation is suggested and substantiated, attempts to estimate inflow pressures from the length of the annulus fluid column will remain, at best, poor approximations.

REFERENCES

1. Gilbert, W. E.: "An Oil-Well Pump Dynagraph," *API Drill. Prod. Practice,* 1936, p. 94.
2. Marsh, H. N., and E. V. Watts: "Practical Dynamometer Tests," *API Drill. Prod. Practice,* 1938, p. 162.
3. Zaba, Joseph, and W. T. Doherty: *Practical Petroleum Engineers' Handbook,* 5th ed., Gulf Publishing Company, Houston, Tex., 1970.
4. *Sucker Rod Handbook,* Bethlehem Steel Company, Bethlehem, Pa., 1958.

5. Hosford, Eugene, and Emory Kemler: "Typical Dynamometer Cards and Their Application," *API Drill. Prod. Practice,* 1939, p. 81.
6. Gibbs, S. G.: "Predicting the Behavior of Sucker-Rod Pumping Systems," *J. Petrol. Technol.,* **15**(7):769 (1963).
7. Gibbs, S. G., and A. B. Neely: "Computer Diagnosis of Down-Hole Conditions in Sucker Rod Pumping Wells," *J. Petrol. Technol.,* **18**(1):91 (1966).
8. Eickmeier, J. R.: "Diagnostic Analysis of Dynamometer Cards," *J. Petrol. Technol.,* **19**(1):97 (1967).
9. Gibbs, S. G.: "A General Method for Predicting Rod Pumping System Performance," *SPE Paper No. 6850, 52d Annual Meeting of SPE and AIME,* Denver, Colo., October 1977.
10. Herbert, W. F.: "Sucker Rod Pumps Now Analyzed with Digital Computer," *Pet. Equipt.,* January-February 1966.
11. Clark, Ned, H. F. Dangberg, and P. L. Kartzke: "The Study of Pumping-Well Problems," *API Drill. Prod. Practice,* 1938, p. 209.
12. Langer, B. F., and E. H. Lamberger: "Calculation of Load and Stroke in Oil-Well Pump Rods," *J. Appl. Mech.,* March 1943, p. 1.
13. Ayre, R. S., L. S. Jacobsen, and A. Phillips: "Steady Forced Vibration of a Non-Conservative System with Variable Mass: A Pumping System," *J. Franklin Inst.,* **250**:315 (1950).
14. Agnew, Bob G.: "The Dynamometer as a Production Tool," *API Drill. Prod. Practice,* 1957, p. 161.

Special Problems—Pumping Wells

11

11-1 INTRODUCTION

Much of the material to which this chapter is devoted is qualitative and is, to some extent, speculative. It is included in the belief that sucker rod pumping has for too long been taken to be a lift mechanism that—for all its challenges in the realm of equipment development—is fundamentally simple and fully understood. This attitude has, perhaps, been fostered by the very real achievements in the design of mechanical systems that operate with ever-increasing efficiency in increasingly demanding conditions of well depth, hole deviation, sand production, wax formation, exploitation of high-viscosity crudes, and so on. The attitude has been reinforced by shortages of personnel and by the need to devote time and energy to ensuring continuity in the pumping operation. Moreover, it is not easy to obtain—because of the presence of the sucker rod string—the basic information required to analyze formation and well performance, although there have been some major advances in this field, as outlined in Sec. 10-10. Finally, pumping wells, individually, are not spectacular producers—they are generally the older wells and, provided that their production rates are holding up within reason, engineering effort and expertise and management interest are centered not on them but on newer wells with higher potentials.

But pumping production remains the heart of the industry and it is hoped that the discussion of the topics that follows will stimulate renewed interest in some of the questions and possibilities that remain unanswered and unresolved. It should be stressed that final solutions lie in field experimentation, field testing, and in the analysis of field results.

11-2 EFFECT OF FREE GAS ON PUMP EFFICIENCY: GAS ANCHORS

A glance at the pumping cycle illustrated in Fig. 9-1 shows that the presence of free gas in the pump barrel cuts pumping efficiency. On the upstroke, the SV does not open until the pressure from below (that is, from the formation) exceeds the pressure in the space between the SV and the plunger. If some fraction of this latter volume is occupied by free gas, the pressure falls gradually as the plunger moves upward, so that the SV does not open until part of the upstroke is lost.[1] Similarly, when the plunger is moving downward, there may be a considerable loss in effective downward stroke before the TV opens as a result of free gas present in the working barrel. Defining *pump volumetric efficiency* E_v as the liquid volume drawn into the pump expressed as a fraction of the volume swept by the pump plunger, it follows that pump volumetric efficiencies decrease as the pumped free-gas/liquid ratios increase, other factors remaining the same.

One method of improving pumping efficiency, then, is to divert the free gas up the annulus of the well in question and bleed this gas off at the casinghead; that is, to pump with the casing open and with some device installed at the bottom of the hole to separate the free gas from the liquid. The free gas produced from the casing annulus may be used to power the prime mover, it may be fed back into the well's flow line, or it may go into a separate gas-gathering system.

The down-the-hole liquid and gas separator is known as a *gas anchor,* and three common types are illustrated in Fig. 11-1. The theory behind the *poor-man anchor* (Fig. 11-1*a*) is that during the downward motion of the liquid and free gas within the anchor, the free gas bubbles tend to segregate and to move back out of the anchor ports and up the casing-tubing annulus. The *stinger,* or *tail pipe,* within the anchor is usually about 5 ft long and made of 1-in. pipe or something similar, while the outer casing of the anchor is frequently of the same diameter as the tubing in the hole. These dimensions, although based to some extent on experience, are to a high degree arbitrary, and considerable work needs to be carried out to determine the best anchor geometry in any particular well. If, for instance, the tail pipe is too long or of too small a diameter, there is a considerable pressure drop over its length before fluid enters the pump, which is undesirable from the point of view of well

[1] This effect is not very marked in general, owing to the small clearance volume between the TV and SV at the bottom of the stroke and to the fact that during the downstroke the free gas between the TV and SV is liable to gravity segregation and so tends to move upward through the traveling valve.

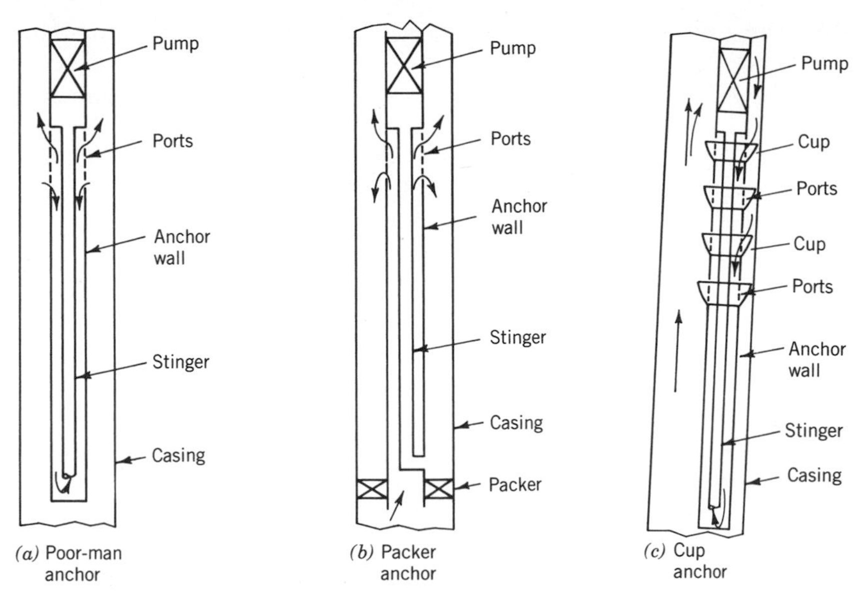

(a) Poor-man anchor *(b)* Packer anchor *(c)* Cup anchor

Fig. 11-1 Three common types of gas anchor.

productivity. If the tail pipe is too short, the gas has little time during which to segregate and to escape while the fluids are moving down to the bottom of the anchor. If the tail pipe is of too large a diameter, the downpass annular area becomes small, with the result that the liquid velocity in this region may become so great that the gas bubbles are entrained and carried along with the liquid. If the outside anchor diameter is made too large, the casing-anchor annulus may become so small that considerable pressure loss may be incurred during movement of fluids from the formation to the anchor intake, again reducing formation productivity. Moreover, the optimum anchor geometry depends upon the productivity of the well: the greater the production rate, the greater the chance of gas entrainment in the anchor downpass area. Most of these questions remain unresolved, but it is known, as might be expected from the anchor geometry, that the efficiency[2] of the poor-man gas anchor decreases rapidly as the liquid throughput rate increases.

The *packer-type gas anchor,* illustrated in Fig. 11-1*b*, has the advantages of a gravity feed to the pump intake and also of a large downpass area, namely, the casing-tubing annulus. Its chief disadvantage lies in the packer installation itself, which may at times prove awkward to set in or remove from the hole and can be troublesome if the well produces sand; in this case, sand may collect on top of the packer. Once again, many unresolved questions arise about optimum anchor geometry in relation to a particular formation's producing

[2] The efficiency of a gas anchor may be defined as the ratio that the volume of free gas diverted up the annulus bears to the volume of free gas present at the anchor intake (see below).

characteristics and about the effect of operating pressure on anchor efficiency.

The *Hague* or *cup anchor* (Fig. 11-1*c*) is essentially a poor-man anchor with a metal "eavestrough" welded around the circumference of the anchor wall immediately below each set of ports. Anchor action is dependent on the facts that no well bore is absolutely vertical and that the anchor will rest against the low side of the hole (Fig. 11-1*c*). It follows that there is no advantage in running a cup anchor if tubing centralizers or a tubing anchor is to be used.

The theory behind this anchor is that the liquid and gas, as they are produced from the formation, move up the high side of the hole, as this is the area that has the better hydraulic radius (Fig. 11-2). The cups divert the fluid flow past the anchor intake ports. But the upward movement of the liquid must eventually stop (otherwise the well would flow), and the liquid must drain back down the low side of the hole. During this downward movement, free gas tends to escape upward—as in the packer anchor—and the relatively gas-free liquid is picked up by the eavestroughs on the way down and diverted into the anchor itself.

It may be noted that the advantage claimed for the packer-type and cup-type anchors is that the gas has an opportunity to separate from the oil in the casing-tubing annulus and, particularly, while the oil is moving downward in this annulus. This advantage is maximized if well, fluid, and formation conditions are such that it is possible to "sump the pump," that is, to locate the pump intake below the bottom perforations. In the relatively few wells in which this is a practical option, placing the pump below the bottom of the producing formation is a simple and effective method for achieving good down-hole separation of gas without recourse to any mechanical device.

In analyzing the effectiveness of the down-hole pump, reference is made to the *volumetric efficiency* E_v of the pump, defined by

$$E_v = \frac{\text{liquid rate pumped from the well}}{\text{down-hole pump displacement rate}} \tag{11-1}$$

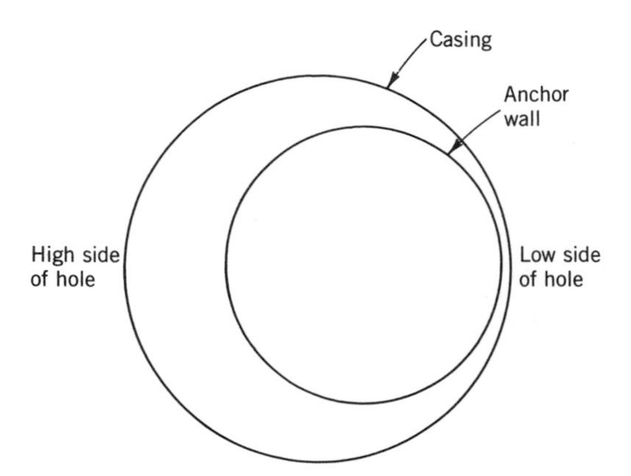

Fig. 11-2 Cup anchor: tubing lies against low side of hole giving good hydraulic radius on the high side.

or, in symbols,

$$E_v = \frac{q}{D} \tag{11-2}$$

The effectiveness of the gas anchor may best be expressed in terms of a *gas-anchor efficiency* E_g, defined as the proportion of the free gas present at the anchor intake that is diverted up the annulus. If, for the sake of simplicity, it is assumed that only negligible volumes of gas come out of solution as the liquid is produced up the hole, and if the effects of temperature on gas volume are omitted from the equations, then the volume rate of gas (measured at the intake pressure p) present at the anchor intake is $qR\, p_a/p$, where R is the producing GLR and p_a is the atmospheric pressure. Of this volume, a quantity qRp_aE_g/p is diverted up the annulus, so that the volume rate of gas entering the pump is

$$D - q = qRp_a\,(1 - E_g)/p \tag{11-3}$$

Dividing Eq. (11-3) by q and using Eq. (11-2) gives

$$\frac{1}{E_v} = 1 + Rp_a\,(1 - E_g)/p$$

$$E_v = \frac{1}{1 + Rp_a\,(1 - E_g)/p} \tag{11-4}$$

Equation (11-4) shows that as p increases, the pump volumetric efficiency, and hence the liquid production rate DE_v, increases.

However, as p—that is, the producing BHP—increases, the production rate from the formation decreases. It follows that pumping production may, in certain circumstances at least, represent a compromise between the performance of the formation and that of the equipment in the hole.

Besides neglecting gas solubility and temperature variation, the analysis given above begs the question of the dependence of E_g on rate and on pressure. Both will clearly influence E_g, because both will affect the entrainment of gas bubbles in the liquid stream. Liquid viscosity and surface tension will also influence the gas-anchor efficiency.

In general terms, various possibilities may be illustrated as shown in Fig. 11-3. Curve (1) of that figure is what might be anticipated if the well were pumped with the annulus closed so that all the production from the formation is passed through the pump. The shape of curve 1 is derived from Eq. (11-4) with E_g set equal to zero. If a bleed line were attached to the casinghead, then some gas would be produced via the annulus, and a curve such as curve 2 would result. Curves (3) and (4) are examples of what might occur with gas-anchor operation. In particular, curve (4) might be expected with either a cup-type or a packer-type anchor. The reason is that at the lower pressures, higher resultant volumes of gas would be moving upward while the liquid was draining back down the hole to the anchor intake, and the volume of gas entrained with the liquid would decrease.

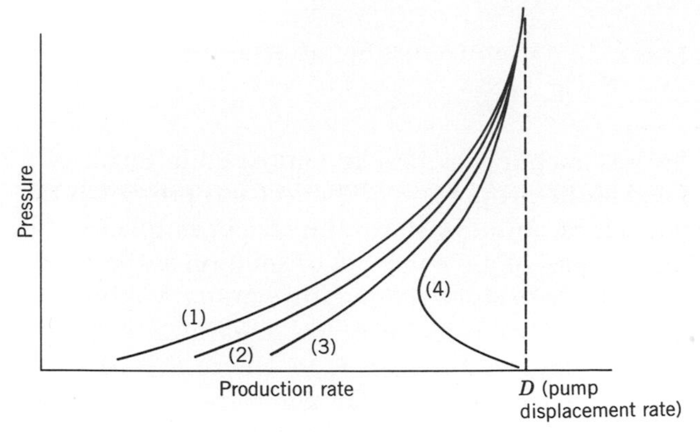

Fig. 11-3 Liquid rate to pump as function of pressure: (1) annulus closed; (2) annulus open; (3) and (4) illustrate possible gas-anchor action.

11-3 AGITATING

Certain pumping wells show no pumping action when tested with a dynamometer (Sec. 10-4) but nevertheless give steady production. If the unit is shut down, production quickly ceases, only to recommence when the reciprocating motion is started again but still without any sign of pumping action. Such a well flows through both the SV and the TV (which are consequently both open throughout the pump cycle) but only as long as the plunger motion is continued. This mode of production, which may continue for many months or may occur intermittently with short periods of true pumping interspersed between the flowing spells, is known as *agitating*. The reason for it seems to be that the action of the rods and plunger causes sufficient gas to come out of solution to permit the well to flow. However, as soon as the unit is stopped, the free-gas/liquid ratio in the tubing falls below the level at which natural flow can be sustained in the well and so production ceases.

11-4 FLUMPING

As the name implies, *flumping* is a combination of flowing and pumping; it differs from agitating in that the pumping action is a true one. The simplest and commonest form of flumping is when a well that is being pumped through the tubing is flowing at the same time from the casing-tubing annulus. There is no question here of a dual completion, and both the annulus and the tubing production come from the same horizon.

Although many wells today are flumping, this condition is, in general, fortuitous and no conscious effort has been made to bring it about. In many pumping wells, however, flumping, if it could be started, would lead to increased production rates. It is therefore worth inquiring into the nature of flumping in the hope that it may be possible to engineer such a production method in certain types of well. As an introduction, an example will be presented to outline the method of approach suggested here and to help emphasize certain points that need to be considered before flumping production is attempted in any well or group of wells.

The method used is to suppose that there is some way of producing liquid from the annulus and then, at various assumed flowing BHPs, to calculate separately the rates at which liquid is produced through the tubing and through the annulus. In this way an annulus IPR is obtained, and the methods for analyzing flowing well performance (Chaps. 4 and 5) may be used to study the presumed annulus flow.

Example 11-1 A well producing from a section from 5004 to 5017 ft is completed with $5^1/_2$-in. perforated casing and 5000 ft of $2^3/_8$-in. tubing. The static BHP in the well at 5000 ft is 1800 psi, and the well's pumped-off rate is 600 bbl/day. The producing GOR is 550 cu ft/bbl, and the well is water-free. From PVT data it is known that there is 200 cu ft/bbl of gas in solution in the oil at 1800 psi and reservoir temperature. The well is being pumped against a line pressure of 60 psig, and the displacement rate (swept volume) of the plunger is 350 bbl/day. Will the well flump, and if so, how may the flumping action be started?

It will be assumed for the purposes of this example that the flowing pressure-distribution curves for a $2^3/_8$-in., $5^1/_2$-in. annulus are identical to those for a $2^7/_8$-in. tubing.

In solving the problem three assumptions will be made: when the well produces on the tubing and the annulus simultaneously, the GOR in each is the same; the producing GOR is independent of the offtake rate and remains constant at 550 cu ft/bbl; and temperature variations and the gas supercompressibility factor may be neglected.

The first step is to draw in the (oil) IPR for the formation. On the assumption that it is a straight line, this may be done as shown in Fig. 11-4 (line 1); the plunger capacity (line 2) is shown on the same figure.

To determine the volume of oil pumped at any particular BHP, it is necessary to find the free-gas volume associated with the oil at that pressure. To do this despite the scarcity of information in this example, use will be made of the fact that the curve of gas solubility against pressure is roughly straight except at very low pressures, and a gas-solubility curve as shown in Fig. 11-5, based on the single known value of 200 cu ft/bbl at 1800 psi, will be assumed.

Using the three assumptions listed above, it is now a straightforward calculation to obtain curves of oil production rate versus pressure for the tubing and for the annulus. The calculations are listed in Table 11-1, and in Fig. 11-4 the total (oil plus free gas) formation IPR is shown (curve 3), together with the curves of oil production into the tubing and into the annulus (curves 4 and 5). The broken line (labeled 6) defines what the situation would be if the well were pumped with the casing closed so that all the output from the formation had to go through the pump (point *A*). Under such circumstances the pumping rate from the well would be some 188 bbl/day of oil.

If it is supposed that the well's casing is open, the question is whether natural

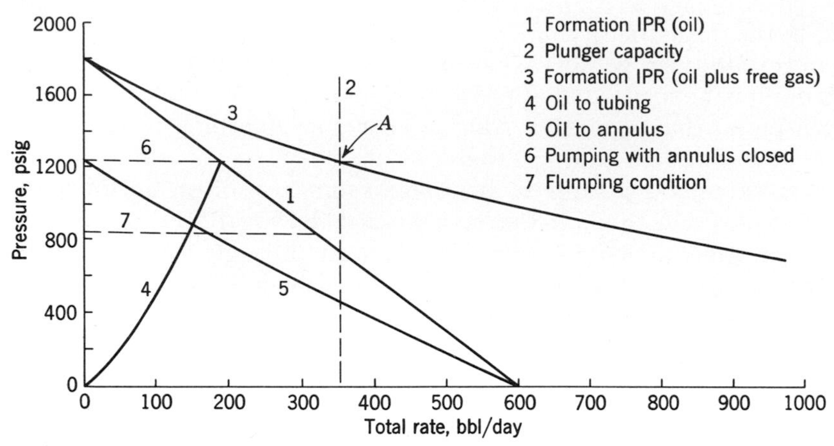

Fig. 11-4 Example 11-1: Distribution of production between tubing and annulus at various assumed flowing BHPs.

flow can take place in the annulus, given the annulus GOR of 550 cu ft/bbl and the annulus IPR typified by line 5 of Fig. 11-4. This is a flowing well problem and may be approached by the methods of Chaps. 4 and 5. Since the line pressure is 60 psig, a reasonable minimum CHP for controlled annular flow would be 100 psig. Using this value for the CHP, the flowing BHP at different rates is determined in Table 11-2 (the pressure-distribution curves for $2^7/_8$-in. tubing being used, as stipulated at the beginning of this example; see Figs. 4-11 through 4-15). The results are plotted in Fig. 11-6 together with the annulus IPR (curve 5 of Fig. 11-4). From Fig. 11-6 it is apparent that if the flowing BHP can be reduced to 840 psig, the well is capable of sustained flow at 170 bbl/day through the annulus against a CHP of 100 psig. Returning to Fig. 11-4, the broken line labeled 7 has been drawn at the pressure of 840 psig and shows that while the annulus is

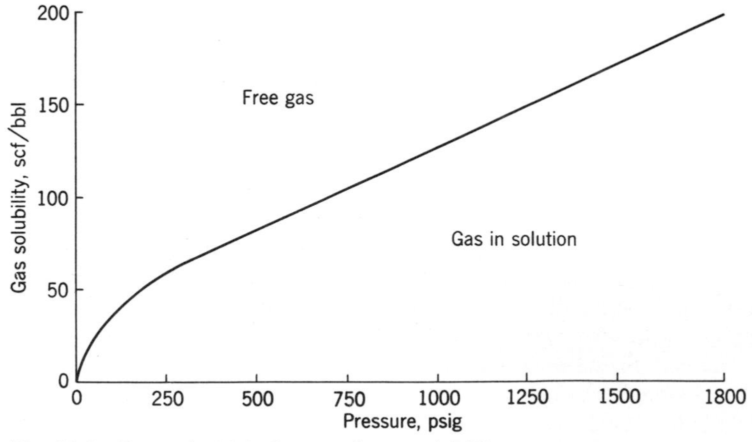

Fig. 11-5 Example 11-1: Assumed gas-solubility curve.

TABLE 11-1 EXAMPLE 11-1: Daily Oil Rates to Pump and to Annulus at Various Intake Pressures

Pressure, psig	Drawdown, psig	Oil Rate, bbl/day	Free-Gas/Oil Ratio, cu ft/bbl*	Free Gas at Standard Conditions, bbl/day†	Free Gas at Pump Conditions, bbl/day‡	Total Vol. from Formation, bbl/day	Oil to Pump, bbl/day§	Oil to Annulus, bbl/day¶
1800	0	0	(350)	0	0	0	0	0
1600	200	67	368	4,392	41	108	67	0
1400	400	133	386	9,145	97	230	133	0
1200	600	200	404	14,393	178	378	185	15
1000	800	267	422	20,070	297	564	166	101
800	1000	333	440	26,099	480	813	144	189
600	1200	400	458	32,633	796	1,196	117	283
400	1400	467	476	39,596	1,431	1,898	86	381
200	1600	533	496	47,091	3,285	3,818	49	484
0	1800	600	550	58,782	58,782	59,382	4	596

* Free-gas/oil ratio equals 550 less gas in solution (from Fig. 11-5).
† Free-gas/oil ratio multiplied by daily oil rate and converted into barrels.
‡ Converted making allowance for pressure only.
§ Oil to pump is equal to plunger displacement multiplied by ratio of oil to total rate, on assumption that GOR in both tubing and annulus is the same.
¶ Oil to annulus is oil rate from formation less oil to pump.

TABLE 11-2 EXAMPLE 11-1: Flowing Performance in the Annulus (CHP 100 psig; GOR 0.55 mcf/bbl; Pressure-Distribution Curves for 2⅜-in., 5½-in. Annulus Assumed to Be the Same as Those for 2⅞-in. Tubing)

Assumed Oil Rate, bbl/day	Equiv. Depth of 100-psi CHP, ft	Equiv. Depth of p_{wf}, ft	p_{wf}, psi
50	750	5750	1050
100	950	5950	925
200	1000	6000	820
400	950	5950	760

flowing at the rate of 170 bbl/day, the tubing will be pumping at the rate of 150 bbl/day, for a total production rate of 320 bbl/day. Thus flumping will substantially increase the offtake rate from the well over the rate obtainable by pumping with the casing closed.

The question remains of how the well can be brought into this flumping state. The first step evidently is to place the correct size of choke in the line from the casinghead. This may be determined from Fig. 5-1 and turns out to be 28/64 in. Once this size bean has been placed in the annulus line, the next step is to reduce the flowing BHP to 840 psig. If the well has been previously pumping with the casing closed, the casing can be opened up on the 28/64-in. choke. If the well has been previously pumping with the casing open, the flowing BHP under the pumping operation can be determined from the well's production rate and IPR. This pressure will presumably be greater than 840 psig (otherwise, there would be nothing to be gained by trying to flump the well). Suppose it is 1150 psig; this implies a steady pumping rate of 215 bbl/day (Fig. 11-4) with some free gas being diverted into the open annulus. The flowing BHP can be reduced to 840 psig by closing in the casing and permitting the CHP to build up by the difference between 1150 and 840 psig, that is, by 310 psi (the increase in CHP is due to free gas accumu-

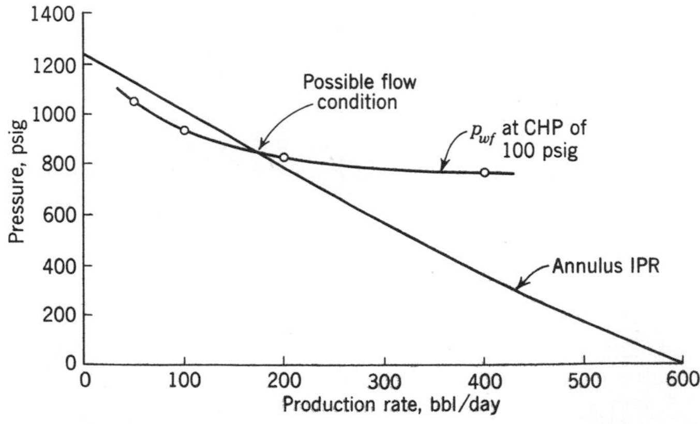

Fig. 11-6 Example 11-1: Establishing annulus flow conditions.

lating in the annulus and reducing the liquid head there). When the required buildup of 310 psi in the CHP has taken place, the casing can be opened up once more, this time on the $^{28}/_{64}$-in. bean, and the flowing BHP will drop to 840 psig.

The final step in the above example may have to be carried out two or three times—perhaps with slightly different choke sizes—before the well can be induced to flow on the annulus.

One apparent shortcoming of the argument illustrated by Example 11-1 is that if a well will flump, then it should flow through the tubing (the natural GLR is evidently sufficiently high to sustain natural flow through the annulus); it might be questioned whether flumping is ever a practical option. The answer lies in gas anchor performance; with an efficient bottom-hole separator, the GLR in the annulus is higher than the natural formation GLR and that in the tubing is lower. This not only improves pumping efficiencies[3] but may permit flow to take place in the annulus, whereas without pump or gas anchor the well would be unable to produce. It is therefore necessary to have a full quantitative understanding of gas anchor operation in order to choose those wells in which flumping is a possibility, to forecast the flumping production rate, to install the correct size of choke in the line from the casinghead, and to achieve the BHP needed before flumping will commence.

What type of well will flump? The answer to this question depends largely on the efficiency of the particular gas anchor in the hole, but as a practical expedient it seems reasonable to try to induce annulus flow in those wells which, when pumping with the casing open, have a high fluid level in the annulus (Secs. 10-8 and 10-9). As was mentioned in Sec. 1-7, the formation GORs in a depletion-drive field tend to rise with cumulative production after an initial period when they are relatively low. Pumps and units might be installed upon completion of the wells in a field with a low initial GOR. Some years later GORs might have risen to such an extent that flumping has become a very real possibility in many of the wells. It is in this type of situation that the wells should be examined with a view to the practicability of inducing flow in the annulus simultaneously with pumping production up the tubing.

11-5 OPTIMUM PUMP SETTING DEPTH: ANNULUS INTERMITTING

Superposition of the IPR grid, that is, pressure versus rate curves at various depths (Fig. 4-24), on the pump/gas anchor (P/GA) curves of Fig. 11-3 shows that for curves of type 1, 2, or 3 the pump should be set at low in the hole as

[3] An attempt to illustrate this is made in Prob. 29, where an arbitrarily assumed gas anchor efficiency is superimposed on the data of Example 11-1. On this assumption it is found that the total production from the well is slightly in excess of the plunger displacement rate, which would imply a pump volumetric efficiency greater than 100 percent if all the production were incorrectly supposed to be taking place through the tubing.

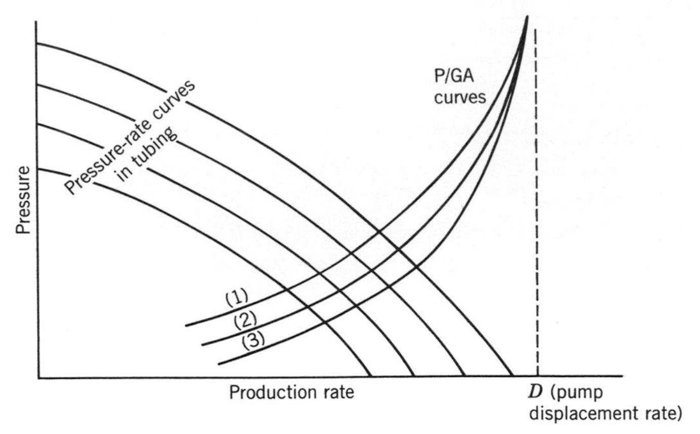

Fig. 11-7 Pump/gas anchor curves superimposed on pressure-rate-depth grid for curves 1, 2, and 3 of Fig. 11-3.

possible in order to achieve maximum production rates (Fig. 11-7). If the gas anchor curve is of type 4, the situation is more complex, and some interesting possibilities arise.

Case *a*: Pump displacement rate D greater than formation potential q'.

Figure 11-8 illustrates the case in which the liquid rate that the P/GA combination can handle exceeds the liquid rate from the formation at all flowing BHPs. In such circumstances the well pumps off and the P/GA should be placed as low in the hole as possible.

In the case illustrated in Fig. 11-9 the formation produces more liquid than the P/GA can handle over the range R_1. If the well were pumped down after being shut in for a period, the production rate would stabilize at the state corresponding to the point S. This is a stable production rate, for if the BHP were decreased somewhat, the formation would produce at a greater liquid rate than the P/GA can accommodate, and the surplus liquid would be pro-

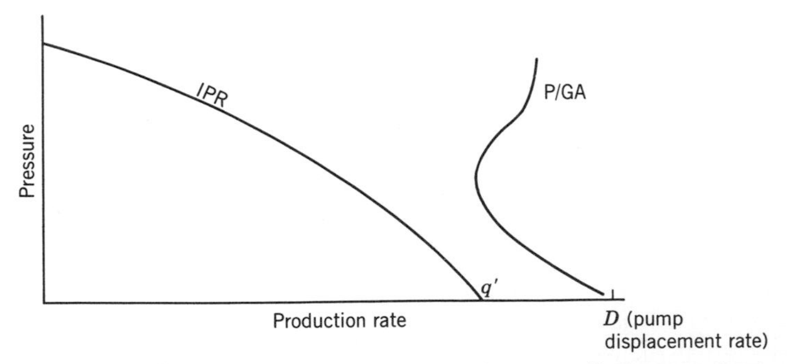

Fig. 11-8 P/GA curve of type 4 (Fig. 11-3); pump displacement rate greater than formation potential.

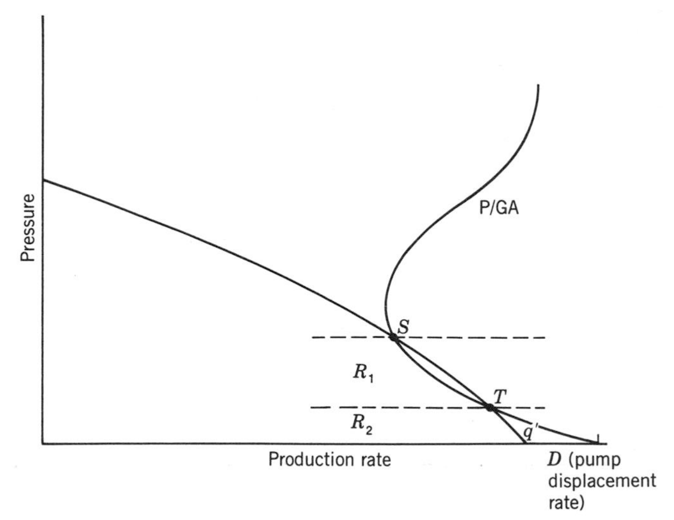

Fig. 11-9 P/GA curve of type 4 (Fig. 11-3); pump displacement rate greater than formation potential, but unstable zone present.

duced into the annulus, increasing the length of the liquid column there, and so increasing the intake pressure until the point *S* was reached. Similarly, any increase in pressure would reduce the rate from the formation, and the P/GA would draw liquid from the annulus, so decreasing the back pressure against the formation, and conditions corresponding to *S* would again be attained.

The annulus could be closed during the pumping operation to force the annulus liquid level down. If the annulus were reopened at a stage corresponding to an intake pressure in the range R_1, liquid would build up once again in the annulus to bring conditions back to *S*. On the other hand, if the annulus were left closed in until the liquid level had been forced down to, or close to, the pump intake and then opened up, bottom-hole conditions might then fall into the range R_2, and the well could be pumped off satisfactorily.

Case *b*: Pump displacement rate less than or about equal to formation potential.

If the P/GA is placed at, or just above, the top perforations, then *S* (Fig. 11-10) is the stable pumping position. There are then three ways in which the production rate might be increased:

1. The pumping equipment might be replaced in order to obtain a larger displacement rate.
2. The well might be kept in an unstable state by intermitting the annulus by means of a timing device at the wellhead. As soon as the stable condition *S* is attained, the annulus would be closed in, and the annulus fluid level forced down. Just before gas blow-round, the casing would be opened,

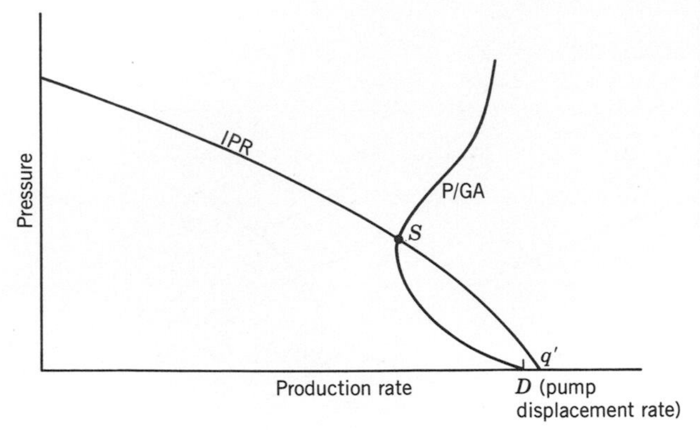

Fig. 11-10 P/GA curve of type 4 (Fig. 11-3); pump displacement rate less than formation potential.

and the formation would produce at a higher rate, the excess liquid moving into the annulus, and so on.

3. The P/GA assembly might be raised off bottom. It is apparent from Fig. 11-11 that the depth at which the pressure versus rate curve (in the casing) is tangential to the P/GA curve is optimal. Since the static pressure can be expected to decline with time (unless the well is subject to pressure maintenance), a pump setting depth somewhat lower than the optimal depth might be advisable. It should be noted that locating the P/GA some way off bottom may lead to reduced production rates even if the gas anchor is operating effectively, and this is illustrated by the curve marked "intermediate depth" on Fig. 11-11.

It is clear that locating the pump off bottom in order to increase the production rate is, at best, a chancy business and—until more is known about gas-

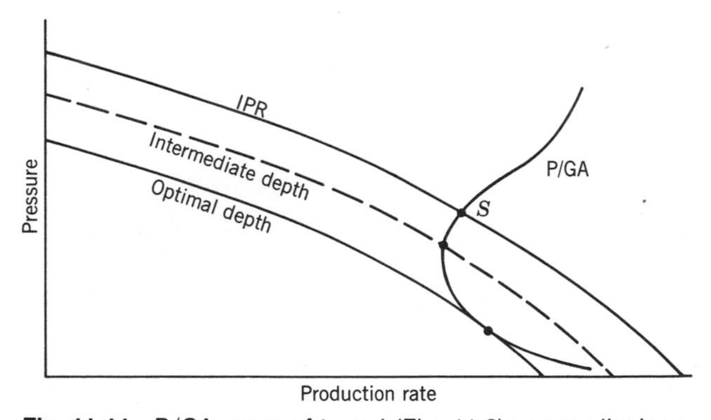

Fig. 11-11 P/GA curve of type 4 (Fig. 11-3); pump displacement rate less than formation potential. Possibility of optimum pump setting depth above the perforations.

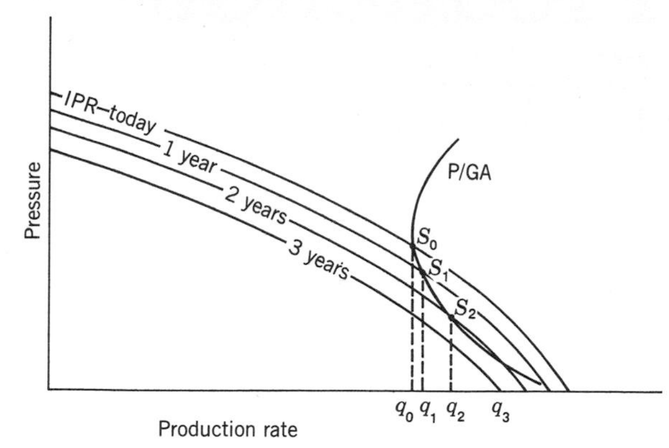

Fig. 11-12 P/GA curve of type 4 (Fig. 11-3); possibility of increased production rate with time.

anchor performance—not something that should be tried on a hit-and-miss basis. However, there are field conditions that make it a practical option and the possibility should not be ignored.

Finally, before undertaking any of the three options summarized here under case *b*, due consideration should be given to the fact that—unless there is a pressure-maintenance project operating in the field—the formation's IPR can be expected to deteriorate with time. As this decrease in IPR occurs, it is possible that the pump will "take hold" over a period of months, or even of years, and that there may be an increase in pumping rate with time (Fig. 11-12).

Production Economics 12

12-1 INTRODUCTION

The ultimate goal underlying the development of the science of reservoir engineering has been defined by Muskat (Ref. 1, p. 27) as the attainment of a maximum efficiency in the exploitation of oil-bearing reservoirs, where the phrase *maximum efficiency* is taken to imply the maximum recovery of oil at a minimum cost. Within these limitations, the goal in the development of production engineering might be said to be *the attainment of maximum efficiency in the operation of those producing wells drilled into an oil-bearing reservoir;* this implies the realization of the maximum profit from each and every such well.

In order to achieve this end, the production engineer must not only know how to analyze and interpret well performance but also be able to translate the resulting interpretation and suggested course of action into dollars and cents to ensure that the recommendation made or the action taken is the one that leads to the maximum profit.

In the first half of this chapter some of the concepts pertaining to the profitability of oil field projects are discussed in broad terms, and in the latter half simple expressions are suggested for making rapid—although admittedly rough—profitability calculations. The advantage of such rapid determinations is that they may eliminate at least some of the possibilities under consideration, leaving perhaps just one or two for a more detailed study.

It will be noted that in the second half of the chapter, specifically in Secs. 12-10, 12-12, and 12-14 to 12-16, a straight-line production rate decline (see

Sec. 2-2) has been assumed in the formulas suggested. As discussed in Sec. 5-5, such a production rate decline is not realistic; it certainly should not be used as the basis for planning future equipment requirements. However, its use for preliminary profitability estimates can be justified because it is *simple to use,* bearing in mind that the many unknowns on the economic side, such as the value of the discount rate applicable to the company, the net value of a barrel of oil, and the changes in these quantities with time, together with the many unknowns on the engineering side, particularly when planning a new project, make a detailed calculation an unrewarding venture. Indeed, such a calculation may at times serve to obscure the main factors and make it difficult to compare the economic possibilities of one project with those of another.[1]

12-2 SOME DEFINITIONS

Gross oil value v is the price received by the field for the oil sold. It is measured in dollars per barrel.

Per-barrel costs c are the costs chargeable against a barrel of oil that depend directly upon the production of that barrel of oil. If the oil is not produced, the per-barrel costs do not arise. Examples of per-barrel costs are royalties, taxes, and certain lifting costs (for example, the cost of the power needed to lift each barrel of oil to the surface).

Oncosts F are costs that go on unchanged when production is reduced or even stopped for a short period. They include the share allotted to the field of overhead costs such as head office costs, research costs, and company insurance and retirement plan costs, in addition to production costs such as production department personnel salaries. Oncosts of this type are measured in dollars per unit time, and the greater part of such oncosts is not directly influenced by the production from a particular field or well. For the majority of successful companies these oncosts remain roughly constant, and for that reason no attempt will be made here to place them on any sort of per-barrel basis. Instead, the net oil value will be defined as follows:

Net oil value u is the gross oil value less the per-barrel costs. When this definition of net oil value is used, the production personnel in some particular area or zone will not be concerned with the oncost attributed to that area by the head office. Rather it will be the duty of the head office to ensure that each producing area or zone in the company is running efficiently in the sense that the oncost attributable to that area is not becoming disproportionately high. As an illustration, suppose that the oil from a certain field is being sold to the refinery for $12.50/bbl. If the costs charged against the field drop by $2.50

[1] Much of what follows in this chapter is taken from a paper entitled "Profitability of Oilfield Projects," presented at the Western Technical Conference of the Engineering Institute of Canada, 1959, and subsequently published in *Southam-MacLean's Oil/Gas World* (Ref. 2).

for every barrel of oil that is not produced, then the net oil value is \$10.00/bbl. The fact that there is a company oncost of, say, \$10 million/year is of no direct concern to the operating departments, since efforts in the field will not affect this flat charge. What it does mean is that, in order to break even on its operations, the company must produce 1 million bbl/year of oil and on every barrel produced in excess of 1 million the net profit will be \$10.

12-3 NEW-INCOME PROJECTS

Consider a typical new project in which an initial capital outlay is required for plant and other facilities, for example, the building of a pipeline between an oil field and a refinery. The financial history of such a project may be illustrated graphically as in Fig. 12-1, by plotting the cumulative profit, or loss, as a function of time. Referring to the figure, from time A to time B' money is being invested in the project until, at its maximum, the cumulative loss reaches the proportions illustrated by BB'.[2] From time B' on, until the project proves uneconomical to continue (time D') and is therefore abandoned, it is supposed that annual revenue exceeds expenditure.[3]

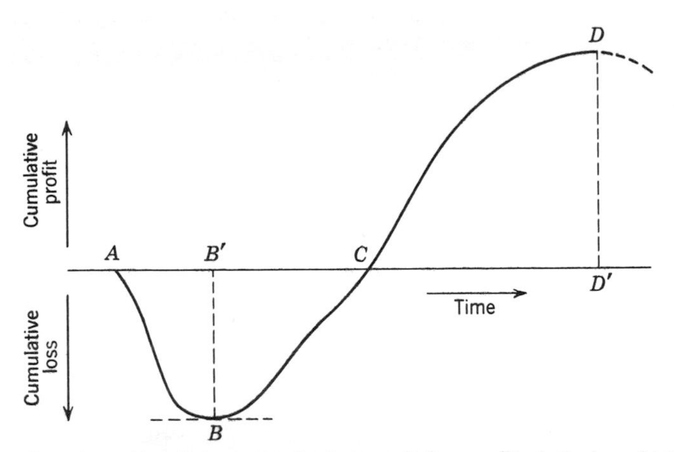

Fig. 12-1 New-income project: cumulative profit plotted against time.

[2] Note that the project might well have started to earn some money prior to B', but until B' was reached, the annual outlay exceeded the income so there was a net loss.

[3] This assumption does not create a serious restriction in the generality of what follows and is brought in merely to eliminate the need to use a series of qualifying phrases in almost every paragraph.

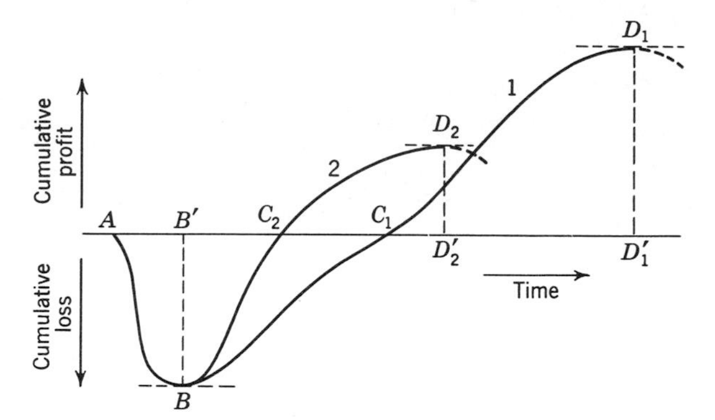

Fig. 12-2 Two new-income projects between which choice must be made.

Definitions

In Fig. 12-1,

BB' = *total capital investment*
AB' = *investment period*
DD' = *final profit,* or *profit* for short (the profit expressed as a percentage of the total capital investment is the *percent profit*)
$B'C$ = *payout* (or *payback*) time
$B'D'$ = *earning life* of the project

It should be noted in passing that AD' is sometimes referred to as the *life* of the project. In the present treatment the project is considered as starting on the day on which it began to record (positive) net incomes. Reasons why the investment period and the earning life of the project are considered as separate entities will be discussed in Sec. 12-8.

Evidently, the value of each of the five quantities mentioned above should be included by the engineer in any proposal for a new-income project that he puts forward to management. Each of these quantities supplies an answer to a very pertinent question:

Can we afford it? *Total capital investment*
How long before we have our money back? *Payout time plus investment period*
How much money do we make? *Profit or percent profit*
How long does it take to make this profit? *Earning life*

Knowledge of these factors is essential to management, but it is by no means the whole story. Consider, for example, the two projects illustrated in Fig. 12-2. At first glance it might appear that project 1 would—obviously—be better for the company than project 2, since the profit made on the same capital would be greater. However, this factor alone does not guarantee the superiority of project 1. Project 2 pays back the initial investment before project 1 (at time C_2 as opposed to time C_1), and it might be that this money becoming available at C_2 would be sufficient to enable an excellent venture to be undertaken but that by time C_1 this opportunity would be lost. Moreover, although project 1 makes more profit than project 2, it has a longer life, which may in fact be disadvantageous. Finally, it is well to remember that $100 invested at 12 percent/year will grow to $176 in 5 years, while $100 invested at 10 percent/year will grow to $177 in 6 years—but very few investors would choose the latter in preference to the former investment.

One method of deciding between two or more projects of types 1 and 2 of Fig. 12-2 is by means of the *rate of return* (for example, Refs. 3–5; other names for methods that are either exactly or virtually the same are the *earning interest rate,* the *investor's cash-flow method,* the *average annual earning power,* and the *discounted cash-flow method*). Before examining this concept in further detail it is necessary to introduce the idea of *present-day value* (*PDV*).

12-4 PRESENT-DAY VALUE (PDV)

If someone owed you $100, it would be to your advantage to be paid today rather than in 10 years' time. For even if you had no immediate need for the money, you could invest it at, say, 8 percent/year and in 10 years' time the sum would have grown to $216. On the other hand, if you were offered the alternative of $100 today or $216 in 10 years' time, then your decision—assuming that you felt that 8 percent was a fair interest rate—would be based on whether you needed the additional capital urgently or whether, for some reason, you preferred to let someone else hold it for you in a form of savings fund.

In this case $100 is said to be the *PDV* or *present worth* of $216 payable in 10 years' time. Evidently, the PDV of a sum of money is dependent on the interest rate used. For instance, the PDV of $200 payable in 10 years' time is $77 at an interest rate of 10 percent/year, while that of the same sum of money payable in 10 years' time is $164 at 2 percent/year.

In symbols, if a sum of money P is invested for n years at an annual interest rate r (expressed as a decimal), then by the end of the period it will have grown to $P(1 + r)^n$. Thus P may be said to be the PDV of $P(1 + r)^n$ payable in n years' time. Or, in other words, the PDV of a sum of money I_n, payable in n years' time, is

$$\text{PDV} = \frac{I_n}{(1 + r)^n} \tag{12-1}$$

where r is the annual interest rate, expressed as a decimal. The quantity $1/(1 + r)^n$ is called the *discount* or *deferment factor.*

12-5 EFFECT OF INFLATION

On Investments

Suppose inflation to be proceeding at an annual rate f (expressed as a decimal). This is taken to mean that a sum of money P now will have, in 1 year's time, a buying power of only $P/(1 + f)$. It follows that a sum of money P invested today at an interest rate r will have a buying power in 1 year's time defined by

$$\text{Buying power} = P\frac{1 + r}{1 + f}$$

The difference between the expression $(1 + r - f)$ and the ratio $(1 + r)/(1 + f)$ is equal to $f(r - f)/(1 + f)$ so that this difference, expressed as a fraction of $(1 + r)/(1 + f)$, is

$$\frac{(1 + r - f) - (1 + r)/(1 + f)}{(1 + r)/(1 + f)} = \frac{f(r - f)}{1 + r}$$

But $f(r - f)/(1 + r)$ is less than x whenever

$$f(r - f) < x(1 + r)$$

or whenever

$$r(f - x) < x + f^2$$

which will always be true, no matter what the value of r, provided the inflation rate f is less than or equal to x.

Hence the expression

$$\text{Buying power} = P(1 + r - f) \qquad (12\text{-}2)$$

may be used as an approximation, the degree of error being at most equal to f. For example, if the annual inflation rate were 8 percent, Eq. (12-2) would be a measure of the actual buying power, accurate to within 8 percent no matter what the value of the interest rate r. Similarly, if the annual inflation rate were 15 percent, Eq. (12-2) would be accurate to within 15 percent, and so on.

Thus, in general, it may be said that inflation can be taken into account in investment calculations by the simple expedient of using the interest rate less the inflation rate in place of the interest rate. This would not hold, of course, to any reasonable degree of accuracy if the inflation rate were to become excessive.

On PDV

Suppose that \$100 were invested at an annual interest rate of 12 percent but that inflation were proceeding at a rate of 8 percent/year. Then in 1 year the sum of \$100 would grow to \$112, but the buying power in present-day dollars of \$112 in 1 year's time would only be \$104, from Eq. (12-2). Thus the sum of \$100 invested now would have a buying power of \$104 (present-day) in 1 year's time.

But \$112 paid in 1 year's time will also have a buying power of \$104 (present-day), so from the point of view of the purchasing power at the time the money will be paid, the PDV of \$112 payable in 1 year's time is \$100. That is,

$$\text{PDV} = \frac{100}{1 + r}$$

and generally

$$\text{PDV} = \frac{P}{(1 + r)^n}$$

It follows that inflation does not affect the PDV calculations.

12-6 EFFECT OF PDV ON PROJECT ANALYSIS

In the discussion of Sec. 12-3 and in Fig. 12-1, no allowance was made for *discounting,* that is, finding the PDV of, future incomes or expenditures.

For the purposes of the present argument, let it be supposed that the investment period (AB' of Fig. 12-1) is short, say, less than 1 year, so that the problem of discounting over this period does not arise (it will be discussed later, Sec. 12-8).

In order to summarize and compare the capital investments and the profits of a venture, it is useful to express all investments and revenues in their equivalent values at one common instant of time. A convenient time base is point B' (Fig. 12-1), at which the project starts to make a (positive) net income (see Sec. 12-8); this point will be used in the present discussion.

Figure 12-3 illustrates the curves of cumulative profit against time that might be obtained for a project in four different cases, namely, incomes undiscounted and incomes discounted at 1, 5, and 10 percent/year. The order in which the curves appear on the graph is self-evident when it is recalled that in discounting future incomes, the income in the nth year has to be multiplied by a factor $1/(1 + r)^n$, which is always less than 1 and which decreases as r increases.

In Sec. 12-3 it was stated that a knowledge of the total capital investment, investment period, payout time, profit, and life of a project is an essential minimum required by management for decision making of the type under review. Although the total capital investment and the life of the project will not be affected by the considerations of the present section, it is obvious from Fig. 12-3 that the discounted profit will decrease, and the discounted payout time will increase, as the discount rate is increased.

Although use of undiscounted values will lead to results unduly favorable

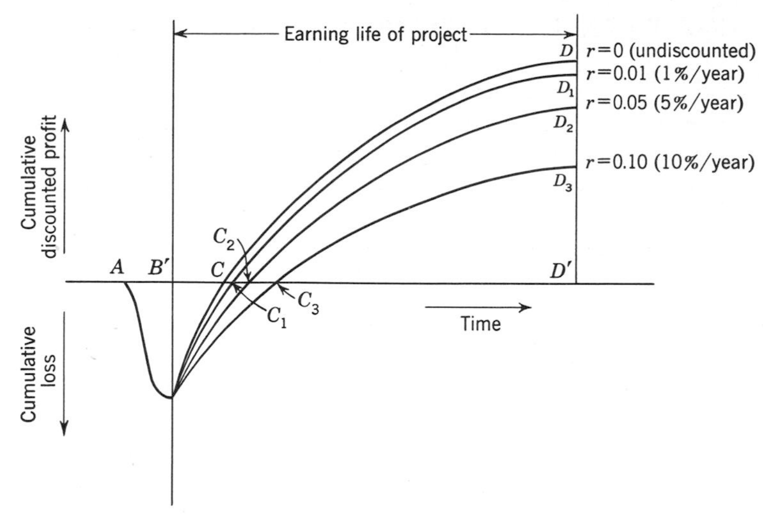

Fig. 12-3 Effect of discount rate on cumulative profit versus time plot: new-income project.

to the project (highest profit, shortest payout time), there is also a major drawback to the use of discounted values. This is provided by the question: What discount rate should be used? If the company has raised the greater part of the capital required by means of a loan, then common sense points to the use of the interest rate on the loan as the discount rate. If, however, as so frequently is the case in the oil industry, the company generates its own capital, then the discount rate should reflect the earning power of the current capital investments of the company. But this is not an easy figure to find with any degree of certainty, and frequently within a company there will be considerable difference of opinion concerning the discount rate that should be used.

In whatever manner these items of contention are resolved in a particular company, what management does require is consistent sets of figures, the set for a particular project outlining the project's financial possibilities. These figures should be based on simple assumptions and be such that the figures for one proposed project can be compared directly with those for another. One possibility, of course, is to present the undiscounted payout time and profit, as in this way any uncertainty as to the most realistic discount rate or its possible variations is avoided. In order to indicate in just how favorable a light the project is being shown by use of the undiscounted method, it might be worthwhile to present the figures in some such way as the following:

Payout time, 2.6 years; if discounted 10 percent/year, 3.3 years
Profit 370 percent; if discounted 10 percent/year, 275 percent

The discount rate used in this type of presentation should, of course, be the one considered to be the most realistic for the company, and the same value should be used in every project presented for management's consideration.

12-7 RATE OF RETURN

To see how rate of return is defined, consider Fig. 12-3 once more. As the discount rate r increases above 10 percent/year, point D will approach D', and for some value R of the discount rate point D will coincide with D', as illustrated in Fig. 12-4. This value R is defined to be the *rate of return* of the project. It can be shown that the rate of return is the rate of interest realized on the capital invested. To put it another way, if the capital for a certain project is borrowed from a bank and if profits from the venture are used to reduce the outstanding debt, then the rate of return of the project will equal the rate of interest the bank charges on the outstanding loan if the final profit from the venture, made at the end of the earning life, is just sufficient to cover the final payment to the bank.

It is evident that, other things being equal, the higher the rate of return, the more attractive the project.

Many advocates of the use of rate of return appear to regard it as the only decision criterion that management needs: proposed projects would be listed in order of decreasing rate of return, and the projects undertaken starting with the top one and working on down the list. But rate of return does not supply sufficient information on its own. For instance, it gives no idea as to the *amounts* of money involved, so that the estimated capital investment and the ultimate (or percent) profit should always be reported; it gives no clear picture of how long it will be before the capital expenditure will be returned to the company by way of operating profit (payout time); and it gives no idea of the life of the project or the time for which money will have to be invested without sufficient compensating returns. This information is just as essential to man-

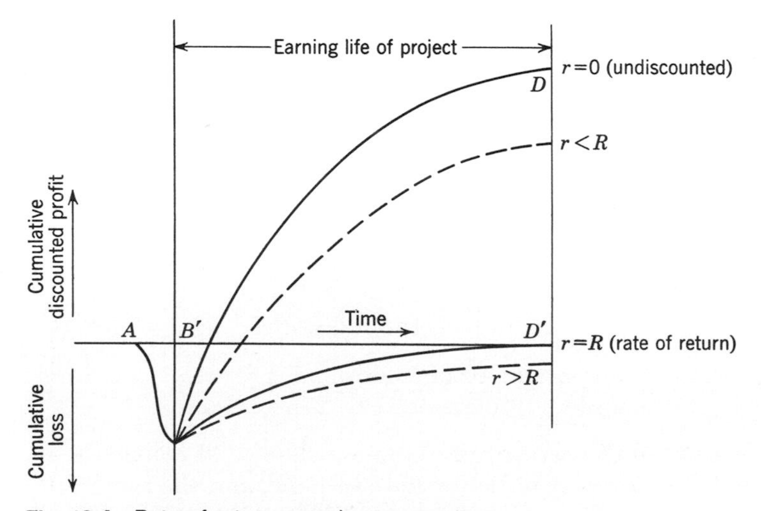

Fig. 12-4 Rate of return: new-income project.

agement as is the ranking criterion, rate of return, in making final decisions as to which projects to put into effect and which to defer or reject.

It thus appears that, in order to decide between projects, management requires, at the minimum, the following six figures:

Total capital investment
Investment period
Profit (or percent profit)
Payout time
Earning life of the project
Rate of return

In addition, the engineers preparing the proposals to be submitted to management might want to include a *risk factor,* which, although necessarily subjective, will give an indication of the chances that the project under consideration might prove a complete or partial failure.

Example 12-1 To illustrate the discussion to this point, consider the following example. It is estimated that the capital cost of a certain project will be \$31,000, that the earning life of the project will be 6 years, and that the net incomes in these 6 years will be \$5000, \$12,000, \$13,000, \$12,000, \$12,000, and \$8000, respectively. Calculate the undiscounted percent profit and payout time, the discounted values based on a discount rate of 10 percent/year, and the rate of return.

For purposes of this example, it will be assumed that the income for any year is received over the course of that year (rather than as a lump sum paid on some particular date). Making this assumption, a reasonably good simplifying approximation is that the entire income for a particular year is paid as a lump sum at the midpoint of the year. This approximation was used in preparing Table 12-1 and Fig. 12-5.

From Table 12-1 it is evident that the undiscounted profit is

$$\$62{,}000 - \$31{,}000 = \$31{,}000$$

so that the undiscounted percent profit is

$$\frac{31{,}000}{31{,}000} \times 100 = 100 \text{ percent}$$

Similarly, the discounted profit is

$$\$46{,}554 - \$31{,}000 = \$15{,}554$$

so that the discounted percent profit is

$$\frac{15{,}554}{31{,}000} \times 100 = 50.2 \text{ percent}$$

From Fig. 12-5, the undiscounted payout time is a little less than 3.1 years, and the discounted payout time is 3.65 years.

A convenient way of finding the rate of return is to plot the cumulative discounted net income at various assumed discount rates against the discount rate. The point at which this curve is intersected by the capital cost of the project will

TABLE 12-1 EXAMPLE 12-1: Discounted Profit as a Function of Time

Year	Undiscounted Net Income	Cumulative Undiscounted Net Income	Undiscounted Profit	Discount Factor 10%/year	Discounted Net Income	Cumulative Discounted Net Income	Discounted Profit
1	\$ 5,000	\$ 5,000	\$−26,000	$\frac{1}{(1,1)^{1/2}} = 0.9535$	\$ 4,768	\$ 4,768	\$−26,232
2	12,000	17,000	−14,000	$\frac{1}{(1.1)^{3/2}} = 0.8668$	10,402	15,170	−15,830
3	13,000	30,000	−1,000	$\frac{1}{(1.1)^{5/2}} = 0.788$	10,244	25,414	−5,586
4	12,000	42,000	+11,000	$\frac{1}{(1.1)^{7/2}} = 0.716$	8,592	34,006	+3,006
5	12,000	54,000	+23,000	$\frac{1}{(1.1)^{9/2}} = 0.651$	7,812	41,818	+10,818
6	8,000	62,000	+31,000	$\frac{1}{(1.1)^{11/2}} = 0.592$	4,736	46,554	+15,554

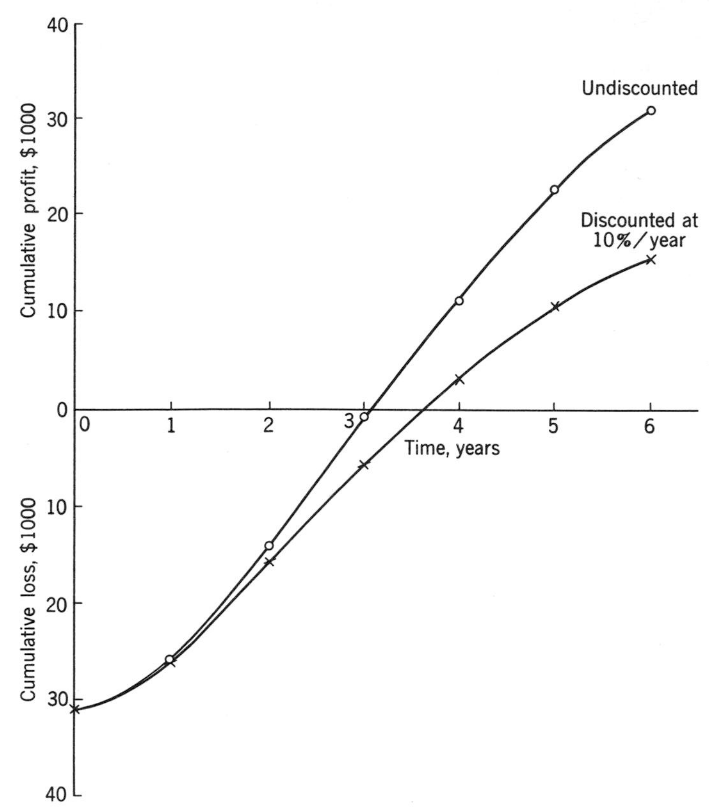

Fig. 12-5 Example 12-1: Cumulative profit versus time plot.

TABLE 12-2 EXAMPLE 12-1: Cumulative Discounted Net Income as a Function of Discount Rate

		Discount Rate					
		10 percent/year		20 percent/year		30 percent/year	
Year	Undiscounted Net Income	Discount Factor	Discounted Net Income	Discount Factor	Discounted Net Income	Discount Factor	Discounted Net Income
1	$ 5,000	0.9535	$ 4,768	0.9130	$ 4,565	0.8771	$ 4,386
2	12,000	0.8668	10,402	0.761	9,132	0.675	8,100
3	13,000	0.788	10,244	0.634	8,242	0.519	6,747
4	12,000	0.716	8,592	0.529	6,348	0.400	4,800
5	12,000	0.651	7,812	0.441	5,292	0.307	3,684
6	8,000	0.592	4,736	0.367	2,936	0.236	1,888
Total	$62,000		$46,554		$36,515		$29,605

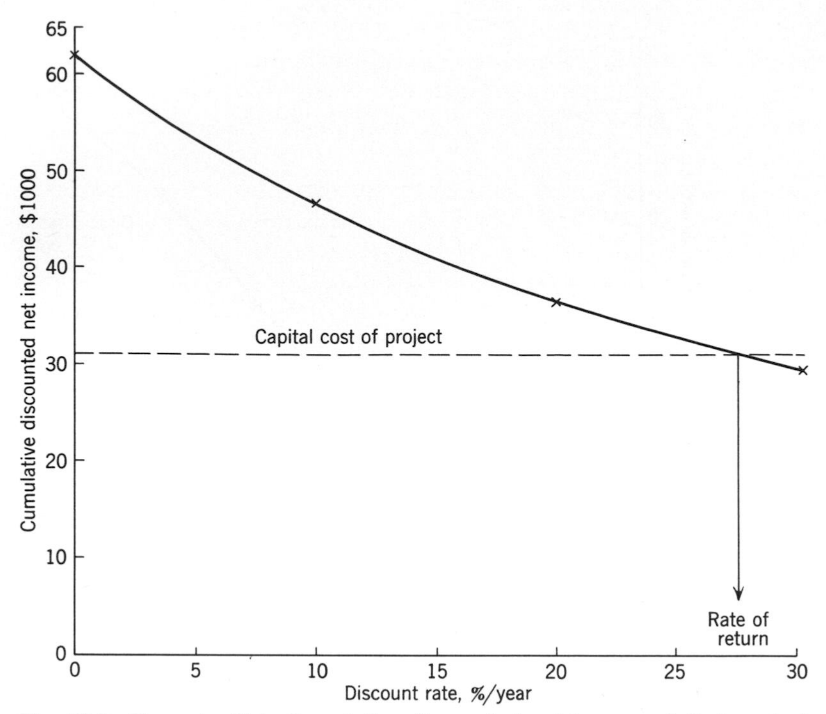

Fig. 12-6 Example 12-1: Cumulative discounted net income plotted against discount rate.

define the rate of return. Table 12-2 and Fig. 12-6 illustrate the procedure, and from Fig. 12-6 a rate of return of 27.6 percent/year for the project under consideration is obtained.

12-8 LONG-TERM CAPITAL EXPENDITURE

In what has gone before, the capital expenditure necessary to set up the project under consideration has been supposed to have been incurred in a short period of time, say, less than 1 year. If, as is frequently the case for any large-scale project and is particularly true of the exploration period preceding the discovery of an oil field, the expenditure is incurred over a considerable length of time, the question arises as to when the project should be considered as starting. There are three ready candidates for the position (Fig. 12-7), namely, the time at which the first expenditure was incurred (point *A*), the time at which the project made its first positive net income (point *B'*), and the time at which the capital costs were paid off and the project started to make a profit (point *C*). These points will be considered in reverse order.

To use point *C* would be unrealistic on two counts: the payout time *B'C* is a function of the discount rate, so that the utilization of *C* would involve a starting point of the project that would vary with the discount rate; the project might never make any profit. In such circumstances point C would never be reached and the project would never be officially started, despite the fact that for many years capital had been poured into machinery, which had been used to turn out salable products.

The use of B' also appears susceptible to the second of these criticisms. The project might never have had even a day on which the income exceeded the expenditure. Projects of this type are indeed commonplace in the oil industry: the outstanding example is of course an exploration venture that is unsuccessful. This seems to indicate that A should be taken as the starting point, but before drawing such a definite conclusion it would be well to examine whether in fact the choice of the starting point is of any importance.

The total capital investment is independent of the starting point, and the PDV of the total capital investment can always be determined whatever *present day* might be chosen, once a discount rate has been agreed upon. Similarly, the investment period, the profit, the payout time, and the earning life of the project are all independent of the starting point.

The rate of return, however, does depend on the starting point used in the calculation. To illustrate this by a simple example, suppose that \$1 is invested today in a project that in 1 year's time yields \$1 net income and in a further year's time yields another \$1. This example is shown graphically in Fig. 12-8; Fig. 12-8*a* shows the yearly expenditures and incomes, and Fig. 12-8*b* is drawn to conform to the pattern of Fig. 12-1.

If A is taken as the starting point, the rate of return R is derived from the equation

$$1 = \frac{1}{1 + R} + \frac{1}{(1 + R)^2}$$

where the left-hand side is the PDV of the expenditure and the right-hand side is the PDV of the net incomes, the present day being taken as the starting point A and R being used as the discount rate.

This equation leads to a value for R of 0.618, or 61.8 percent/year.

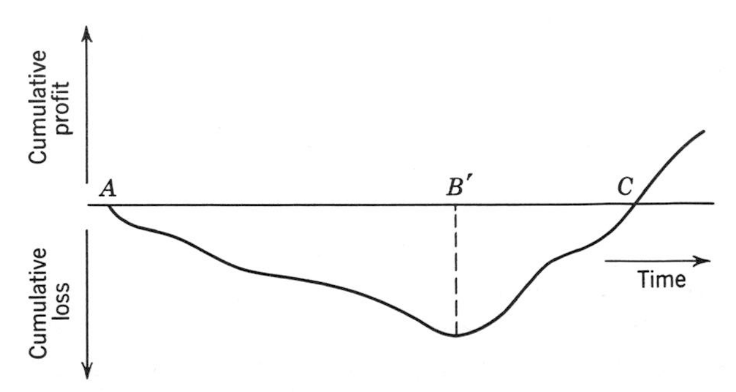

Fig. 12-7 Long-term capital expenditure: three possibilities for the starting point of the project.

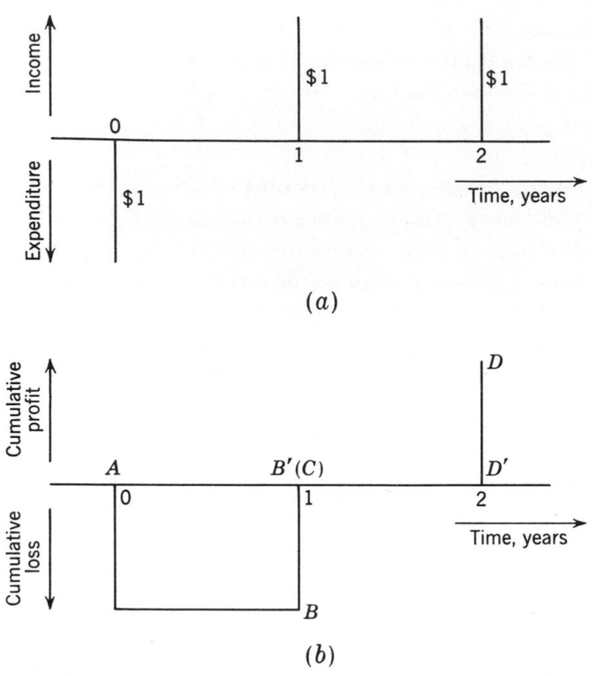

Fig. 12-8 Long-term capital expenditure: example illustrating influence of starting point on rate of return.

On the other hand, if B' is taken as starting point, the expenditure will have been made 1 year ago, so the PDV of that expenditure at the time B' will be $1(1 + r)$, where r is the interest rate. This must be equal to the PDV of the net incomes discounted at the rate of return R. The equation for R becomes

$$1 + r = 1 + \frac{1}{1 + R}$$

If r is taken as, say, 0.1 (10 percent/year), R from this equation will be 9, or 900 percent/year.

Thus the choice of starting point can affect the calculated rate of return to a considerable extent. If A were taken as the start of the project, all capital expenditures would be charged at the rate of return in the rate-of-return calculation. Is this, in fact, a fair way of assessing the project?

Suppose a company contemplates purchasing a machine costing \$2200 and, after estimating the profits this machine will render, calculates that the rate of return is 20 percent/year. Does it make any difference to the attractiveness of the project if the \$2200 is paid out of a current capital account or out of a savings account into which \$200 a year has been paid for the last 9 years, the interest on the capital in the savings account being 4 percent/year? Clearly, the answer is no.

It follows that if the PDV of the total capital investment in two projects is the same and if the earning history of those two projects is also the same, then the two projects should have the same rate of return. The investment history does not enter the picture, except insofar as it determines the PDV of the total capital investment. It would appear logical, therefore, to use the point B' as the starting point of the project, to accumulate all expenditures prior to this point at the *earning power* of the company in order to obtain the PDV of the capital investment at the point B', and then to calculate the rate of return as outlined in Sec. 12-7.

It should be noted that this proposal is at variance with the general formulas given for the calculation of rate of return in much of the literature (see Refs. 5, 7, and 8, for example). These formulas are based on the (generally undiscussed) assumption that point A of Fig. 12-7, that is, the date investment begins, is the starting point of the project, with the result that both investments and incomes are discounted at the rate of return.

12-9 ACCELERATION PROJECTS

An acceleration project may be defined as a project applied to an already existing profitable enterprise in order to bring future net incomes forward in time. A basic assumption made is that the acceleration project causes no alteration in the cumulative undiscounted net income to be received.

Such a definition is, of course, an oversimplification. On the producing side of the oil and gas industries, the assumption would imply not only that no change occurs in the ultimate cumulative production but also that older wells are no more expensive to operate than newer wells, since one result of an acceleration project in a producing field is that the operating life of each well is reduced. Neither of these conditions is likely to be realized in practice. Many projects—for instance, the drilling of infill wells—are justified on the assumption that no increase in ultimate recovery from the well or pool will occur, but such an assumption is generally made because it is, if anything, conservative and because it is virtually impossible to assign a figure to any increase in cumulative oil production that may result. The second point (that is, that of a shorter well life) is hardly ever taken into consideration, but it is one of the factors in favor of an acceleration project, particularly in a pool where conditions are so corrosive that serious trouble may be expected to develop with tubing and possibly with casing strings over an extended producing life.

The justification for using the restricting assumption in the definition of an acceleration project is that if a project should appear profitable when analyzed on the basis of such a restriction, provided it is not a technical failure, it is assured of being profitable in fact. Moreover, predictions made on the basis of the definition quoted have been found by experience to be sufficiently good approximations to be of some practical significance.

Under the definition given, the cumulative profit–cumulative loss history of a typical acceleration project is as shown in Fig. 12-9, in which C_a is the capi-

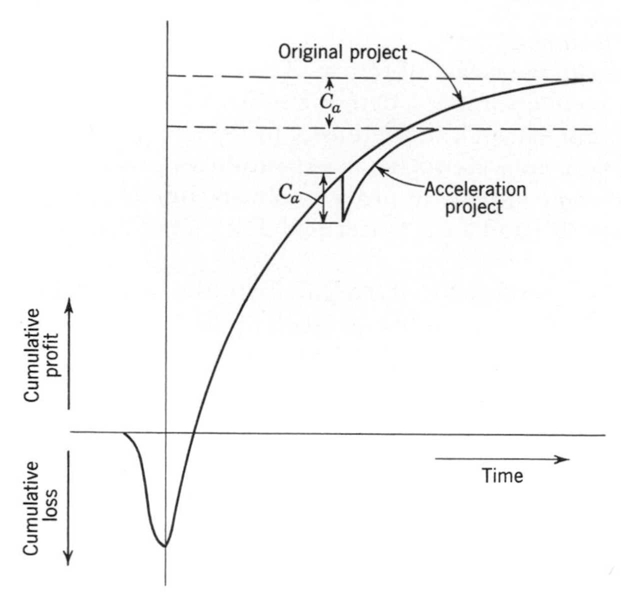

Fig. 12-9 Acceleration project: undiscounted cumulative profit versus time plot.

tal cost of the project. It should be noted that the undeferred profit of the original enterprise exceeds that of the accelerated project by an amount equal to C_a. The economic justification for accelerating lies in the fact that the accelerated income is earned in fewer years than the unaccelerated, so that discounting will have less effect on the former than on the latter. It then becomes conceivable that the discounted profit from the accelerated project might exceed that from the original project. If, in Fig. 12-10, the capital cost of the acceleration project is less than the difference between the cumulative discounted net income (accelerated) at 10 percent and the cumulative discounted net income (unaccelerated) at 10 percent, then the acceleration is profitable with discounting at 10 percent.

Turning now to the decision criteria discussed in Secs. 12-3, 12-6, and 12-7, it is evident that there is no particular difficulty in assigning values to the capital investment, investment period, profit (or percent profit), and future life (perhaps compared with the unaccelerated future life) for an acceleration project. A distinction may be made between the *payback* and *payout* times, the payback time being defined as the time at which the acceleration project has paid back its own capital cost (*OA* of Fig. 12-11) and the payout time as the time at which the difference between the cumulative net income derived from the accelerated project and the cumulative net income that would have been derived from the original (unaccelerated) enterprise if left undisturbed is equal to the capital cost of the acceleration project. In Fig. 12-11, *OB* is the payout time. Both payout and payback times may be calculated on either an undiscounted or a discounted basis.

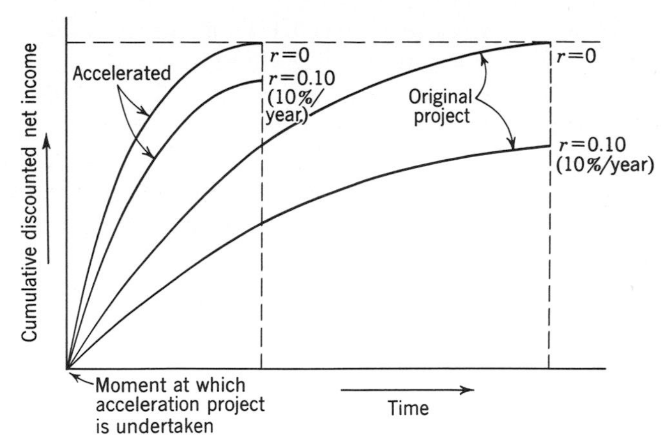

Fig. 12-10 Effect of discounting on accelerated versus original project.

The standard method of determining the rate of return of an acceleration project is to apply a discount rate r to both the accelerated and the unaccelerated projects (Ref. 5). This discount rate is said to be the *rate of return* R when the difference between the PDV of the cumulative net incomes, discounted at the rate R, is C_a.

Figure 12-10 shows that such a procedure will lead, in general, to two answers or none. For when r is zero, the difference in the cumulative net incomes is zero. As r increases, the difference in the cumulative discounted net incomes will also increase for a period; but it must begin to decrease again when r becomes very large, for when r is infinitely large, both the accelerated and unaccelerated curves will coincide with the abscissa, since the discount factors will all be zero. Thus either the difference in PDVs will never be as

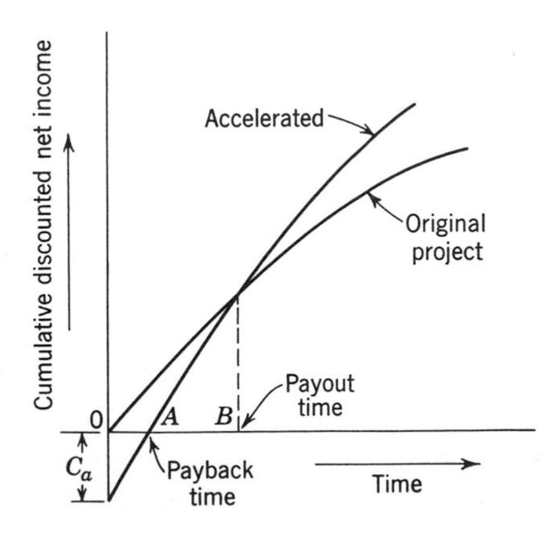

Fig. 12-11 Acceleration project: payout and payback times.

much as C_a, or if this difference takes on the value C_a once as r increases, it will take it on a second time as r continues to increase (Fig. 12-12).

The real significance of Fig. 12-12 is that it defines the limits within which the earning power of the company must fall if the acceleration project under consideration is to be profitable for that company. For if the company earning power is less than r_1 or greater than r_2 (Fig. 12-12), the project will clearly be unprofitable; for a company with an earning power between r_1 and r_2, the project will be worthwhile (Ref. 9).

A more serious criticism may be leveled at the method mentioned above for determining the rate of return of an acceleration project, namely, that the *calculated rate of return is independent of the discount rate.*

At first sight this might appear to be a property strengthening the usefulness of R as a decision criterion for acceleration projects. But consider two companies X and Y, each having exactly the same capital, projects, and so on. Suppose that X has recently been taken over by a new management that is extremely efficient and can invest money to much greater advantage than can the management of Y. Now let both X and Y undertake the same acceleration project; the incomes from the previously existing project come in sooner than they would have, and the company (X) that can reap the greater benefits from that earlier influx of revenue will have gained more benefit from the acceleration project. This implies that the *rate of return of an acceleration project should depend on the earning power of the company:* the greater the earning power, the greater the rate of return.

A definition can in fact be given by which the rate of return has only one value. Suppose that r is the earning power of the current capital investments of the company. Construct graphs for the accelerated and unaccelerated cumulative net incomes of the project under consideration, discounted at rate r (see Fig. 12-13*a*). Next construct the difference curve, accelerated less unaccelerated cumulative discounted net income, as shown in Fig. 12-13*b*. In this fig-

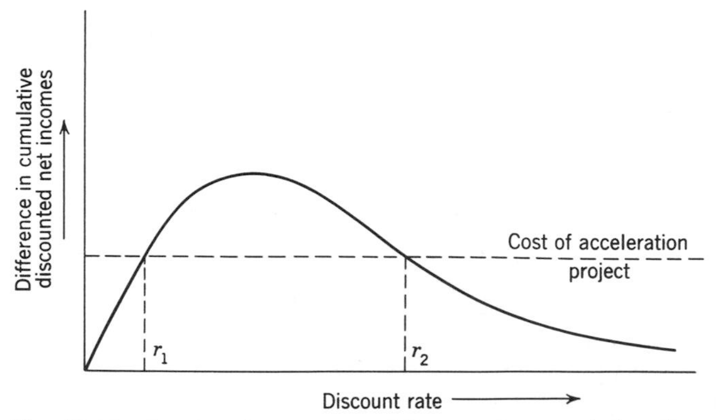

Fig. 12-12 Acceleration project: difference in cumulative discounted net incomes plotted against discount rate.

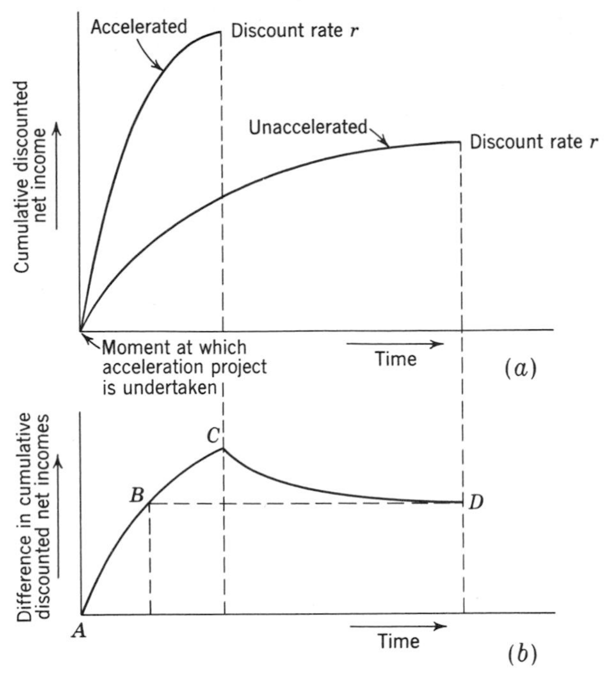

Fig. 12-13 Acceleration project: a method for determining the rate of return.

ure, at point B the difference in the cumulative discounted net incomes is equal to the ultimate difference; that is, from point B on there is no need to worry about the accelerated project vis-à-vis the unaccelerated. Whatever money is made due to the acceleration will, when invested in the general company activities, offset the income that is not realized by virtue of forgoing the unaccelerated earning period CD. Thus the rate of return may be based on the period AB alone and defined to be that discount rate which, when applied to the net income attributable to acceleration over the period AB, results in a cumulative discounted net income (for the period AB) equal to the cost of the acceleration project. Although not immediately obvious, it can be shown that the rate of return calculated in this way increases (as it should; see above) as the earning power of the company increases. Moreover, this method results in a unique value for the rate of return.

For acceleration projects with lives of more than, say, 10 years, the larger of the two values resulting from the rate-of-return calculation outlined first is roughly the same as the value for the rate of return defined by the second method; but for shorter projects the answer derived from the second method lies between the two answers given by the first.

In general, there is no direct comparison between the rate of return of an acceleration project and that of a new-income project. Indeed, closer analysis

TABLE 12-3 EXAMPLE 12-2: Basic Information

Year	Annual Net Income, Dollars: Original Project	If Accelerated
1	4,000	7,000
2	5,000	15,000
3	4,000	6,000
4	6,000	2,000
5	3,000	
6	3,000	
7	2,000	
8	1,000	
9	1,000	
10	1,000	
Total	30,000	30,000

leads to the questioning of the utility of rate of return as one of the decision criteria for acceleration projects. As yet there appears to be no clear-cut answer one way or the other on this point, and the best that can be said is that the results of calculations for the rate of return of acceleration projects should be treated with considerable caution.

At this stage it is appropriate to illustrate some of the ideas relating to acceleration projects by means of an example.

Example 12-2 It is estimated that the future net incomes for a certain project will be as shown in Table 12-3. This project is under consideration for acceleration, and the estimated future net incomes in this eventuality are also shown. The

TABLE 12-4 EXAMPLE 12-2: Difference in Cumulative Discounted Net Incomes as a Function of Time

Year	Discount Factor, 10%/year	Discounted Annual Net Income: Original Project	If Accelerated	Cumulative Discounted Net Income: Original Project	If Accelerated	Difference in Cumulative Discounted Net Incomes
1	0.9535	$3814	$ 6,675	$ 3,814	$ 6,675	$ 2,861
2	0.8668	4334	13,002	8,148	19,677	11,529
3	0.788	3152	4,728	11,300	24,405	13,105
4	0.716	4296	1,432	15,596	25,837	10,241
5	0.651	1953		17,549	25,837	8,288
6	0.592	1776		19,325	25,837	6,512
7	0.538	1076		20,401	25,837	5,436
8	0.489	489		20,890	25,837	4,947
9	0.445	445		21,335	25,837	4,502
10	0.404	404		21,739	25,837	4,098

capital cost involved in undertaking the acceleration is $3000. Determine the discounted profit and payback and payout times, draw a curve of difference in cumulative discounted net incomes (accelerated less unaccelerated) against discount rate, and find the rate of return if the earning power of the company is 10 percent/year. *Note:* It will be assumed that returns come in regularly, but in order to keep the discount calculations within reasonable proportions in this illustrative example, an approximation will be used in determining the discount factors, namely, that the income in any period (taken as 1 year, with the exception of one step in the rate-of-return calculation) will be discounted as if it were paid as a lump sum at the midpoint of that period.

In Table 12-4 the future annual net incomes for both the original (unaccelerated) and the accelerated project are discounted at the company earning power of 10 percent/year. The final difference in cumulative discounted net incomes is $4098, so allowing for the capital cost of $3000 needed to set the acceleration project in motion, it is seen that the discounted profit is $1098, or 36.6 percent.

The cumulative discounted net incomes for both the unaccelerated and accelerated cases are shown in Fig. 12-14 as functions of time, and it is apparent that the payback time of the proposed project is about 0.45 year, or $5^1/_2$ months.

In Fig. 12-15 the difference in cumulative discounted net incomes is shown on an annual basis, and it can be seen that the payout time is a little over 1 year. Also, from Fig. 12-15 it appears that the rate-of-return calculation may be carried out using the difference in income over the initial 1.15 years only.

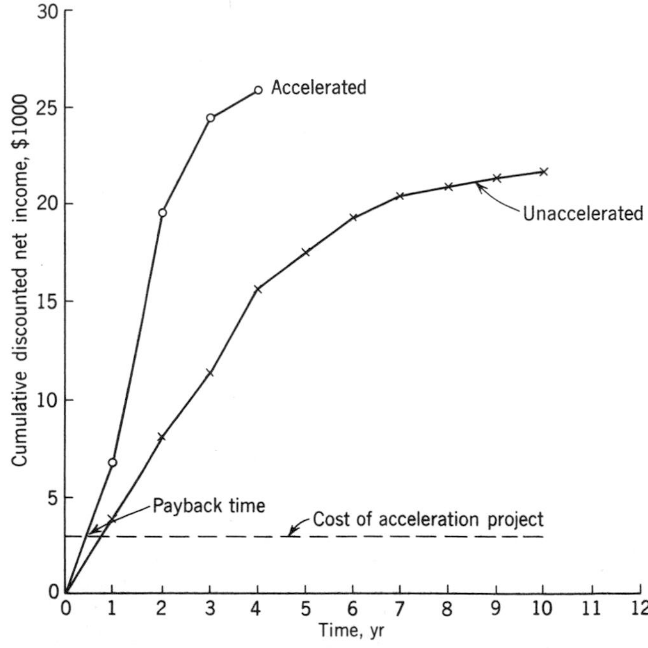

Fig. 12-14 Example 12-2: Cumulative discounted net incomes plotted against time.

TABLE 12-5 EXAMPLE 12-2: Difference in Cumulative Discounted Net Incomes as a Function of the Discount Rate

Year	Discount Rate														
	30%/Year			50%/Year			100%/Year			200%/Year			400%/Year		
	Discount Factor	Discounted Annual Income		Discount Factor	Discounted Annual Income		Discount Factor	Discounted Annual Income		Discount Factor	Discounted Annual Income		Discount Factor	Discounted Annual Income	
		Unacc.	Acc.		Unacc.	Acc.		Unacc.	Acc.		Unacc.	Acc.		Unacc.	Acc.
1	0.877	$ 3,508	$ 6,139	0.817	$3,268	$ 5,719	0.707	$2,828	$ 4,949	0.578	$2,312	$4,046	0.448	$1,792	$3,136
2	0.675	3,375	10,125	0.545	2,725	8,175	0.354	1,770	5,310	0.193	965	2,895	0.090	450	1,350
3	0.519	2,076	3,114	0.363	1,452	2,178	0.177	708	1,062	0.064	256	384	0.018	72	108
4	0.400	2,400	800	0.242	1,452	484	0.089	534	178	0.021	126	42	0.004	24	8
5	0.308	924		0.161	483		0.044	132		0.007	21		0.001	3	
6	0.236	708		0.108	324		0.022	66		0.002	6		0.000	0	
7	0.182	364		0.072	144		0.011	22		0.001	2		0.000	0	
8	0.140	140		0.048	48		0.006	6		0.000	0		0.000	0	
9	0.108	108		0.032	32		0.003	3		0.000	0		0.000	0	
10	0.083	83		0.021	21		0.001	1		0.000	0		0.000	0	
Total		$13,686	$20,178		$9,949	$16,556		$6,070	$11,499		$3,688	$7,367		$2,341	$4,602
Difference (accelerated less unaccelerated)		$6,492			$6,607			$5,429			$3,679			$2,261	

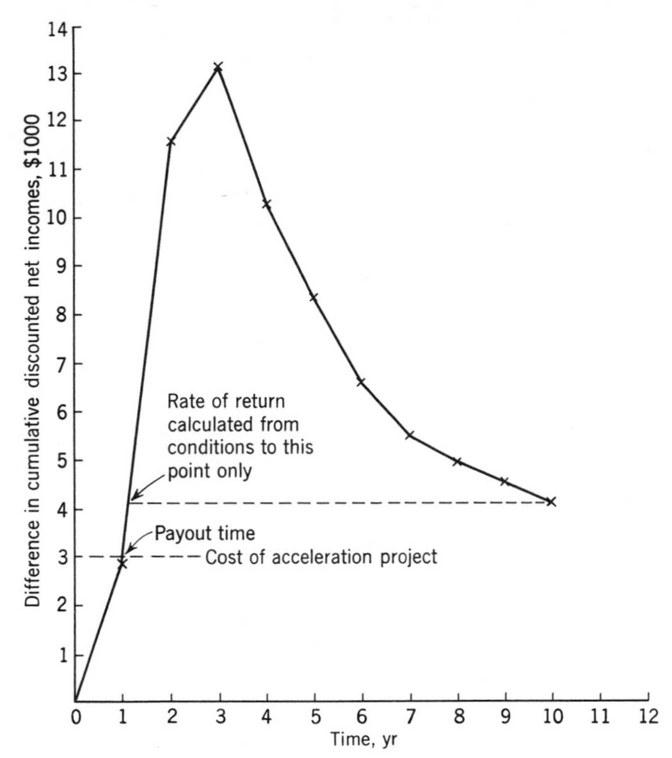

Fig. 12-15 Example 12-2: Difference in cumulative discounted net incomes plotted against time.

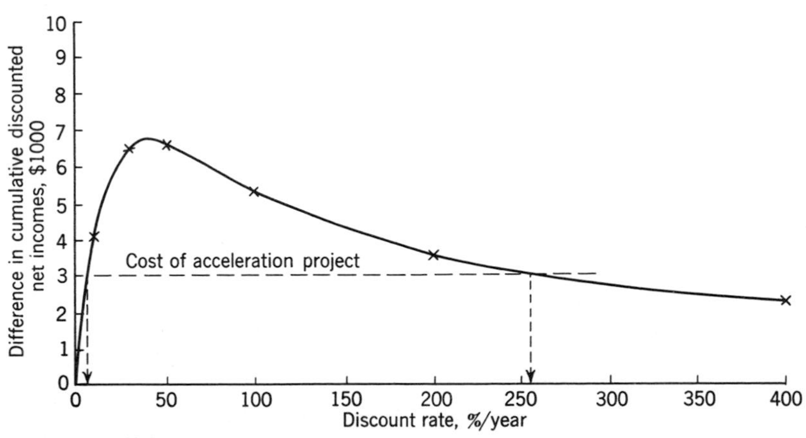

Fig. 12-16 Example 12-2: Difference in cumulative discounted net incomes plotted against discount rate.

TABLE 12-6 EXAMPLE 12-2: Difference in Cumulative Discounted Net Incomes over First 1.15 Years as a Function of the Discount Rate

Year	Undiscounted Net Income			Discount Rate									
				20%/Year		50%/Year		80%/Year		100%/Year		150%/Year	
	Unacc.	Acc.	Diff.	Discount Factor	Discounted Income Difference	Discount Factor	Discounted Income Difference	Discount Factor	Discounted Income Difference	Discount Factor	Discounted Income Difference	Discount Factor	Discounted Income Difference
1	$4000	$7000	$3000	0.913	$2739	0.817	$2451	0.745	$2235	0.707	$2121	0.632	$1896
1 to 1.15	750*	2250*	1500	0.822†	1233	0.647	971	0.532	798	0.475	713	0.374	561
Total			$4500		$3972		$3422		$3033		$2834		$2457

* These figures assume a regular income rate during the second year.

† This is $\frac{1}{(1.2)^{1.075}}$.

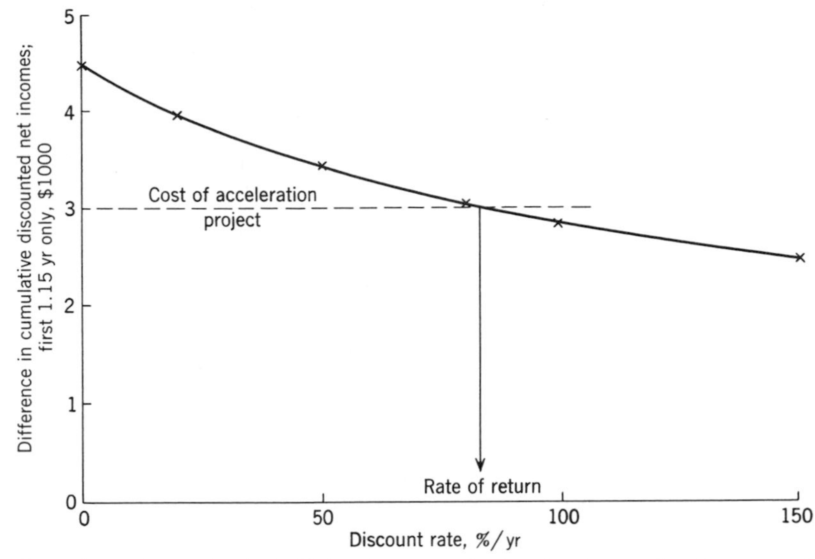

Fig. 12-17 Example 12-2: Determination of rate of return using the second method.

From the calculations set forth in Table 12-5, a plot of difference in cumulative discounted net incomes against discount rate may be made (Fig. 12-16), and this shows that, provided the earning power of the company lies between 6 and 255 percent/year, the acceleration project will be profitable.

Finally, using the differences in income over the initial 1.15 years only, Table 12-6 has been drawn up and Fig. 12-17 (which is in all respects similar to Fig. 12-6 for a new-income project) prepared. This figure shows that the rate of return for the acceleration project is about 83 percent/year, which, it should be noted, differs considerably from either of the limits obtained from Fig. 12-16.

12-10 SHORT-TERM REDUCTIONS OR INCREASES IN PRODUCTION RATE

It is frequently of interest to operating personnel to know the cost of taking a well off production for a short period[4] (for example, in order to run a BHP survey or because there is a delay in bringing a work-over hoist on to the site) or the cost of a temporary restriction in a field's production rate (because, for instance, of a partial power failure causing some pumping wells to go off production). Alternatively, a simple method of estimating the profit resulting from a short-lived increase in production rate may be useful. An example of a technique that frequently results in short-term production rate increases from

[4] The term *short* in this section will refer to a time interval of less than 3 months' duration. For such values the approximations of Sec. 12-16 are reasonably good.

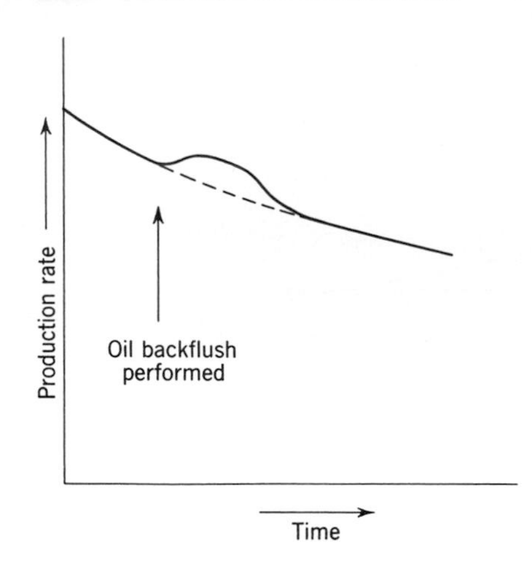

Fig. 12-18 Short-term increase in production rate resulting from oil backflush.

a well is the *oil backflush* procedure (Ref. 10). The behavior of production rate with time resulting from such a treatment might be as shown in Fig. 12-18.

In none of these cases would it prove really worthwhile, even if it were possible, to run a full-scale acceleration (or deceleration) project analysis. What is required is to be able to assign, with reasonable accuracy, a value to each barrel of delayed or accelerated oil, so that the total volume of oil in the hump of Fig. 12-18, for instance, can be converted directly to a gain in net income.

It will be shown later (Sec. 12-16) that to a good degree of approximation in many practical instances, the loss (gain) incurred per barrel of delayed (accelerated) oil is given by:

$$\frac{j}{b + j} u \tag{12-3}$$

where j = the continuous (or nominal) discount rate defined by $\exp(j) = 1 + r$ (see Sec. 12-12)
r = annual discount rate expressed as a decimal
b = the continuous (or nominal) decline rate defined by $\exp(-b) = 1 - d$ (see Sec. 2-2)
d = annual production rate decline expressed as a decimal
u = net oil value as defined in Sec. 12-2, that is, the gross oil value less the per-barrel costs

It will be noted that expression (12-3) implies that the loss incurred by deferred production is equal to the full net oil value when the production rate is normally constant (for example, prorated production), for in this case the decline rate is zero and the factor $j/(b + j)$ reduces to unity. Such a result will be realistic when the normal[5] life is of such a length that the PDV of any produc-

[5] The word *normal* in this context is used to imply that there are no deliberate attempts made to increase, but, on the other hand, no unexpected and unusual reductions in, the production rate.

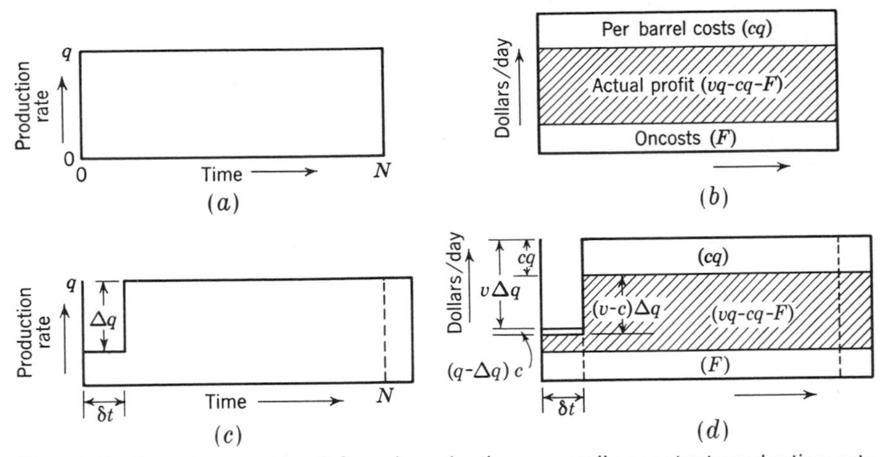

Fig. 12-19 Loss incurred by deferred production: normally constant production rate.

tion taken at the end of this normal life is negligibly small (Fig. 12-19).

In Fig. 12-20, the factor $j/(b + j)$ is shown as a function of the annual production decline rate (expressed as a percentage) for various discount rates, and it will be noted that the value to be assigned to delayed (accelerated) oil steadily decreases as the production decline rate increases. Also shown in Fig. 12-20 are curves of $r/(d + r)$ at various discount rates, and it will be seen that for discount rates of less than, say, 10 percent/year, this simpler expression, which does not involve use of nominal (continuous) rates, is of sufficient accuracy.

Example 12-3 A well pumping at capacity at 250 bbl/day and declining at 8 percent/year is closed in for 1 week while waiting for repairs. Assuming that there is

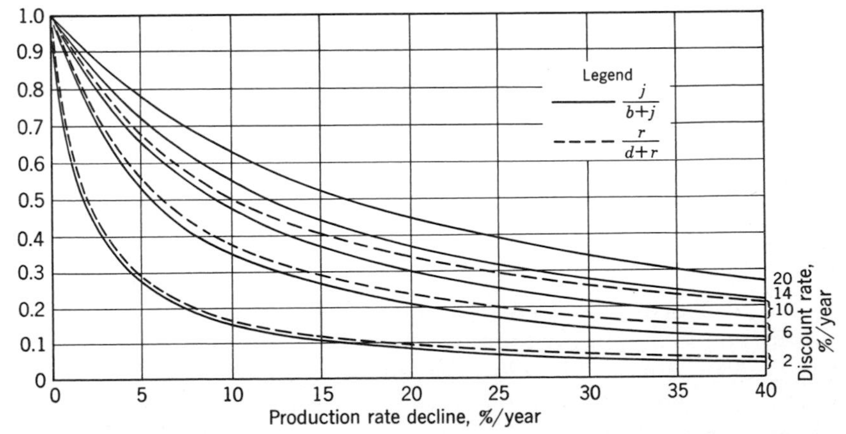

Fig. 12-20 Factors $j/(b + j)$ and $r/(d + r)$ plotted as functions of discount rate and production rate decline.

no possibility for transferring the well's production to another well and that the earning power of the company is 10 percent/year, determine the loss due to the delay. The gross oil value (after payment of royalties and taxes) is \$12.25/bbl and the per-barrel costs are \$2.00.

The total delayed oil is 250 bbl/day for 1 week, or 1750 bbl, and the value of $r/(d + r)$ is $10/(8 + 10)$, or 0.556. Thus the loss per barrel of delayed oil is $\$0.556 \times (12.25 - 2.00)$, or \$5.7, and the total loss due to the delay is $\$5.7 \times 1750$, that is, \$9975.

If use is made of the expression $j/(b + j)$, it is necessary to determine j and b. Since

$$\exp(j) = 1 + r = 1.10$$

in this case j is 0.0953. Similarly, b comes from

$$\exp(-b) = 1 - d = 0.92$$

so that b is equal to 0.0834. Thus,

$$\frac{j}{b + j} = \frac{0.0953}{0.0834 + 0.0953} = 0.533$$

and the total loss due to the delay is

$$\$0.533(12.25 - 2.00) \times 1750 = \$9550$$

It must be emphasized that this is a real loss in income, which can never be recovered.

In order to illustrate the use of expression (12-3), three practical applications will now be discussed.

Optimum Number of Work-over Rigs

Suppose that a field is operating with three work-over strings in constant use but, nevertheless, with an average *low-and-off* production of 250 bbl/day. It is assumed that, within reasonable limits, markets are available for produced oil. Let the selling price of the oil be \$12/bbl, the costs that vary with the production rate amount to \$2/bbl, the current decline rate be 10 percent/year, and the value of money to the company concerned be 8 percent/year. Is three the optimum number of work-over rigs (for maximum profit), or should two or four be used?

The value of the factor $j/(b + j)$ is 0.422, so that the loss per barrel of delayed oil is

$$\$0.422(12.0 - 2.0) = \$4.22$$

If it is estimated that an additional hoist will reduce the average low-and-off production to 50 bbl/day, the reduction in the loss due to delayed oil will be $\$4.22(250 - 50)$, that is, \$844/day, or about \$25,650/month.

If the monthly cost for the work-over rig were more than this figure, then the operation of a fourth hoist would not be economically attractive. On the other hand, if the monthly work-over rig cost were less than $25,650, a more detailed analysis would be justified, based on the thought that if the *normal* production rate were raised by some 200 bbl/day, the field decline rate would probably be altered to some extent, so that the simplified analysis would no longer be applicable.

A similar analysis based on the anticipated increase in low-and-off production resulting from a reduction from three to two hoists will enable a decision to be made as to whether or not the possibility of such a reduction should be studied in greater detail.

Sucker-Rod-String Replacement Program

Each breakage in a sucker rod string involves certain costs, namely, the cost of the hoist work involved, the cost of the replacement rod, and the loss incurred from the delay in oil production. In order to determine the third of these items, a value must be assigned to each barrel of delayed oil. With these figures available, a graph of the type illustrated in Fig. 12-21 may be prepared. Each time breakage occurs, the total cost of the breakage is calculated and plotted as shown. The objective is to minimize the cost per unit time (or cost per barrel), and the time at which the string should have been replaced can be identified from the graph shortly after this time has passed.

Such a plot gives a ready method for comparing different grades of sucker rod. That grade which results in the minimum slope of the tangent line *AO* will be the most economical, regardless of the initial cost of the string. It will be noted that the optimum grade of string to use will depend on the speed with which hoist work can be put into effect. A company that is in a position to work over a well soon after it has gone off production can stand a relatively inferior grade of rod, since the delay loss attendant on each break is small. On

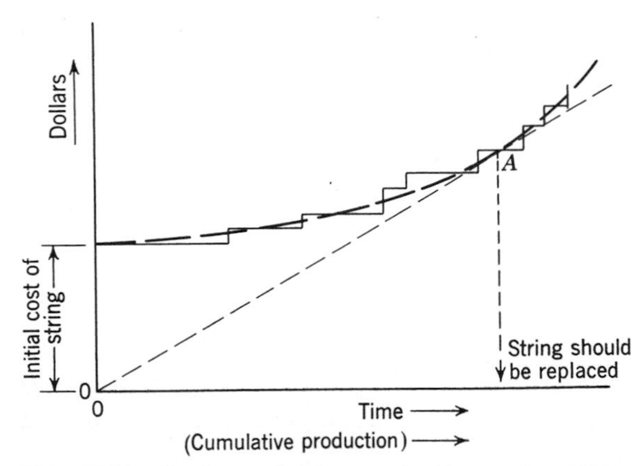

Fig. 12-21 Sucker-rod-string evaluation and replacement.

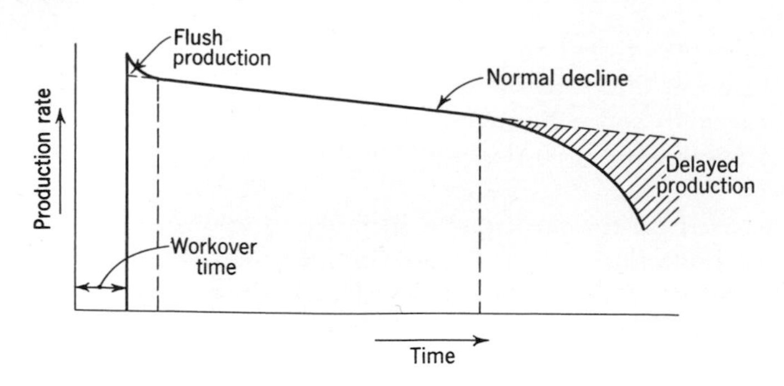

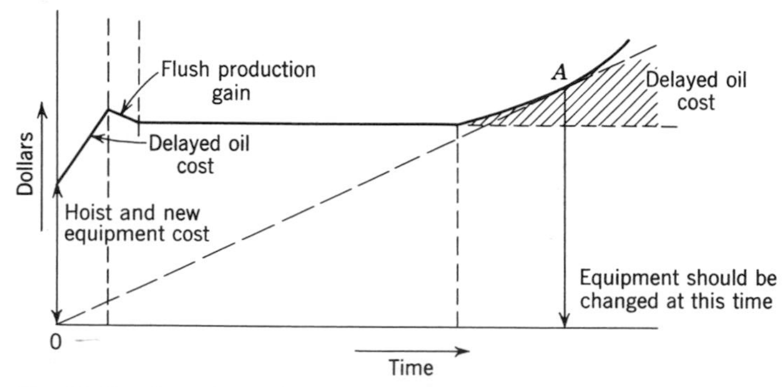

Fig. 12-22 Pumping equipment evaluation and replacement.

the other hand, a company that is operating within a minimum work-over effort should use a higher-grade rod string, since each breakage may involve a considerable loss due to delayed oil production.

Pumping Equipment Evaluation and Replacement

This problem can be approached by a method similar to that used in setting up a sucker-rod-string replacement program. Figure 12-22, which is self-explanatory, outlines one possible approach. Once again the slope of the tangent line *AO* should be minimized.

12-11 THE QUESTION OF TAXES

Those taxes, such as royalties, which are charged on a per-barrel basis may, and in fact should, be included in the per-barrel cost defined in Sec. 12-2. But such a simple approach cannot take into account all facets of a company's tax position, which may become extremely complicated in view of depletion allowances, for example, and decisions taken to ensure the minimum tax liability. In any full-scale profitability analysis, estimates of the tax must be made on a year-to-year basis and deducted from the gross income for that year. There are several discussions in the literature (Refs. 11–13) of the effects of taxation on economic evaluations, and no attempt will be made here to analyze this complex subject. It should be emphasized that the methods suggested in the subsequent sections of this chapter are shortcuts to obtain quick estimates as to the profitabilities of various projects under consideration; the more profitable of these projects must then be subjected to full and searching economic analyses run on the basis of anticipated yearly gross profits and expenses, including taxation, before final decisions can be taken as to their attractiveness to the company.

12-12 CONTINUOUS (NOMINAL) DISCOUNT RATES

Because it will greatly simplify much of the subsequent discussion, the concept of continuous (or nominal) discount rates will be introduced (Ref. 6, 14). If r is the discount rate expressed as a decimal, the *continuous* or *nominal* discount rate j is defined by the expression

$$\exp(j) = (1 + r) \tag{12-4}$$

With this definition, the PDV of an income I_n occurring n years in the future is $I_n/(1 + r)^n$, or $I_n \exp(-jn)$.

In order to illustrate the way in which the nominal discount rate may be used, two cases will be considered: a constant-income project and a project having an income that declines at a constant rate (Sec. 2-2).

Constant-Income Project

Suppose a project is such that the annual net income is a constant (for example, a pipeline, a refinery, or a tanker). Let the annual net income be I dollars, and let the earning life of the project be N years. Then, referring to Fig. 12-23, the PDV (at the start of the project, as defined in Sec. 12-8) of the income earned during the time dt is found to be

$$I\frac{dt}{(1 + r)^t} = I \exp(-jt)\, dt$$

Thus the PDV of the total net income from the project is

$$\int_0^N I \exp(-jt)\, dt = I\frac{1 - \exp(-jN)}{j} \tag{12-5}$$

If C is the PDV of the capital cost of the project (referred to the starting point) and if the *continuous rate of return* J is defined by the equation

$$\exp(J) = 1 + R \tag{12-6}$$

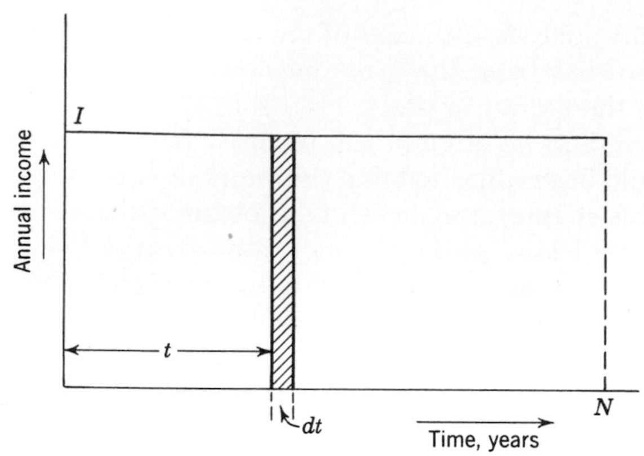

Fig. 12-23 Constant annual income.

where R is the rate of return, then J will be determined from the equation

$$C = I \frac{1 - \exp(-JN)}{J}$$

since the expression on the right-hand side is the net income discounted at the rate of return.

Provided JN is large, $\exp(-JN)$ may be neglected and

$$J = \frac{I}{C}$$

But the payout time t_p of the project is, by definition, such that

$$t_p = \frac{C}{I}$$

Thus for a constant-income project the continuous rate of return is the reciprocal of the payout time, provided that the life of the project is sufficiently long to make $\exp(-JN)$ very small. In practice, the result will usually hold if N is greater than 15 years.

It should be stressed at this point that in order to obtain the annual rate of return, recourse must be made to Eq. (12-6). Continuous discount rates and rates of return are introduced in order to simplify the mathematical calculations, but they must always be reconverted to actual discount rates and rates of return at the end of the calculation.

Income Declining at a Constant Rate

In this section reference will be made to oil production assumed to be declining at a constant rate. If the production rate at the start of the project is q_0 bbl per unit time, if the continuous decline rate is b, and if the life of the field

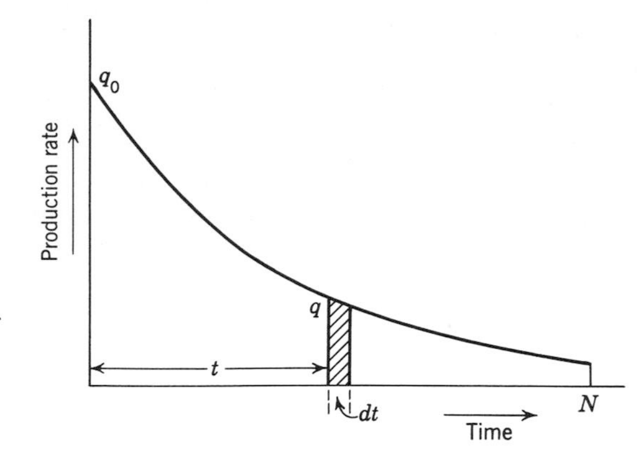

Fig. 12-24 Declining production rate.

(project) is N time units, then the PDV of an element of production $q\ dt$ occurring at the time t after the start is

$$\frac{uq\ dt}{(1+r)^t} = uq \exp(-jt)\ dt$$

where u is the net value of a barrel of oil (Fig. 12-24). But

$$q = q_0(1-d)^t = q_0 \exp(-bt)$$

so the PDV of the element of production is

$$uq_0 \exp[-(b+j)t]\ dt$$

Thus the PDV of the total field production is

$$\int_0^N uq_0 \exp[-(b+j)t]\ dt = uq_0 \frac{1-\exp[-(b+j)N]}{b+j} \tag{12-7}$$

The similarity between this expression and Eq. (12-5) should be noted. In fact, Eq. (12-7) reduces to Eq. (12-5) if b is equal to zero and I is written in place of uq_0.

Phillips (Ref. 6) has termed the factor $(b+j)$ the *composite decline-discount factor.*

12-13 THE FUNCTION [1 − exp (−x)]/x

It was shown in Sec. 2-2 that the cumulative oil production over a certain period under constant production-rate-decline conditions is equal to the difference between the production rates at the beginning and at the end of the period divided by the continuous decline rate. This result was stated in Eq. (2-7) as

$$Q - Q_0 = \frac{q_0 - q}{b}$$

But, from Eq. (2-10)

$$q = q_0 \exp[-b(t - t_0)]$$

and so

$$Q - Q_0 = \frac{q_0 - q_0 \exp[-b(t - t_0)]}{b}$$

If, for simplicity, both time and cumulative production are measured from the condition when the production rate was q_0, this equation becomes

$$Q = \frac{q_0[1 - \exp(-bt)]}{b}$$

or

$$Q = q_0 t \frac{1 - \exp(-bt)}{bt} \tag{12-8}$$

From Eq. (12-5), the PDV of the net income from a constant-income project is

$$\text{PDV} = IN \frac{1 - \exp(-jN)}{jN} \tag{12-9}$$

while from Eq. (12-7), the PDV of the net income from a well or field with a steadily declining rate is

$$\text{PDV} = q_0 uN \frac{1 - \exp[-(b + j)N]}{(b + j)N} \tag{12-10}$$

It will be noted that Eqs. (12-8), (12-9), and (12-10), respectively, may be written in the forms

$$Q = q_0 t \psi(x) \qquad x = bt$$
$$\text{PDV} = IN\psi(x) \qquad x = jN$$

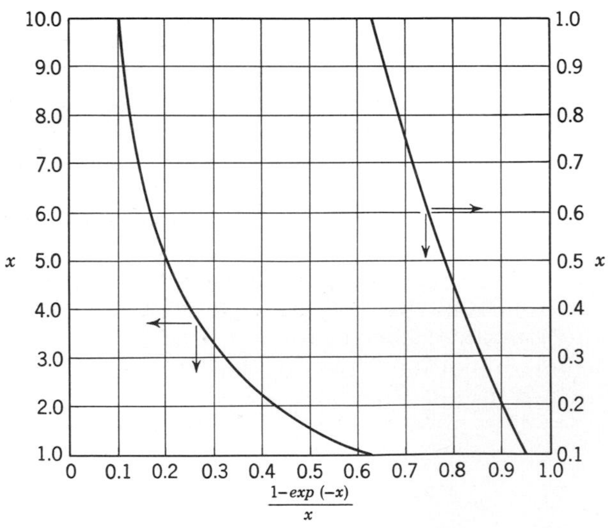

Fig. 12-25 Plot of $[1 - \exp(-x)]/x$ against x.

and $$\text{PDV} = q_0 u N \psi(x) \qquad x = (b + j)N$$

where $$\psi(x) = \frac{1 - \exp(-x)}{x} \tag{12-11}$$

Calculations will therefore be simplified by using a curve of

$$\frac{1 - \exp(-x)}{x}$$

plotted as a function of x, so that the value of the factor $\psi(x)$ can be read once the value of x is known.

This curve, for values of x lying between 0.1 and 10.0, is plotted in Fig. 12-25; for values of x greater than 10.0, $\psi(x)$ is equal to $1/x$ to sufficient accuracy, and for values of x less than 0.1, $\psi(x)$ is equal to $(1 - x/2)$ to sufficient accuracy.

12-14 DETERMINATION OF DECISION CRITERIA: NEW-INCOME PROJECT

The initial production rate q_0, the net value u of a barrel of oil, the earning life N of the project, the decline rate d and hence the continuous decline rate b, and the value of the capital investment C at the start of the project, as well as the discount rate to be used, will all have to be determined as well as possible at the commencement of the study of the project.

Profit

The PDV of the total field production is given by Eq. (12-10). Since C is the PDV of the capital expenditure at the start of the project, the profit is

$$q_0 u N \psi(x) - C$$

where x is equal to $(b + j)N$ and the profit, expressed as a percentage of the capital investment, is

$$\left[\frac{q_0 u N}{C} \psi(x) - 1\right] 100$$

Once the value of $(b + j)N$ has been determined, the corresponding value of $\psi(x)$ may either be calculated or read directly from Fig. 12-25. Thus the percent profit may be obtained. If the undiscounted profit is required, put j equal to zero.

Payout Time

Let t_p be the payout time. Then from Eq. (12-7), the PDV of the field production from the start of the project to payout time t_p is

$$q_0 u \frac{1 - \exp\left[-(b + j)t_p\right]}{b + j}$$

By definition of payout time, this quantity must equal C, so t_p is given by

$$C = q_0 u \frac{1 - \exp\left[-(b + j)t_p\right]}{b + j}$$

which simplifies to

$$t_p = -\frac{1}{(b + j)} \ln \left[1 - \frac{C(b + j)}{q_0 u}\right]$$

If the undiscounted payout time is required, put j equal to zero.

Rate of Return

Let J be the continuous rate of return as defined by Eq. (12-6). Then if the oil revenues are discounted at the rate J, the PDV (referred to the start of the project) will be $q_0 u N \psi(y)$, where y is equal to $(b + J)N$. By the definition of rate of return, this must equal C, so

$$\psi(y) = \frac{C}{q_0 u N}$$

The term on the right-hand side may be estimated, and so the value of $\psi(y)$ is known. This enables y to be determined, either by a trial-and-error calculation or directly from Fig. 12-25 by entering the known value of $\psi(y)$ on the abscissa and then reading the value of y from the ordinate. Since b and N are known and y is equal to $(b + J)N$, J may be determined. The rate of return R is readily computed from Eq. (12-6).

Example 12-4 A successful wildcat well in northern Alberta flowed at an initial 100 bbl/day on test, apparently declining at 15 percent/year. It is estimated that for a capital expenditure of \$6 million, an additional 12 holes could be drilled, of which 3 would be dry, and the necessary gathering and storage facilities could be built. Moreover, it appears that the net operating profit would be \$8/bbl and that the earning life of the field would be 15 years.

The undiscounted payout time, the percent profit (discounted at 9 percent/year), and the rate of return to be expected may be calculated as follows:

There will be 10 producing wells in all, so taking unit time to be 1 year, q_0 is $365 \times 100 \times 10$ bbl/year, while

$$\begin{aligned} u &= \$8 \\ N &= 15 \\ d &= 0.15 \\ \exp(-b) &= 1 - d = 0.85 \qquad b = 0.1626 \\ r &= 0.09 \\ \exp(j) &= 1 + r = 1.09 \qquad j = 0.0862 \end{aligned}$$

Percent Profit (Discounted)
This is

$$\left[\frac{q_0 u N}{C} \psi(x) - 1\right] 100$$

where $$x = (b + j)N = 0.2488 \times 15 = 3.73$$

From Eq. (12-11), $\psi(3.73)$ is 0.262, and the percent profit is seen to be 91 percent.
Payout Time (Undiscounted)

$$t_p = -\frac{1}{(b + j)} \ln \left[1 - \frac{C(b + j)}{q_0 u}\right]$$

and on substituting, remembering that in this case j is zero, it is seen that t_p is 2.9 years.
Rate of Return
This is given by

$$\psi(y) = \frac{C}{q_0 u N}$$

where $$y = (b + J)N$$

Substituting the values given above, it appears that $\psi(y)$ is 0.137. From Fig. 12-25, y is equal to 7.3 so

$$b + J = 0.486$$

and $$J = 0.3234$$

On using $$\exp(J) = 1 + R$$

$$R = 38.2 \text{ percent/year}$$

12-15 DETERMINATION OF DECISION CRITERIA: ACCELERATION PROJECT

Profit

The PDV of the future net incomes of the original project is

$$\text{PDV} = q_0 u N \psi(x)$$

where $$x = (b + j)N$$

The PDV of the future net incomes of the accelerated project is

$$\text{PDV}^{(a)} = q_0^{(a)} u N^{(a)} \psi(x^{(a)})$$

where $$x^{(a)} = (b^{(a)} + j)N^{(a)}$$

and $b^{(a)}$ and $N^{(a)}$ are derived from Eqs. (2-15) and (2-16), respectively. The profit is equal to $\text{PDV}^{(a)}$ less PDV less the capital cost of the acceleration project.

Payback Time

If t_b is the payback time and $C^{(a)}$ is the capital cost of the acceleration project, then referring to the definition of payback time given in Sec. 12-9, t_b may be calculated from the equation

$$C^{(a)} = q_0^{(a)} u \frac{1 - \exp[-(b^{(a)} + j)t_p]}{b^{(a)} + j}$$

Payout Time

If t_p is the payout time, then, referring to the definition of payout time given in Sec. 12-9, it may be calculated from the equation

$$C^{(a)} = q_0^{(a)} u \frac{1 - \exp[-(b_0^{(a)} + j)t_p]}{b^{(a)} + j} - q_0 u \frac{1 - \exp[-(b + j)t_p]}{b + j}$$

Rate of Return

In order to calculate the rate of return, it is necessary first to determine the cutoff time, say t_c, defined in Sec. 12-9 (see, for example, Fig. 12-15). If P is the profit as calculated above, then the time t_c may be found from the equation

$$C^{(a)} + P = q_0^{(a)} u \frac{1 - \exp[-(b^{(a)} + j)t_c]}{b^{(a)} + j} - q_0 u \frac{1 - \exp[-(b + j)t_c]}{b + j}$$

Using this value t_c, the continuous rate of return J is then defined by

$$C^{(a)} = q_0^{(a)} u \frac{1 - \exp[-(b^{(a)} + J)t_c]}{b^{(a)} + J} - q_0 u \frac{1 - \exp[-(b + J)t_c]}{b + J}$$

Discussion

It will be noted that all the equations of this section are unwieldy and that the determinations of payout time and rate of return require trial-and-error or graphical methods. It would appear therefore that the prime objective of obtaining a quick answer has been lost, and it is suggested that each acceleration project be treated on its own merits, utilizing the methods outlined in Sec. 12-9.

12-16 SHORT-TERM REDUCTIONS OR INCREASES IN PRODUCTION RATE: MATHEMATICAL ANALYSIS

It will be assumed that the *normal*[6] production rate is a function of the cumulative production.

[6] A discussion of the word *normal* as used here is given in a footnote in Sec. 12-10.

Suppose that at time zero the *normal* production rate is q_0 bbl/day but that for a period of time δt (years) the production rate is restricted by an amount Δq, so that the actual production rate over this period is $q_0 - \Delta q$ bbl/day (Fig. 12-26). Referring to this figure, it is seen that the area *AHFO*, which is equal to the *normal* cumulative production in time *OF*, must equal the area *DEGO*, which is the actual cumulative production, that is, $365(q_0 - \Delta q)\,\delta t$ bbl, from the assumption that the *normal* production rate is a function of the cumulative.

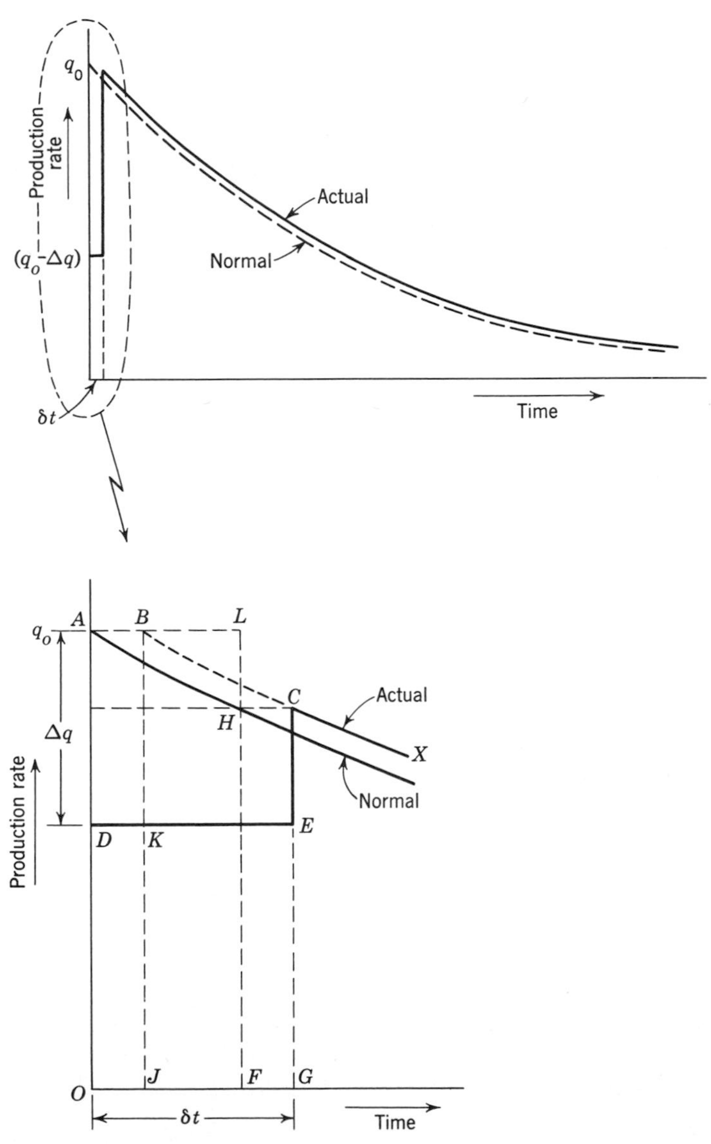

Fig. 12-26 Short-term reduction in production rate: actual and normal production rate histories.

Since area *AHFO* equals area *ALFO* approximately (δt is small), it follows that

$$365\, q_0(\text{time } OF) = 365\, (q_0 - \Delta q)\, \delta t$$

or

$$OF = \frac{q_0 - \Delta q}{q_0}\, \delta t$$

Thus the time *FG* is equal to $\Delta q\, \delta t / q_0$.

It follows that the new decline curve is shifted through a time

$$HC = FG = \frac{\Delta q}{q_0}\, \delta t$$

compared with the old (that is, the one unaffected by any production rate restriction).

If $\overline{V}$ is the PDV of the gross value of the unaffected production, then the PDV of the production under the curve *JBCX* will be $\overline{V}$ *at the time J*, or $\overline{V}/(1 + r)^{OJ}$ *now.* This equals

$$\overline{V} \exp\,(-j \times OJ) \qquad \text{or} \qquad \overline{V} \exp\left(-j \frac{\Delta q}{q_0}\, \delta t\right) \tag{12-12}$$

where j is defined by the equation $\exp\,(j) = 1 + r$

The actual production is

$$ODECX = JBCX + ODKJ - KBCE \tag{12-13}$$

but

PDV of

$$\text{production } ODKJ = \frac{365(q_0 - \Delta q)OJ}{(1 + r)^{1/2 OJ}}$$

$$= 365(q_0 - \Delta q)\frac{\Delta q}{q_0}\, \delta t \exp\left(-\frac{1}{2} j \frac{\Delta q}{q_0}\, \delta t\right)$$

$$= 365(q_0 - \Delta q)\frac{\Delta q}{q_0}\, \delta t \left(1 - \frac{1}{2} j \frac{\Delta q}{q_0}\, \delta t\right) \qquad \text{approximately}$$

$$= 365(q_0 - \Delta q)\frac{\Delta q}{q_0}\, \delta t \qquad \text{approximately} \tag{12-14}$$

neglecting terms in $j(\delta t)^2$.

$$\text{PDV of production } KBCE = 365\, \Delta q \frac{JG}{(1 + r)^{OJ + 1/2 JG}} \qquad \text{approximately}$$

where

$$JG = \delta t - OJ = \delta t - \frac{\Delta q}{q_0}\, \delta t = \frac{q_0 - \Delta q}{q_0}\, \delta t$$

and

$$OJ + \frac{1}{2} JG = \frac{\Delta q}{q_0}\, \delta t + \frac{1}{2} \frac{q_0 - \Delta q}{q_0}\, \delta t = \frac{1}{2} \frac{1}{q_0}(q_0 + \Delta q)\, \delta t$$

so

$$\text{PDV of production } KBCE = 365\,\Delta q \left(\frac{q_0 - \Delta q}{q_0}\right) \delta t \exp\left[-\frac{j}{2q_0}(q_0 + \Delta q)\,\delta t\right]$$

$$= 365\,\Delta q \left(\frac{q_0 - \Delta q}{q_0}\right) \delta t \qquad (12\text{-}15)$$

approximately, neglecting terms in $j(\delta t)^2$.

Thus, from Eqs. (12-14) and (12-15), the PDV of the production *ODKJ* is equal to the PDV of the production *KBCE* to a good degree of approximation, provided δt is small.

Hence, from expression (12-12) and Eq. (12-13), the PDV of the actual production equals $\overline{V} \exp(-j\,\Delta q\,\delta t/q_0)$, where $\overline{V}$ is the PDV of the gross value of the unaffected *normal* production.

Thus the loss in PDV due to the delay is

$$\overline{V} - \overline{V} \exp\left(-j\frac{\Delta q}{q_0}\delta t\right) = \overline{V}j\left(\frac{\Delta q}{q_0}\right)\delta t \qquad \text{approximately}$$

so that the loss in gross income per barrel of delayed oil is

$$\frac{\overline{V}j}{365q_0} \qquad (12\text{-}16)$$

There is, similarly, a saving in the PDV of the money paid out in per-barrel costs. If $\overline{C}$ is the PDV of the total costs chargeable on a unit of production basis, this saving per barrel of delayed oil will amount to

$$\frac{\overline{C}j}{365q_0} \qquad (12\text{-}17)$$

Equations (12-16) and (12-17) taken together show that there is a net loss per barrel of delayed oil of amount

$$\frac{(\overline{V} - \overline{C})j}{365q_0} \qquad (12\text{-}18)$$

A similar expression gives the net gain per barrel of accelerated oil under short-term production rate increases.

Equation (12-18) is of little practical significance unless values can be assigned to $\overline{V}$ and $\overline{C}$. Considering only the case of a straight-line production rate decline, the PDV of the gross income is

$$\overline{V} = 365q_0v\frac{1 - \exp[-(b + j)N]}{b + j}$$

and the PDV of the total costs chargeable on a unit of production basis is

$$\overline{C} = 365q_0c\frac{1 - \exp[-(b + j)N]}{b + j}$$

Substituting these expressions in Eq. (12-18) leads to the result that the net loss per barrel of delayed oil is

$$(v - c)\frac{j}{b + j}\{1 - \exp[-(b + j)N]\} \tag{12-19}$$

But $v - c$ is the net value u, and if it is assumed that the future life N is long so that $\exp[-(b + j)N]$ is negligibly small, this expression reduces to

$$\frac{ju}{b + j}$$

which is expression (12-3) and was used in the discussion and examples of Sec. 12-10.

REFERENCES

1. Muskat, Morris: *Physical Principles of Oil Production,* McGraw-Hill Book Company, Inc., New York, 1949.
2. Nind, T. E. W.: "Profitability of Oilfield Projects," *Southam-MacLean's Oil/Gas World,* December 1959, p. 14.
3. Dean, Joel: "Measuring the Productivity of Capital," *Harvard Buisness Rev.,* January-February 1954, p. 120.
4. Newendorp, Paul D.: *Decision Analysis for Petroleum Exploration,* The Petroleum Publishing Company, Tulsa, Okla., 1975.
5. Glanville, James W.: "Rate of Return Calculations as a Measure of Investment Opportunities," *J. Petrol. Technol.,* **9**(6):12 (1957).
6. Phillips, Charles E.: "The Relationship between Rate of Return, Payout, and Ultimate Return in Oil and Gas Properties," *J. Petrol. Technol.,* **10**(9):26 (1958).
7. Horner, W. L., and I. F. Roebuck: "Economics and Prediction of Oil Recovery by Fluid Injection Operations," in *Improving Oil Recovery,* Department of Petroleum Engineering, University of Texas, Austin, Tex., 1957.
8. Lefkovits, H. C., H. Kanner, and R. B. Harbottle: "On Multiple Rates of Return," *Proc. Fifth World Petrol. Congr., New York,* Sec. IX: 67 (1959).
9. Woody, L. D., Jr., and T. D. Capshaw: "Investment Evaluation by Present-Value Profile," *J. Petrol. Technol.,* **12**(6):15 (1960).
10. Ghauri, W. K.: "Results of Well Stimulation by Hydraulic Fracturing and High Rate Oil Backflush," *J. Petrol. Technol.,* **12**(6):19 (1960).
11. Bullion, J. Waddy: "Tax Considerations in Oil Transactions," *J. Petrol. Technol.,* **8**(8):12 (1956).
12. Watkins, P. B.: "Economic Evaluations," *J. Petrol. Technol.,* **11**(11):20 (1959).
13. Breeding, Clark W., and John R. Herzfield: "Effect of Taxation on Valuation and Production Engineering," *J. Petrol. Technol.,* **10**(9):21 (1958).
14. Brons, F., and M. W. McGarry: "Methods for Calculating Profitabilities," *Paper* 870-*G. 32d Annual Fall Meeting, Soc. Petrol. Engineers, AIME, Dallas, Tex.,* 1957.

Conversion Factors

To Convert From	To	Multiply By
meter (m)	foot (ft)	3.281
foot (ft)	meter (m)	0.305
kilometer (km)	mile (mi)	0.621
mile (mi)	kilometer (km)	1.609
acre	square meter (m^2)	4,046.87
square centimeter (sq cm)	square inch (sq. in.)	0.155
square inch (sq in.)	square centimeter (sq cm)	6.452
barrel (bbl)	cubic foot (cu ft)	5.614
barrel (bbl)	cubic meter (m^3)	0.159
barrel (bbl)	gallon (gal; U.S. liquid)	42
cubic foot (cu ft)	barrel (bbl)	0.178
cubic foot (cu ft)	cubic meter (m^3)	2.832×10^{-2}
cubic meter (m^3)	barrel (bbl)	6.290
cubic foot per barrel (cu ft/bbl)	volume per volume (vol/vol)	0.178
volume per volume (vol/vol)	cubic foot per barrel (cu ft/bbl)	5.614
barrel per day (bbl/day)	cubic meter per hour (m^3/hr)	6.625×10^{-3}
cubic meter per hour (m^3/hr)	barrel per day (bbl/day)	150.96

To Convert From	To	Multiply By
pound (lb)	kilogram (kg)	0.454
kilogram (kg)	pound (lb)	2.205
kilogram (kg)	ton (metric)	1×10^{-3}
atmosphere (atm)	pound per square inch (psi)	14.696
atmosphere (atm)	kilogram per square centimeter (kg/cm^2)	1.033
pounds per square inch (psi)	kilogram per square centimeter (kg/cm^2)	7.031×10^{-2}
pounds per square inch (psi)	megapascals	6.895×10^{-3}
foot-pound (ft-lb)	British thermal unit (Btu)	1.284×10^{-3}
foot-pound (ft-lb)	kilogram-meter (kg-m)	0.138

Note: In SI units, volume is measured in cubic meters, distance is measured in meters, and pressure is measured in megapascals.

Problems

1. The production rate of the Crazy Dog pool was held constant at its allowable of 2500 bbl/day for some years. Recent production rate figures, expressed in terms of the cumulative oil production from the pool, have been as follows:

Cumulative Production from Pool, 10^6 bbl	Oil Production Rate, bbl/day
3.20	1920
3.45	1645
3.70	1650
4.10	1230
4.40*	1095

* Present-day conditions.

Assuming a pool economic limit of 200 bbl/day, determine the future life of the pool, the total life of the pool, the future production to be expected, and the ultimate cumulative production.

If an infilling program were carried out now to increase the present production rate to 1400 bbl/day, what would be the future life of the pool?

Answers: Future life, 6.4 years; total life, 12.3 years; future production, 1,200,000 bbl; ultimate cumulative, 5,600,000 bbl; if infilling program carried out, future life is 5.7 years.[1]

2. A new oil field is thought to have recoverable reserves of 10 million bbl, and from information on a similar neighboring pool it is estimated that the average well will come in at 100 bbl/day, flowing, with a decline rate of 17 percent/year. It is expected that the wells will flow down to a production rate of 40 bbl/day and that they will then be put to the pump for an average initial pumping rate of 60 bbl/day. Taking the economic limit for pumping wells to be 3 bbl/day, how many producing wells should be drilled to exploit the field and what will be its life?

Answers: Number of wells, 52; life of field, 15.6 years.

3. A pool that is drilled on a 40-acre spacing has a productive horizon 15 ft thick with an average permeability of 10 md. The oil has a viscosity at reservoir conditions of 6 cP, and the oil formation volume factor is 1.17. If the wells are completed with 5-in. casing, estimate the average PI. What is the potential of the average well if the static reservoir pressure is 1250 psig?

Answers: PI, 0.0188 bbl/(day)(psi); potential, 23.5 bbl/day.

4. Reservoir pressure in the High Hope pool has declined linearly with cumulative oil withdrawal since the pool was first put into production, the rate of pressure drop being 1 psi per 5000 bbl of oil produced. The pool is drained by 12 wells, the average PI being 0.3 bbl/(day)(psi) per well, and as a matter of policy, these wells are produced at a drawdown equal to 50 percent of the prevailing static reservoir pressure. Show that the wells have a straight-line decline, the annual decline rate being 12.3 percent.

It is estimated that the reservoir pressure will have fallen to 300 psi after a total producing life of 16 years. What was the initial reservoir pressure?

Answer: 2460 psi.

5. A well came in at 116 bbl/day and by the end of the first year had declined to 94 bbl/day. It was found that, at a cumulative production of 143,500 bbl, the PI was 0.053 bbl/(day)(psi) and the static pressure of

[1] Answers are, in most cases, approximate in that they depend on subjective interpretations of data and, in those problems that involve the use of pressure-distribution curves, on the particular set of curves.

the formation was 770 psi. What was the flowing BHP at that time? Assuming that the same flowing BHP will prevail at abandonment, what will be the static pressure of the formation at abandonment if the well's economic limit is taken as 2 bbl/day?

Answers: Flowing BHP, 138 psig; static BHP, 176 psig.

6. A series of tests is made on a certain well with the following results:

Oil Rate bbl/day	WOR	Flowing BHP, psi
40	0	2360
56	0.785	1950
61	1.440	1524
70	1.855	1000

Draw the curve of water cut versus gross production rate and the oil, water, and gross IPRs. Might any harm result from shutting this well in for a few days?

7. A flowing well with 5540 ft of tubing in the hole is completed without a casing-tubing packer. The CHP is 480 psig when the production rate is 750 bbl/day and 760 psig when the production rate is 525 bbl/day. What are the PI, static pressure, and potential of the well?

Answers: PI, 0.71 bbl/(day)(psi); static pressure, 1600 psig; potential, 1135 bbl/day.

8. Well 1121 South Block is flowing at 1120 bbl/day through $2^7/_8$-in. tubing. There is zero water cut, and the GLR is 820 cu ft/bbl. A pressure survey on the well shows that the flowing pressure at 6470 ft (the foot of the tubing) is 675 psig, while the pressure buildup survey gives a static pressure of 2080 psig at a datum level of 6500 ft.

Using Vogel's method, draw the IPR curve, and estimate the well's potential.

Reservoir analysis indicates that the ratio of the value of $k_{ro}/B_o\mu_o$ now to its value at the static pressure of 1500 psig is 1.57. Estimate what the well's potential rate will be when the static pressure has dropped to 1500 psig.

Answers: Present potential, 1322 bbl/day; potential at static pressure of 1500 psig, 609 bbl/day.

9. Use the data of Prob. 8 and Fetkovich's method to draw the present-day

IPR (that is, at a static pressure of 2080 psig), and determine the well's potential. What will be the well's potential at the static pressure of 1500 psig under Fetkovich's method?

Answers: Present potential, 1255 bbl/day; potential at static pressure of 1500 psig, 475 bbl/day.

10. The data shown in Table P-1 were obtained from a pressure survey on a well in Main Pass Block 35. Find the value of M in Eq. (4-15) that gives the best fit between the calculated and the measured results.

Answer: $M = 0.2$ gives a good fit.

11. A well completed over the interval 2994 to 3032 ft (below kelly bushing) has $2^3/_8$-in. tubing hung at 3000 ft. The well is flowing 320 bbl/day, zero water cut, at a GOR of 400 cu ft/bbl with a THP of 500 psig. The static pressure is 1850 psig at 3000 ft. What would be the effect of changing the choke size to $^1/_2$ in.?

Answers: Flow rate, 450 bbl/day; THP, 170 psig.

12. A certain well is completed with 7500 ft of $3^1/_2$-in. tubing in the hole, the tubing shoe being located just above the top perforations. The well is

TABLE P-1 PROBLEM 10: Data from Pressure Survey on Well in Main Pass Block 35

Depth, ft	Pressure, psig
0	96
500	161
1000	287
1500	433
2000	580
2500	767
3000	991
3500	1227
4000	1466
5000	1937
6000	2403
7000	2794

Tubing $2^3/_8$ in., 4.7 lb/ft
Oil rate 26 bbl/day, 26.0°API
Water rate 104 bbl/day, sp gr 1.15
GLR, 138 cu ft/bbl

flowing 130 bbl/day of oil with a water cut of 25 percent and a GOR of 1200 cu ft/bbl. If the well's static pressure is 2800 psig and its gross PI is 0.32 bbl/(day)(psi), estimate the size of choke in the flow line. At what oil rate would the well flow if a $^1/_2$-in. bean were substituted for the current one?

Answers: Choke size, $^3/_{16}$ in. On $^1/_2$-in. choke: flow rate, 450 bbl/day gross.

13. A well was completed with 7-in. casing perforated (2 shots/ft) from 7216 to 7253 ft with 7000 ft of $2^3/_8$-in. tubing in the hole. The well was flowing steadily at 320 bbl/day of clean oil, GOR 800 cu ft/bbl, against an $^{11}/_{16}$-in. choke when a smaller choke was accidentally inserted in the flow line. When the well stabilized against the new choke, it was flowing with a THP of 300 psig and a CHP of 993 psig. Determine the new bean size, the well's static pressure, and the PI. (Assume no annulus packer.)

Answers: Bean size, $^7/_{32}$ in.; static pressure, 1425 psig; PI, 0.52 bbl/(day)(psi).

14. A new flowing well completed with $2^7/_8$-in. tubing hung at the top of the perforations at 5500 ft was initially produced on $^1/_4$-in. choke, the THP stabilizing at 400 psig. After a few days' production the choke size was increased to $^1/_2$ in., and the THP stabilized at 270 psig. One week later the choke size was again increased, and the well then gauged at 600 bbl/day of clean oil, GOR 800 cu ft/bbl, THP 140 psig. Estimate the well's static pressure and its pumped-off potential.

Answers: Static pressure, 1550 psig; potential, 850 bbl/day.

15. A flowing well is completed with 7332 ft of $2^3/_8$-in. tubing, the tubing shoe being located opposite the top of the perforations. At what rate will the well flow against a $^1/_4$-in. choke under each of the following conditions?

a. Static pressure, 3000 psig; PI, 0.42 bbl/(day)(psi); GOR, 200 cu ft/bbl.
b. Static pressure, 2500 psig; PI, 0.33 bbl/(day)(psi); GOR, 330 cu ft/bbl.
c. Static pressure, 2000 psig; PI, 0.29 bbl/(day)(psi); GOR, 500 cu ft/bbl.
d. Static pressure, 1500 psig; PI, 0.27 bbl/(day)(psi); GOR, 1000 cu ft/bbl.

On the same sheet of graph paper plot the pumped-off potential and the actual flow rate against the static BHP.

TABLE P-2 PROBLEM 16: Data Resulting from Tests on Wells, A, B, C, and D

Well	Cumulative Oil Production: Well, bbl	Cumulative Oil Production: Field, bbl	Static Pressure at Datum, psig	Oil Rate, bbl/day	Flowing BHP, psig	GOR, cu ft/bbl	Water Cut, %
				630	1470	200	2.0
A	100,000	1,482,000	2120	460	1720	189	2.2
				380	1880	212	1.7
				460	1200	1180	0.0
B	200,000	1,973,000	1950	320	1530	1180	0.0
				235	1700	1140	0.2
				520	2720	121	10.3
C	621	621	3000	420	2820	193	8.7
				360	2880	210	4.5
				180	820	1116	0.2
C	330,000	3,426,000	1520	120	1160	1141	0.2
				90	1240	1132	0.3
				340	1190	1863	0.0
D	170,000	2,471,000	1790	200	1390	1869	0.0
				220	1470	1855	0.0
				115	650	622	0.2
D	370,000	4,600,000	1300	90	840	641	1.3
				60	960	629	0.2

Answers: (*a*) Flow rate, 410 bbl/day; (*b*) flow rate, 270 bbl/day; (*c*) flow rate, 170 bbl/day; (*d*) flow rate, 115 bbl/day.

***16.** The data shown in Table P-2 have been obtained from a series of tests on four wells in a certain field. By analogy with gas well performance a reasonable assumption might be that the production rate q is related to the drawdown $p_s - p_{wf}$ by an equation of the form

$$q = k(p_s - p_{wf})^n$$

where k and n are constants in any particular test but may vary from test to test. It is further postulated that there is a relationship between the values of k and the values of n.

Using a log-log plot of production rate against drawdown to determine k and n values for each of the six tests reported in Table P-2, construct a graph of log k as a function of n and hence construct a regular grid on log-log paper of production rate against drawdown for values of n equal to 0.4, 0.5, 0.6, 0.7, and 0.8.

* The more difficult problems are marked with an asterisk.

TABLE P-3 PROBLEM 16: Additional Data on Well E*

Well's Cumulative Production, bbl	Static BHP, psig	Oil Rate, bbl/day	GOR, cu ft/bbl	CHP, psig
0	3100			
150,000		440	550	1947
160,000		300	700	1925
200,000	2440			
260,000		350	2000	1620

* Well E was completed without a tubing-casing packer in the hole.

Well E is currently flowing on $2^3/_8$-in. tubing at 200 bbl/day of clean oil, GOR 700 cu ft/bbl, through $^1/_4$-in. choke. This well is perforated from 8003 to 8021 ft, and the tubing is hung at 8000 ft. The well's cumulative production to date is 460,000 bbl, and the current static pressure at the datum of 8000 ft is 1750 psig. The initial flowing BHP on well E was 2910 psig at a production rate of 540 bbl/day, GOR 200 cu ft/bbl. Some additional data from well E are listed in Table P-3.

Plot the current IPR for well E. What is the well's potential at the present time?

For well E prepare a graph showing the variation in static BHP and in GOR with cumulative oil production from the well. On the same graph plot the production rate that would have been obtained from the well if it had been produced at a constant drawdown of 100 psi. Extrapolate these three curves to higher cumulatives as well as possible, and use these extrapolated curves to answer the following questions:

What would have been the production rate from the well at a drawdown of 600 psi when its cumulative production was 100,000 bbl?

What will be the future flowing life history of this well on $2^3/_8$-in. tubing, assuming that the THP is maintained at 100 psig?

What will be the well's maximum inflow potential when it ceases to flow, and what percentage of this potential will it actually be making immediately prior to dying?

Answers: Potential of well E now, 700 bbl/day. Production rate at a 600-psi drawdown when well's cumulative was 100,000 bbl, 660 bbl/day. On $2^3/_8$-in. tubing at THP of 100 psig: flowing 390 bbl/day at 460,000 bbl cumulative; flowing 335 bbl/day at 480,000 bbl cumulative; flowing 275 bbl/day at 500,000 bbl cumulative; flowing 200 bbl/day at 525,000 bbl cumulative. Well dies at about 530,000 bbl cumulative, at which time potential is 585 bbl/day. Final flowing production rate is 175 bbl/day, or 30 percent of potential.

***17.** Well data are as follows:

Depth:	9200 ft
7-in. casing:	9100 ft
$4^1/_2$-in. liner:	9050 to 9200 ft
GOR:	450 cu ft/bbl
Water cut:	10 percent
PI:	0.333 bbl/(day)(psi)
Static pressure at 9000 ft:	3000 psig
Flow-line pressure:	60 psig

Available equipment includes 1.9-, $2^3/_8$-, $2^7/_8$-, and $3^1/_2$-in. tubing and a compressor of 135 hp and 2000-psig outlet pressure. Input gas is available at 35 psig.

a. What size tubing will give the maximum rate on natural flow, and what bean size is required?
b. What maximum rate of flow could be obtained by gas lift through $2^7/_8$-in. tubing at 9000 ft? What horsepower would be required? (Take a THP of 60 psig, and assume that the compressor outlet pressure is equal to the flowing BHP.)
c. Using the available compressor, what will the well make on gas lift through 9000 ft of $2^7/_8$-in. tubing?
d. If $3^1/_2$-in. tubing is run to 9000 ft and 1.9-in. tubing is run inside the $3^1/_2$-in. tubing to 4000 ft, what is the maximum gas-lift rate and what horsepower is required?

Answers: (*a*) Assuming a tubing-head pressure of 100 psig, 1.9-in. tubing gives a maximum flow rate of 490 bbl/day through an $^{11}/_{16}$-in. choke. (*b*) 740 bbl/day, 260 hp. (*c*) 650 bbl/day. (*d*) 480 bbl/day, 35 hp.

***18.** Well 132-C is completed over the interval 7152 to 7213 ft below tubing head. The static pressure at the datum level of 7150 ft is 2200 psig. The well is flowing through 7100 ft of $2^7/_8$-in. tubing at 396 bbl/day of clean oil, GLR 400 cu ft/bbl, with a THP of 270 psig. The Vogel-type IPR curve has proved to be reasonably accurate for other wells in the same pool.

The desirability is being investigated of using gas from a gas well in the vicinity to increase the production rate from Well 132-C, and it is estimated that 300,000 scf/day of gas can be delivered to the wellhead at 550 psig.

What maximum gas-lift rate can be anticipated against a THP of 150 psig; and what would be the depth of the lowest gas-injection point?

Answers: **Maximum rate 620 bbl/day; depth of lowest gas-injection point 3200 ft.**

***19.** Design a differential valve string for the following well:

Productive interval:	10,000 to 10,170 ft
Static pressure:	2700 psi at 10,000 ft
PI:	0.333 bbl/(day)(psi)
Formation GLR:	300 cu ft/bbl
Oil gravity:	23° API

Gas of sp gr 0.65 is available in unlimited quantities at a pressure of 700 psi. Differential valves of various choke sizes with a spring setting of 150 psi are available.

Answers: In $3^1/_2$-in. tubing with a 50-psi THP:

Valve 1, depth 4840 ft, choke $^{11}/_{64}$ in.
Valve 2, depth 5550 ft, choke $^{14}/_{64}$ in.
Valve 3, depth 6450 ft, choke $^{17}/_{64}$ in.
Valve 4, depth 7300 ft, choke $^{21}/_{64}$ in.
Valve 5, depth 8150 ft, choke $^{23}/_{64}$ in.
Valve 6, depth 9000 ft, choke $^{27}/_{64}$ in.

With this string, valve 6 is lowest injection point, maximum gas-lift rate is 660 bbl/day, and injection gas required is 2050 mcf/day.

***20.** Design a string of pressure-charged bellows valves for the well of Prob. 19, assuming that the top valve will have a bellows pressure of 650 psi and that the bellows pressure is reduced by 50 psi per valve on moving down the string.

Answers: In $3^1/_2$-in. tubing with a 50-psi THP:

Valve 1, depth 4840 ft, port $^{23}/_{64}$ in.
Valve 2, depth 6950 ft, port $^{25}/_{64}$ in.
Valve 3, depth 7950 ft, port $^{29}/_{64}$ in.
Valve 4, depth 8250 ft, port $^{40}/_{64}$ in.

A fifth valve appears to have little value. With the string shown, valve 4 is the lowest injection point, maximum gas-lift rate is 635 bbl/day, and injection gas required is 2020 mcf/day.

***21.** Consideration is being given to placing well 7 in Block A on intermittent gas lift. The well is completed over the interval 11,032 to 11,071 ft below tubing head with 11,000 ft of $2^7/_8$-in. tubing (internal cross-sectional area 0.0318 sq ft) in the hole. Initial planning is being carried out on the basis of a closed system utilizing the $2^7/_8$-in. tubing, and on the assumption that the upward slug velocity during the injection period will be 750 ft/min. Other well data are as follows:

Average PI:	1.7 bbl/(day)(psi)
Formation static pressure:	1570 psig
Pressure gradient exerted by formation liquid:	0.29 psi/ft
Minimum line pressure in gathering system:	150 psig

Determine the optimum number of cycles per day and calculate the anticipated production rate, the maximum THP, the volume of injection gas required per cycle, and the maximum gas-injection pressure for assumed values of the fallback per cycle of 0, 2, 4, 6, 8, and 10 bbl. Use these calculations to predict probable results.

Answers: Optimum number of cycles per day, 44; production rate, 580 bbl/day (±60); volume of injected gas per cycle, 25.7 mcf (±0.8); volume of injected gas per day, 1130 mcf (±35); maximum gas-injection pressure, 1230 psig (±35).

22. What would be the effect if a chamber-lift installation were made in well 7 of Block A, described in Prob. 21, the installation involving 2000 ft of $4^1/_2$-in. tubing (internal cross-sectional area 0.0850 sq ft) run on 9000 ft of $2^7/_8$-in. tubing? (In this problem it is necessary to assume a value for the THP during production of the slug; the calculations of Prob. 21 would suggest 300 psig as a reasonable figure. The explanation for this point of difference between the solution to Prob. 21 and that to Prob. 22 is that the gas pressure required to move the slug up the $2^7/_8$-in. tubing is considerably greater than the pressure needed to lift the liquid out of the $4^1/_2$-in. chamber because of the length of the slug in the smaller tubing.)

Answers: Optimum number of cycles per day, 31; production rate, 720 bbl/day (±50); volume of injected gas per cycle, 53.5 mcf (±3.5); volume of injected gas per day, 1660 mcf (±110); maximum gas-injection pressure, 1880 psig (±115).

Note: If a figure of 400 psig were assumed for the THP during production of the slug, the answers for the optimum number of cycles per day and for the production rate would not be changed. However, the predicted volume of injected gas per cycle would rise to 56.7 mcf (±3.5) and the maximum gas-injection pressure to 1980 psig (±115).

23. A well completed with 4492 ft of $2^3/_8$-in. tubing has a static pressure of 1100 psig at 4500 ft. It is producing on plunger lift with an average CHP of 220 psig. What are the production rate and efficiency of the operation, if it is assumed that the trap pressure may be neglected?

Answers: 518 bbl/day; 78 percent.

24. Show by an analysis of Eq. (9-21) that when rod stretch and overtravel are taken into account, it remains true that for a given plunger size, maximum plunger displacement rate is obtained by using the longest stroke length available (even at the expense of accepting fewer strokes per minute).

25. Determine the intake pressure in a pumping well in which the pump is set at 3750 ft below the casinghead, given that during a test in which the production rate from the well was maintained at a constant level, the depth to the free fluid level in the annulus was found to be 3610 ft when the CHP was 140 psig and 1780 ft when the CHP was 35 psig.

Answer: 159 psig.

26. SV and TV checks are carried out on a well completed with 2700 ft of $^3/_4$-in. rods and 2-in. diameter plunger. A load of 3810 lb is recorded during the former and a load of 4820 lb during the latter test. Estimate the pressure exerted by the formation at the SV while the well is pumping, if the THP is 100 psig.

Answer: 333 psig.

27. A pumping well is producing 230 bbl/day of oil and 85 bbl/day of water with a tubing GOR of 140 scf/bbl. The pump displacement rate is estimated to be 385 bbl/day, and the PVT data of the oil are as shown in Table P-4. Estimate the pressure at the pump intake.

Answer: 625 psig.

28. A 7000-ft well is completed with $4^1/_2$-in. casing. The static pressure is 1800 psig, the PI 0.4 bbl/(day)(psi), and the GLR 0.3 mcf/bbl. A $2^3/_8$-in.

TABLE P-4 PROBLEM 27: Assumed PVT Data

Pressure, psia	B_o, bbl/bbl	B_g, bbl/scf	R_s, scf/bbl
2500	1.250	0.0015	225
2200	1.225	0.0018	206
1900	1.203	0.0023	189
1600	1.180	0.0030	170
1300	1.156	0.0039	152
1000	1.133	0.0050	134
700	1.108	0.0065	116
400	1.080	0.0085	83
200	1.045	0.0110	48

tubing string, with a P/GA combination on bottom, is used to produce the well.

Assuming that flow gradients in $4^1/_2$-in. casing are identical with those in $3^1/_2$-in. tubing, that the pump volumetric displacement rate is 600 bbl/day, and that the P/GA efficiency is given by

$$E_v = \frac{1}{1 + 0.02\sqrt{p_i}}$$

approximately, when p_i is less than 650 psig, determine the optimum pump setting depth and the maximum (stable) pumping rate.

Answers: Optimum setting depth in the range 4500 to 5000 ft; maximum stable pumping rate in the range 440 to 490 bbl/day.

29. Using the data of Example 11-1, determine the production rate attainable by flumping when a gas anchor is installed immediately below the pump, if the gas-anchor efficiency is taken to be of the form

$$1 - \frac{\text{(oil rate through anchor into pump, bbl/day)}}{1200}$$

Answer: If gas is bled continuously from the annulus but the well is pumped only, a rate of 280 bbl/day is attainable. If the well is flumped against a CHP of 100 psig, 120 bbl/day may be flowed from the annulus while 235 bbl/day is pumped from the tubing, giving an apparent pump volumetric efficiency of a little over 100 percent.

***30.** A well is to be pumped using a conventional unit having a crank/pitman ratio of 0.2 and a maximum polished rod stroke of 48 in. Rods of $^7/_8$-in. diameter and a tubing pump of $2^3/_4$-in. diameter are to be used. The maximum permissible polished rod stress is 30,000 psi, and the downward acceleration of the rod is not to rise above 60 percent of the acceleration due to gravity.

A gas anchor is to be installed immediately below the pump; if it is decided not to set the pump opposite the productive formation but higher up the hole, the production will be flowed up $2^7/_8$-in. tubing to the pump intake. The gas diverted up the annulus by the anchor is to be bled continuously from the casinghead.

Determine the depth at which the pump should be set to achieve the maximum oil production rate, given that: the productive formation is 3500 ft below the surface; the gas in solution in the oil is a straight-line function of the pressure, ranging from 10 cu ft/bbl at atmospheric pressure to 200 cu ft/bbl at 1000 psig; the producing GOR of the well is 300 cu ft/bbl; the static pressure in the formation is 1000 psig; the formation PI is 0.5 bbl/(day)(psi); the oil gravity is 20°API; temperature effects may be neglected; and the gas-anchor efficiency, that is, the ratio that the volume of free gas diverted up the annulus bears to the volume of free gas present at the anchor intake, both volumes being measured at the intake pressure, is believed to be

$$1 - \frac{\text{oil rate in bbl/day}}{1000}$$

Answer: Optimum setting depth is some 400 to 500 ft off bottom, with an attainable rate of 400 to 410 bbl/day.

31. A well that is currently able to produce 100 bbl/day of clean oil and is showing a production rate decline of 5 percent/year is off production for 1 month. Assuming a net oil value of $12.0/bbl and a discount rate of 8 percent/year, how much money is lost as a result of the shutdown?

Answer: $21,925.

32. The drilling of a well on a certain location is under consideration. It is estimated that the initial production rate will be 200 bbl/day, with a decline of 2 percent/month. The drilling cost is estimated at $775,000, the net oil value at $10.0/bbl, and the economic limit at 5 bbl/day. Determine the undiscounted payout time, the discounted percentage profit based on a discount rate of 9 percent/year, and the rate of return.

Answers: Undiscounted payout time, 1.2 years; discounted percentage profit, 185 percent; rate of return, 101 percent/year.

33. A well is shut in for 0.1 year awaiting repairs. The expected future life of the well is 7 years. Taking the oncosts to be $8/bbl and the selling price to be $22/bbl and using a 10 percent/year discount rate:

a. Determine for decline rates of 0, 5, 10, 20, and 30 percent/year the PDV loss per barrel of delayed oil. Express the results as percentages of the net operating profit per barrel. Plot the results.
b. Determine how the PDV loss per barrel would be affected if the expected life of the well were 20 instead of 7 years. Plot the results.

Answers:

a. PDV loss per barrel:

At 0 percent/year decline, 48.8 percent
At 5 percent/year decline, 41.7 percent
At 10 percent/year decline, 35.8 percent
At 20 percent/year decline, 26.7 percent
At 30 percent/year decline, 20.2 percent

b. If the expected life of the well were 20 years, the loss per barrel, as a percentage of the net operating profit per barrel, would be close to $100j/(b+j)$:

At 0 percent decline, 100.0
At 5 percent, 65.0
At 10 percent, 47.4
At 20 percent, 29.9
At 30 percent, 21.1

34. A new-income project is under consideration for a pool, the production history of which is showing a straight-line decline and the future life of which is estimated to be long. It is company policy for such a project that the discounted payout time must be less than Y years and the discounted percentage profit must be greater than X percent. Show that, provided the annual decline rate b is less than

$$\frac{1}{Y}\ln\left(\frac{X+100}{X}\right)-j$$

where j is the continuous discount rate, then if the project satisfies the requirement as to payout time, it will automatically satisfy the requirement as to profitability.

If Y is 1 year and X is 200 percent, find the limiting annual decline rates for discount rates of 0, 5, 10, and 15 percent/year.

Answers: 33.3, 30.0, 26.6, and 23.3 percent/year.

35. A profitability analysis is carried out on a newly drilled well draining a hitherto untapped source of oil. The well is assumed to have a straight-line decline, and it is estimated that the undiscounted profit, expressed as a percentage of the capital investment, will be P percent. Show that the undiscounted payout time x is given by the equation

$$\exp(bx) = 1 + \frac{100}{P}$$

and the continuous rate of return J by the equation

$$J = \frac{bP}{100}$$

where b is the continuous decline rate, provided that $\exp(-bT)$ and similar terms may be neglected (T is the life of the well).

36. In Sec. 12-9 two ways of defining the rate of return for an acceleration project are discussed. Show that for the case of exponential production rate decline and high rate of return on the acceleration project, the two methods result in the same answer for the rate of return, namely, the larger solution of the equation

$$C^{(a)} = \frac{q_0^{(a)} u}{b^{(a)} + J} - \frac{q_0 u}{b + J}$$

where the symbols are those of Secs. 2-2 and 12-15.

***37.** A new well, drilled for a cost of $800,000, has an initial oil production rate of 230 bbl/day with a steady decline of 2.25 percent/month. If the net oil value is $8/bbl, the economic limit is 5 bbl/day, and the company's earning power is 12 percent/year, find the discounted payout time, the discounted percentage profit, and the rate of return.

After 5 years' production the well is considered for acceleration, it being estimated that the then current production rate could be doubled for an expenditure of $60,000. Assuming that the net value of the oil remains the same and that the economic limit and the total recoverable oil are not influenced by the increased production rate, determine the payback and payout times, the percentage profit, and the rate of return applicable to the acceleration project.

Answers: If not accelerated: payout time, 1.59 years; percentage profit, 117 percent; rate of return, 76.3 percent. If accelerated: payback time, 2.2 months; payout time, 5.1 months; percentage profit, 31.3 percent; rate of return, 215 percent/year.

38. In analyzing a particular pool in an attempt to determine the well spacing for maximum profit, the assumptions are made that the ultimate oil recovery is independent of the spacing and that the average well's production rate declines exponentially. It is also assumed that the economic limit may be taken to be zero to a good degree of approximation. Let R be the recoverable oil, bbl; q_0 the average initial production rate per well, bbl/day; n the number of wells; D those development costs unaffected by the number of wells drilled, dollars; C the development costs per well, dollars; L the PDV of the total lifting costs per well, dollars; u the net oil value, dollars/bbl; and j the continuous annual discount rate. Prove that the total PDV profit from the pool is

$$\frac{365 q_0 R n u}{365 q_0 n + R j} - n(C + L) - D \qquad \text{dollars}$$

Show that the number of wells required for maximum PDV profit is hence

$$\frac{R}{365 q_0} \left[\left(\frac{365 j u q_0}{C + L} \right)^{1/2} - j \right]$$

39. Use the formula of Prob. 38 to determine the optimum well spacing under the following conditions:

Recoverable oil:	400 bbl/acre-ft
Net oil sand:	20 ft
Average initial oil production rate per well:	77 bbl/day
Development costs per well:	$720,000
PDV: lifting costs	$90,000/well
Net oil value:	$10.50/bbl
Discount rate:	10 percent/year

Answer: 38.5 acres/well or, say, 40-acre spacing.

***40.** It is estimated that the future net incomes of a certain project will be as shown in Table P-5 and that the capital cost involved in undertaking the acceleration project will be $8000. Determine the discounted profit, payback time, and payout time for the proposed acceleration project. Draw a curve of difference in cumulative discounted net incomes against discount rate, and find the rate of return if the earning power of the company is 15 percent/year.

Answers: Profit, $4280 (53.5 percent); payback time, 5 months; payout time, 7 months; rate of return, 218 percent/year.

***41.** In estimating the profitability of an associated-gas gathering and processing plant, it is assumed that the gas production rate will decline exponentially after the plant goes on stream. Prove that the PDV of the profit (or loss) resulting from the installation of the system is (in dollars)

TABLE P-5 PROBLEM 40: Estimated Future Net Incomes

	Annual Net Income, Dollars	
Year	If Unaccelerated	If Accelerated
1	6000	21,000
2	6000	15,000
3	6000	7,000
4	4000	3,000
5	4000	
6	4000	
7	4000	
8	4000	
9	3000	
10	2000	
11	2000	
12	1000	

$$Q_0 \exp(-b\tau)\left[3.65E(x-c)\frac{1-\exp(-j\tau)}{j} - K\right]$$
$$+ 3.65EQ_0(x-c)\frac{\exp[-(b+j)\tau] - \exp[-(b+j)T]}{b+j} - G$$

where b = annual continuous decline rate of the gas production
c = operating cost per mcf of gas gathered and processed, cents
E = plant efficiency (that is, of the total throughput, only a fraction E is effectively processed)
G = capital cost of the gathering system, dollars
j = annual continuous discount rate
K = capital cost of compressors and plant, dollars per mcf/day of capacity
Q_0 = rate of gas production from the pool when the plant goes on stream, mcf/day
T = time between the plant going on stream and the economic limit of the pool being reached, years
x = sales value of gas plus products, cents/mcf
τ = time during which the plant is operating at capacity, years

Assuming that the capital cost G of the gathering system is, within reasonable limits, independent of the size of plant installed, prove that the size of plant required for maximum PDV profit (or minimum PDV loss) is such that the capital cost of the plant is equal to the PDV of the net operating income during the period when the system is operating at capacity.

Notes on Solutions to the Problems

1. Plot the data on regular graph paper. The best-fit line gives the ultimate cumulative production and hence the future cumulative.

The value of the continuous decline rate b is found to be 0.262 from Eq. (2-7).

The future life is determined from Eq. (2-10). This equation is also used to determine the length of the period to date of the declining production rate. The cumulative production at which the well started to decline (2.4×10^6 bbl) is read from the graph, and so the length of the steady production period is readily calculated. The sum of these three time periods (6.4, 3.25, and 2.63 years) gives the total life of the well.

The continuous decline rate, if the infilling program were carried out, is obtained from Eq. (2-15).

It should be noted that the original data, when plotted on semilog paper, define a reasonably good straight line. It follows that the equations for harmonic decline could be applied. Such an approach gives an ultimate cumulative production of 8 million barrels.

2. Use of Eqs. (2-7) and (2-15) permits determination of both the flowing (117,800 bbl) and pumping (74,600 bbl) productions of a typical well. The total number of wells required is then 10,000,000/192,400 or 52. The life of the well (flowing and pumping phases separately) comes from Eq. (2-10).

3. This problem is solved by direct application of Eq. (1-8), (1-9), or (1-10).

4. $$\frac{dp}{dQ} = -\frac{1}{5000}$$

$$p = p_0 - Q/5000 \qquad \text{(a)}$$

where Q is the cumulative production when the pressure is p.

The producing policy determines that

$$q = 0.3 \times (0.5p)$$

so that $$p = q/0.15 \qquad \text{(b)}$$

Combining Eqs. (a) and (b) results in an expression giving q as a linear function of Q, which is one of the bases for a straight-line decline (Fig. 2-2).

The slope of this line gives b (0.1312), from which, by Eq. (2-12), d is found (12.3 percent/year).

The last part of the question is answered by substituting (b) above in Eq. (2-10).

5. b from Eq. (2-10)
q from Eq. (2-7)
p_{wf} from Eq. (3-2)

Static pressure at abandonment from Eq. (3-2).

6. See Example 3-5. The data given show that the pressure in the oil zone is higher than that in the water zone, so that if the well were shut in for a few days, some oil might be lost to the water zone, but no permanent damage would be done.

7. See Example 4-5.

8.

$$\frac{q}{q'} = \left(1 - \frac{p_{wf}}{\bar{p}}\right)\left(1 + 0.8\frac{p_{wf}}{\bar{p}}\right) \qquad \text{from Eq. (3-8)}$$

$$\frac{1120}{q'} = \left(1 - \frac{690}{2095}\right)\left(1 + 0.8\frac{690}{2095}\right)$$

$$= 0.847$$

$$q' = 1322 \text{ bbl/day}$$

$$q = 1322\left(1 - \frac{p_{wf}}{2095}\right)\left(1 + 0.8\frac{p_{wf}}{2095}\right)$$

This gives the following results:

p_{wf}	q
2095	0
1800	315
1500	591
1200	824
900	1013
690	1120
300	1262
0	1322

$$J_p^* = 1.8 \times 1322/2095 = 1.136 \text{ bbl/(day)(psi)} \quad \text{from Eq. (3-10)}$$
$$J_f^* = 1.136/1.57 = 0.723 \text{ bbl/(day)(psi)} \quad \text{from Eq. (3-12)}$$
$$\frac{1.8q_f'}{\bar{p}_f} = J_f^* \quad \text{from Eq. (3-10)}$$

so, substituting,

$$q_f' = \frac{0.723 \times 1515}{1.8} = 609 \text{ bbl/day}$$

9.

$$q = J_{oi}' (p_i^2 - p_{wf}^2) \quad \text{from Eq. (3-17)}$$

so that

$$J_{oi}' = 1120/(2095^2 - 690^2)$$
$$= 2.86 \times 10^{-4} \text{ bbl/(day)(psia)}^2$$
$$q = 2.86 \times \left[438.9 - \left(\frac{p_{wf}}{100}\right)^2\right]$$

This gives the following results:

p_{wf}	q
2095	0
1800	329
1500	612
1200	843
900	1024
690	1119
300	1230
0	1255

$$J_o' = J_{oi}' \, p_s/p_i \quad \text{from Eq. (3-21)}$$
$$= 2.86 \times 10^{-4} \times 1515/2095 = 2.07 \times 10^{-4}$$
$$q = 2.07 \times 10^{-4} (p_s^2 - p_{wf}^2)$$

so that when

$$p_s = 1515 \quad \text{and} \quad p_{wf} = 0$$
$$q = 475 \text{ bbl/day}$$

10. A plot indicates that the last of the measurements may be in error. Take as the anchor points

Depth = 0, pressure = 96 psig = 111 psia
Depth = 6000 ft, pressure = 2403 psig = 2418 psia

$$\text{Liquid specific gravity} = \frac{(0.898 \times 26) + (1.15 \times 104)}{26 + 104} = 1.099$$

and hence

$$0.433 \, \rho = 0.476$$

Eq. (4-15) becomes

$$0.476 \, H = A \ln\left(\frac{p}{p_{th}}\right) + (1 - M)(p - p_{th})$$

Taking the value of M to be 0.2, and inserting the anchor-point data

$$2856 = A \times 3.08 + 1846$$
$$A = 327.9$$

With $M = 0.2$, $A = 327.9$, the following table results:

p	$327.9 \ln\left(\frac{p}{111}\right)$	$0.8\,(p - 111)$	$0.476H$	H
165	130	43	173	363
315	341	163	504	1060
615	561	403	964	2025
915	692	643	1335	2805
1315	810	963	1773	3725
1715	898	1283	2181	4582
2115	967	1603	2570	5399

11. Use Figs. (4-18) and (4-19) to construct the following table:

q, bbl/day	Equiv. Depth 500 psig, ft	Equiv. Depth Tubing Shoe, ft	p_{wf}, psig
50	3500	6500	1100
100	3500	6500	1050
200	4000	7000	950
400	4000	7000	1000
600	3750	6750	1000

Plot the curve of p_{wf} (assumed THP of 500 psig) against q from this table. Enter at the actual rate of 320 bbl/day to determine the producing point on the IPR, which may now be drawn, assuming it to be straight (a curved IPR could also be put in based on either Vogel, or Fetkovich; see Probs. 8 and 9).

Now construct the THP curve using the actual flowing BHPs [from the IPR and Figs. (4-18) and (4-19)].

Equation (5-1), (5-2), or (5-3) can be used to define the choke performance line for the $^1/_2$-in. choke. Use of Eq. (5-1) leads to

$$p_{tf} = 0.375\,q$$

This line may be plotted, and its intersection with the THP curve gives the required results.

12.

$$q_o = 130 \text{ bbl/day}$$
$$\text{Gross liquid rate} = {}^4/_3 \times 130 = 173 \text{ bbl/day}$$
$$\text{GLR} = {}^3/_4 \times 1200 = 900 \text{ cu ft/bbl}$$

$$J = \frac{q}{2800 - p_{wf}}$$

Use of the pressure distribution curves of Figs. 4-19 and 4-20 gives:

q	p_{wf}	Equivalent Depth p_{wf}	Equivalent Depth THP	THP
50	2640	11500	4000	700
100	2480	12250	4750	650
200	2170	13000	5500	600
400	1550	11000	3500	350
800	920	7500	0	0

From the graph, the value of the THP at the gross liquid rate of 173 bbl/day is 640 psig. Using Eq. (5-1),

$$640 = \frac{600(0.9)^{1/2} \times 173}{S^2}$$

from which $S = 12.4$, or choke size $= {}^{12}/_{64} = {}^{3}/_{16}$ in.

On ${}^{1}/_{2}$-in. choke, using Eq. (5-1),

$$p_{tf} = \frac{600(0.9)^{1/2}\, q}{(32)^2}$$

Plotting this line, it is seen that the intersection occurs when q (gross) = 450 bbl/day.

13. Use the choke performance equation [for example, Eq. (5-1)] to determine the THP (96 psig) at the flow rate of 320 bbl/day. Then use the pressure-distribution curves in $2^3/_8$-in. tubing to plot the pressure at the bottom of the tubing as a function of rate for: (*a*) THP = 100 (96) psig; and (*b*) THP = 300 psig. The rate of 320 bbl/day, THP = 100 psig, locates one point on the IPR. The CHP of 993 psig may be converted to p_{wf} using Eq. (4-18) and this value, used with the 300-psig THP curve, gives a second point on the IPR.

14. Assuming a GLR of 800 cu ft/bbl throughout, curves of pressure at the foot of the tubing against rate may be plotted for THPs of 400, 270, and 140 psig. In each case there is sufficient information given to determine the point at which the well is actually producing on the corresponding curve. Thus three points on the (apparently curved) IPR are obtained.

15. See Example 5-4.

16. The problem is solved by the following steps:

1. Use data of Table P-2 to plot rate against drawdown on double-log paper (see Fig. 3-16). Draw in the straight line that best fits the results from each well, and so determine values of J and n as defined in Eq. (3-14). Plot these results (semilog paper is convenient, with J plotted on the logarithmic scale) and from that plot determine J values at regular n intervals (say $n = 0.4, 0.5, 0.6, 0.7$, and 0.8). Plot the resulting grid of lines (Fig. 3-16) on double-log paper, rate versus drawdown.
2. Determine, from one of the choke-performance equations, the cur-

rent THP, and hence the flowing BHP, of well E. Locate well E on the grid defined in step 1, and so draw in the straight line representing the IPR of well E. Use this line to plot the IPR of well E now on a regular p-q graph. The present-day potential of 700 bbl/day may be read off (see Fig. P16-1).

3. From the data given, prepare plots for well E of static pressure and GOR against the well's cumulative production. Extrapolate these curves as well as possible to higher cumulatives (Fig. P16-2).
4. On Fig. P16-2 plot the production rate at the reference drawdown of 100 psig against cumulative production and extrapolate as well as possible. These points may be obtained, from the data presented, at cumulatives of 0, 150,000, 160,000, and 260,000 bbl. As an example, at 160,000 bbl Table P-3 shows an oil rate of 300 bbl/day and a value of p_{wf} [determined from the CHP adjusted through Eq.

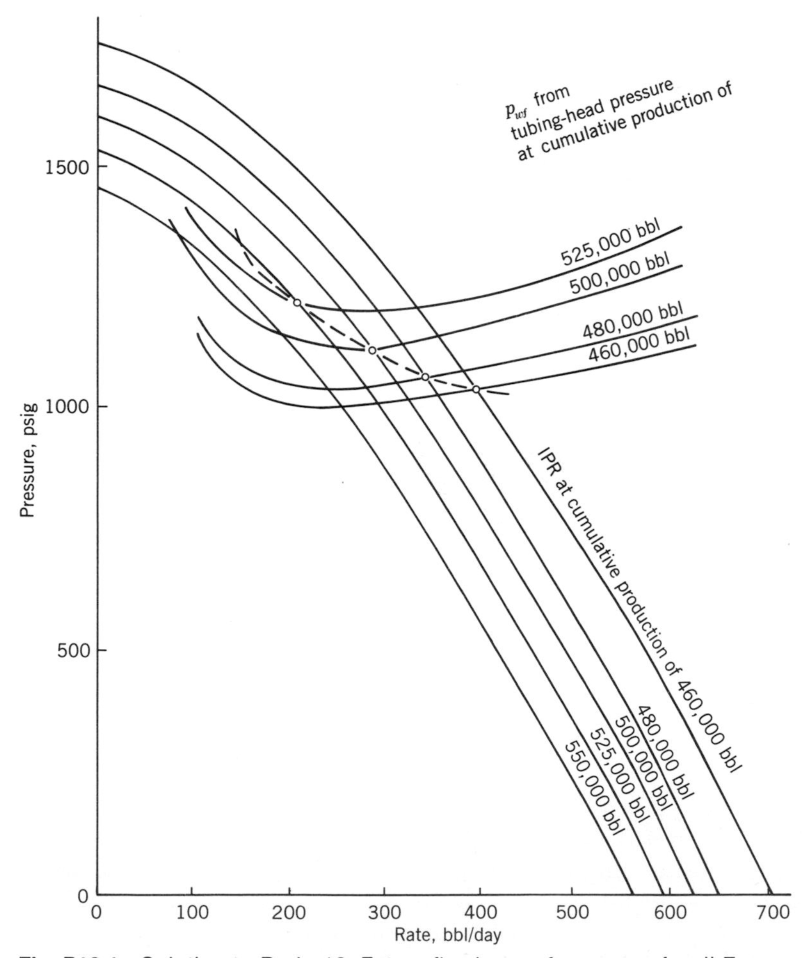

Fig. P16-1 Solution to Prob. 16: Future flowing performance of well E.

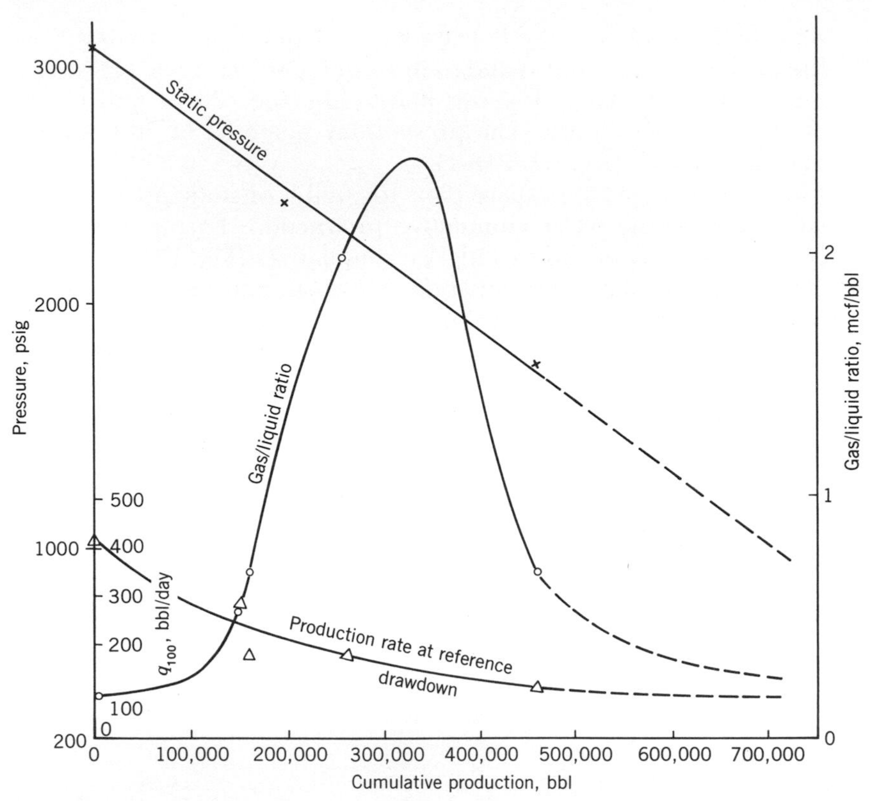

Fig. P16-2 Solution to Prob. 16: Performance of static pressure, GLR, and production rate at reference drawdown, well E.

(4-18)] of 2363 psig. From the static-pressure line of Fig. P16-2, p_s at this cumulative for well E is 2620 psig, so that the drawdown at 300 bbl/day is 257 psi.

Locate this point on the grid defined in step 1, and interpolate the corresponding IPR line. This cuts the $\Delta p = 100$ line at $q =$ 180 bbl/day.

5. To determine the production rate from well E at a drawdown of 600 psi when its cumulative production was 100,000 bbl, read off q_{100} at 100,000 bbl from Fig. P16-2; this is 293 bbl/day. Locate back onto the q-Δp grid on double-log paper. Draw in the corresponding grid line, and determine the value of q (660 bbl/day) when $\Delta p =$ 600 psi.
6. To determine the future flowing life, first choose some future regular cumulative production steps, for example, 480,000 bbl, 500,000 bbl, and so forth. Read off the corresponding values of q_{100} from Fig. P16-2. Locate the points on the grid of step 1, and draw in

the corresponding grid lines. Read off a series of rate versus drawdown values for each line.

Since the value of p_s at each cumulative may be obtained from Fig. P16-2, the IPR curve at each assumed cumulative may be plotted (shown on Fig. P16–1).

Considering now, for example, the situation at a cumulative of 460,000 bbl, the GOR may be obtained from Fig. P16–2, and so the curve of pressure at the tubing shoe (assuming 100 psi THP) may be plotted on Fig. P16–1. The intersection with the IPR gives the flowing production rate (390 bbl/day).

This process is continued at increasingly higher assumed cumulatives until no intersection occurs—this situation is reached at a cumulative slightly in excess of 525,000 bbl (Fig. P16-1)—at which point the well dies.

17. See Examples 6-2, 6-3, 6-5, and 6-4 (part 3).

18. See Prob. 8 and Example 6-6.

19. See Sec. 7-4.

20. See Sec. 7-6.

21.

$$b = 5.614\,Jw/a = 87.03$$
$$T = 11000/750 \text{ min} = 0.0102 \text{ day}$$

From Fig. 8-8

$$n_{\text{opt}} = 44 \text{ cycles/day}$$

From Eq. (8-14)

$$q = \frac{a(p_s - p_m)}{5.614w}\, n \left\{1 - \exp\left[-b\left(\frac{1}{n} - T\right)\right]\right\}$$

so $$q_{\text{opt}} = 0.571(1570 - p_m) \qquad \text{bbl/day}$$

From Eq. (8-5)

$$p_m = P_1 + 5.614\,\frac{w}{a}\,Q_m$$
$$= 150 + 51.20Q_m$$

so that $$q_{\text{opt}} = 810.82 - 29.24Q_m \qquad \text{bbl/day}$$

This leads to the following results:

Q_m	q_{opt}
0	810.8
2	752.6
4	693.9
6	635.4
8	576.9
10	518.4

From Eq. (8-19) and Eq. (8-20)

$$p_g = P_1 + 5.614 \frac{w}{a} \left(\frac{q}{n} + Q_m \right) = P_2 + \frac{7.48wq}{an}$$

$$P_2 = P_1 + \frac{5.614w}{a} Q_m - \frac{1.866w}{a} \frac{q}{n}$$

which gives $P_2 = 150 + 51.20Q_m - 0.387q$

Using the results above, this gives:

Q_m	P_2
0	(negative)
2	(negative)
4	86.3
6	211.3
8	336.3
10	461.4

The tubing volume V = 11,000 × 0.0318 = 349.8 cu ft
Gas volume per cycle = 23.80 $[P_2 - 150 + 1.55q]$ cu ft from Eq. (8-21)

Use of the above results gives:

Q_m	Gas Vol/Cycle (cu ft)
6	24,900
8	25,716
10	26,535

Finally,

$$p_g = P_2 + \frac{7.48wq}{an}$$

$$= P_2 + 1.55q$$

giving

Q_m	p_g
6	1196.2
8	1230.5
10	1264.9

22.

$$b = \frac{5.614Jw}{a} = 32.56$$

$$T = 0.0102$$

From Fig. 8-8,

$$n_{opt} = 31 \text{ cycles/day}$$

As in the Prob. 21 solution,

$$q_{opt} = 0.542\,(1570 - p_m) \qquad \text{bbl/day}$$
$$p_m = 150 + 19.15Q_m$$
$$q_{opt} = 769.64 - 10.38Q_m \qquad \text{bbl/day}$$

leading to

Q_m	q
0	769.6
2	748.9
4	728.1
6	707.4
8	686.6
10	665.8

Assuming P_2 = 300 psig, and using

$$\text{Tubing volume V} = (2000 \times 0.085) + (9000 \times 0.0318)$$
$$= 456 \text{ cu ft}$$
$$\text{Gas volume per cycle} = 31.02(150 + 2.200q) \qquad \text{cu ft}$$
$$p_g = 300 + 2.2q$$

Q_m	Gas Vol/Cycle (cu ft)	p_g
0	57170	1993
2	55774	1948
4	54347	1902
6	52920	1856
8	51524	1811
10	50097	1765

If P_2 is assumed to be 400 psig, the gas volume per cycle is 31.02 $(250 + 2.200q)$, which gives

Q_m	Gas Vol/Cycle (cu ft)
0	60272
2	58876
4	57449
6	56022
8	54626
10	53199

23. Using Eq. (4-18),

$$p_{wf} = 257 \text{ psia} = 242 \text{ psig}$$
$$q = J(p_s - p_{wf}) = 858\,J \qquad \text{bbl/day} \qquad \text{(a)}$$

From Eq. (8-24), neglecting P_1,

$$q = 1100J/(1 + KJ) \qquad \text{(b)}$$

where, from Eq. (8-27)

$$K = \frac{0.00226 \times 4.492}{0.0217}$$
$$= 0.467 \text{ psi/(bbl)(day)}$$

Equating expressions (a) and (b) and using the calculated value of K,

$$J = 0.604 \text{ bbl/(day)(psi)}$$
$$q = 518 \text{ bbl/day}$$

24. From Eq. (9-21)

$$S_p = S - \frac{12D}{A_rE}\left(W_f - \frac{SN^2M}{70{,}500} W_r\right)$$

while plunger displacement rate is S_pN.

From Eq. (9-16)

$$N = \left[\frac{70{,}500L}{S(1 - c/h)}\right]^{1/2} = K/S^{1/2}, \text{ say}$$

Combining these equations gives an expression for plunger displacement rate as a function of S. Differentiating gives the rate of change of the plunger displacement rate with respect to S as

$$\frac{N}{2S}\left[S + \frac{12D}{A_rE}\left(W_f - \frac{SN^2M}{70{,}500} W_r\right)\right]$$

which is positive in all practical circumstances.

25. See Example 10-1.

26. See Example 10-2.

27. See Example 10-3.

28. Prepare a pressure-rate-depth grid (Fig. 4-24) and superimpose the pressure versus rate curve defined by the assumed equation for gas-anchor performance.

29. Prepare a table similar to Table 11-1, but using the assumed gas-anchor efficiency, instead of the assumption made in Example 11-1 that the GLRs in the tubing and in the annulus will be the same.

30. If the gas anchor is assumed to be set at a particular depth, say 3000 ft, it is possible to determine a curve of oil rate from the formation against the total volume rate (that is, the oil rate plus the free-gas rate) that must pass through the anchor if stable producing conditions are to prevail. The balance of the free gas at the anchor intake is diverted up the annulus according to the equation for the anchor's efficiency. The curve will be determined by the IPR of the formation and the anchor efficiency and will not reflect in any way the capacity of the pump.

A separate calculation will give the pump's maximum displacement rate when the pump is set at 3000 ft. Location of this displacement rate on the above curve defines the actual producing situation when the P/GA is set at 3000 ft, and so the oil production rate.

The calculations are then repeated at other setting depths and a curve results that gives the attainable oil production rate in terms of the setting depth. From this curve, the optimum depth and rate may be obtained (Fig. P30-1).

To obtain the first curve, assume oil rates of 50, 100, 200, 400, and 600 bbl/day. The IPR gives the values of p_{wf}. A flowing calculation then determines the corresponding intake pressures at the assumed setting depth of the gas anchor. The information on gas in solution enables the free-gas volumes at the anchor intake at the various oil rates to be calculated. The free-gas volume is apportioned between casing and tubing using the expression for anchor efficiency.

The maximum displacement rate of the pump at the assumed setting depth is determined as in Example 9-1.

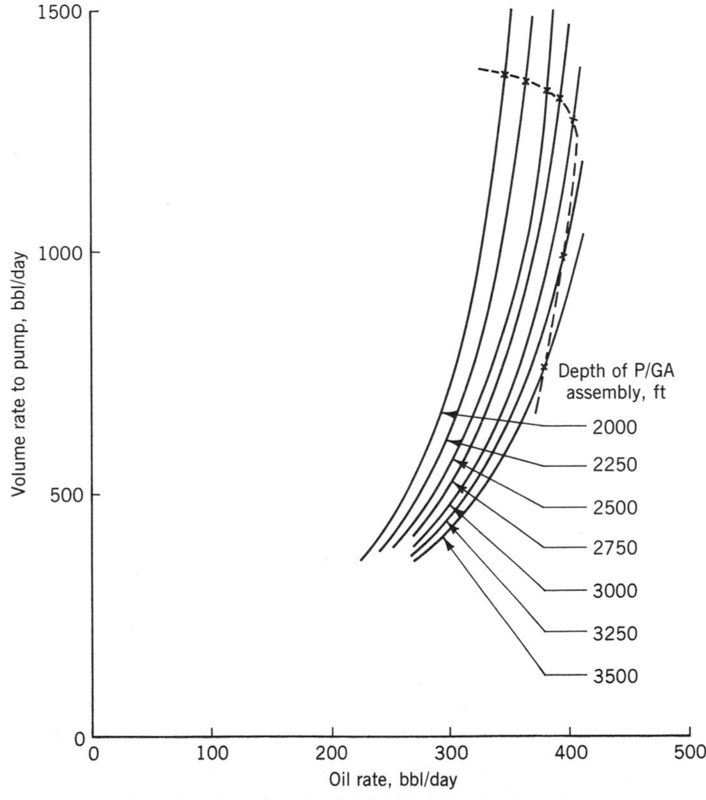

Fig. P30-1 Solution to Prob. 30: Oil production rate plotted against total oil and gas rate entering P/GA assembly at various setting depths.

31. See Example 12-3.
32. See Example 12-4.
33. Use Eq. (12-19).
34. From Sec. 12-14, the payout time t_p is given by

$$\exp\left[-(b+j)t_p\right] = 1 - \frac{C(b+j)}{q_0 u}$$

Since t_p must be less than Y, it follows that

$$1 - \frac{C(b+j)}{q_0 u} > \exp\left[-(b+j)\,Y\right] \qquad \text{(a)}$$

If the decline rate b is such that

$$b < \frac{1}{Y} \ln\left(\frac{X+100}{X}\right) - j$$

it follows that

$$\exp\left[-(b+j)\,Y\right] > \frac{X}{X+100} \qquad \text{(b)}$$

If the project is such that both Eq. (a) and Eq. (b) are satisfied, then

$$1 - \frac{C(b+j)}{q_0 u} > \frac{X}{X+100}$$

which may be written in the form

$$\frac{q_0 u}{C(b+j)} - 1 > \frac{X}{100}$$

The left-hand side of this inequality is the profit, as given by Eq. (12-10), provided N is large.

35. Use the equations of Sec. 12-14, writing $\exp(-bT) = 0$.
36. If the exponential terms are ignored, both methods give J as a root of the equation

$$C^{(a)} = \frac{q_0^{(a)} u}{b^{(a)} + J} - \frac{q_0 u}{b + J}$$

Hence the problem is to show that the second method defines only the larger of the roots of this equation. If J' is the smaller root, then a value of j slightly larger than J' will give a value $C^{(a)} + p$, where p is positive, to the difference

$$\frac{q_0^{(a)} u}{b^{(a)} + j} - \frac{q_0 u}{b + j}$$

This would imply that the value of the discount rate was not yet sufficiently high to be the rate of return as defined by the second method.

37. Use the methods of Example 12-4 in conjunction with Eqs. (2-15) and (2-16) and the formulas of Sec. 12-15.

38. Recovery per well $= \dfrac{R}{n}$

and also $= \dfrac{365q_0}{b}$ from Eq. (2-7)

Hence,
$$b = \frac{365q_0 n}{R} \tag{a}$$

From Eq. (12-10), with neglect of the exponential term,

$$P = \frac{365q_0 un}{b + j} - D - (C + L)\,n$$

where P is PDV profit, so that, from Eq. (a)

$$P = \frac{365q_0 Rnu}{365q_0 n + Rj} - D - (C + L)\,n \tag{b}$$

Differentiating P with respect to n and setting the result equal to zero gives the answer for the number of wells for maximum PDV profit.

39. A direct application of the equation developed in Prob. 38.

40. See Example 12-2.

41. See Fig. P41-1. During the first τ years the volume of gas processed is 365 Q_p mcf/yr, and the volume of gas effectively processed in the interval δt (expressed as a fraction of a year) is 365 $EQ_p\ \delta t$. The income from this is $3.65(x - c)EQ_p\ \delta t$ dollars.

Hence the PDV of the income received while the plant is operating at capacity is

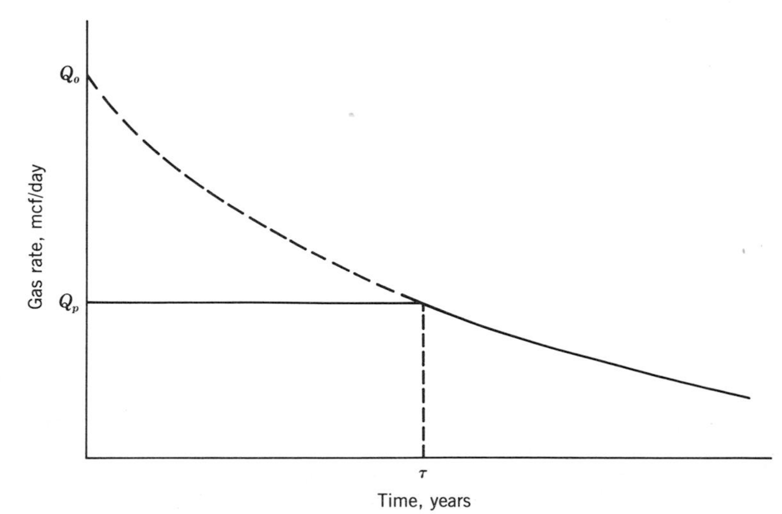

Fig. P41-1 Solution to Prob. 41: Throughput of gas plant.

$$P_1 = 3.65\,(x - c)EQ_p \int_0^{\tau} \exp\,(-jt)\,dt$$

$$= 3.65\,(x - c)EQ_0 \exp\,(-b\tau)\,\frac{1 - \exp\,(-j\tau)}{j}$$

since $Q_p = Q_0 \exp\,(-b\tau)$

Similarly, the PDV of the income received while the plant is operating below capacity is

$$P_2 = 3.65(x - c)EQ_0\,\frac{\exp\,[-(b + j)\tau] - \exp\,[-(b + j)T]}{b + j}$$

and the profit is

$$P = P_1 + P_2 - G - Q_0 \exp\,(-b\tau)\,K$$

which is the first result.

Differentiating P with respect to τ [assuming G to be independent of τ, that is, of the size of plant $Q_0 \exp\,(-b\tau)$] and setting the result equal to zero shows that the optimum plant size occurs when

$$3.65\,(x - c)E\,\frac{1 - \exp\,(-j\tau)}{j} = K$$

or, as is shown by multiplying both sides by $Q_0 \exp\,(-b\tau)$, when P_1 is equal to the capital cost of the plant.

Name Index

Subject Index

About the Author

T. E. W. Nind is professor of mathematics and past president and vice-chancellor, Trent University, Ontario. He was associated for eight years with the Royal Dutch/Shell Group of companies, with production, reservoir, and secondary recovery engineering experience in The Netherlands, Venezuela, and British Borneo. After teaching for eight years in the Department of Geological Sciences, University of Saskatchewan, during which time he was a member of the Provincial Oil and Gas Conservation Board, he joined the faculty of Trent University in 1966. Professor Nind has been a visiting professor at the Universidad Autónoma de México, Mexico City, and at the University of the West Indies, Trinidad. He is a Fellow of the Cambridge Philosophical Society, a member of the Canadian Institute of Mining and Metallurgy and of the Canadian Mathematical Congress, and a registered professional engineer in the province of Ontario.